THE SCIENCE OF LEARNING AND DEVELOPMENT IN EDUCATION

A Research-based Approach to Educational Practice

All teachers need to understand how children and adolescents learn and develop. Traditionally, teachers' understandings were informed by a mix of speculative theory and scientific evidence. However, recent scientific research has provided new insights into how children and adolescents learn and develop that are particularly relevant to teachers. *The Science of Learning and Development in Education* provides teachers with an exciting and comprehensive introduction to this new and emerging field.

This innovative text introduces readers to brain science and the science of complex, dynamic systems as applied to human development. It does not assume or require any prerequisite knowledge and it situates the scientific study of learning and development firmly within the academic discipline of education and its practice in schools and classrooms. Part 1 examines the science of learning and development in the twenty-first century; Part 2 explores the emotional, cultural, moral and empathetic brain; and Part 3 focuses on learning, wellbeing and the ecology of learning environments.

Each of the 11 chapters guides the reader through the subject matter and relevant concepts and terminology. Readers are encouraged to take a critical approach to their study and are provided with opportunities to reflect on key issues as they arise, via questions and prompts interspersed throughout the text.

Written in an engaging style by experts and generously illustrated with colour photographs and diagrams, *The Science of Learning and Development in Education* is an essential resource for pre-service teachers entering the profession, and a valuable aid for all teachers wishing to deepen their scientific understanding of learning and development.

Minkang Kim is Senior Lecturer in Human Development and Education and researches brain science and education in the Faculty of Arts and Social Sciences at The University of Sydney.

Derek Sankey is retired but actively researches brain science and education. He previously taught at the University of London, Institute of Education and the Hong Kong Institute of Education.

Cambridge University Press acknowledges the Australian Aboriginal and Torres Strait Islander peoples of this nation. We acknowledge the traditional custodians of the lands on which our company is located and where we conduct our business. We pay our respects to ancestors and Elders, past and present. Cambridge University Press is committed to honouring Australian Aboriginal and Torres Strait Islander peoples' unique cultural and spiritual relationships to the land, waters and seas and their rich contribution to society.

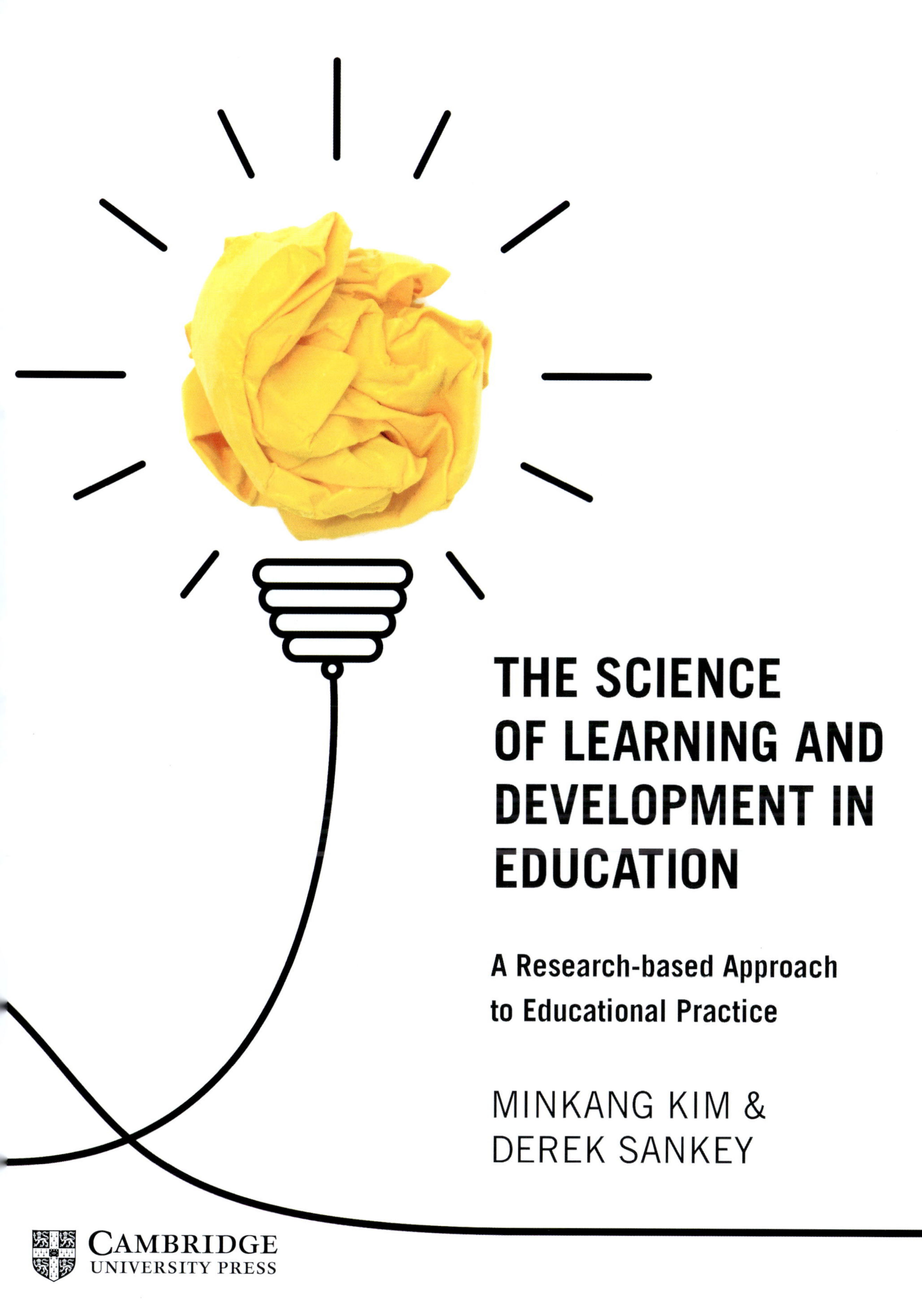

THE SCIENCE OF LEARNING AND DEVELOPMENT IN EDUCATION

A Research-based Approach to Educational Practice

MINKANG KIM &
DEREK SANKEY

CAMBRIDGE
UNIVERSITY PRESS

CAMBRIDGE
UNIVERSITY PRESS

University Printing House, Cambridge CB2 8BS, United Kingdom

One Liberty Plaza, 20th Floor, New York, NY 10006, USA

477 Williamstown Road, Port Melbourne, VIC 3207, Australia

314–321, 3rd Floor, Plot 3, Splendor Forum, Jasola District Centre, New Delhi – 110025, India

103 Penang Road, #05–06/07, Visioncrest Commercial, Singapore 238467

Cambridge University Press is part of the University of Cambridge.

It furthers the University's mission by disseminating knowledge in the pursuit of
education, learning and research at the highest international levels of excellence.

www.cambridge.org
Information on this title: www.cambridge.org/highereducation/isbn/9781108999786

© Cambridge University Press 2022

First published 2022

Cover designed by Cate Furey
Typeset by Integra Software Services Pvt. Ltd
Printed in Singapore by Markono Print Media Pte Ltd, March 2022

A catalogue record for this publication is available from the British Library

A catalogue record for this book is available from the National Library of Australia

ISBN 978-1-108-99978-6 Paperback

Additional resources for this publication at
https://www.cambridge.org/highereducation/isbn/9781108999786/resources

To the future …
For Diana, Markos, Laon and Dawon

FOREWORD

When the findings of projects from the Australian Values Education Program (AVEP) (2003–2010) became apparent, there was much talk of a 'surprise effect'. The surprise was that a program designed to enhance values-based education included such a strong academic achievement effect. From the very first project, the Values Education Study, along with the predictable outcomes of improved behaviour, enhanced relationships and a more settled environment, came evidence of greater student engagement and academic results, along with an allied strengthening of teachers' pedagogical practice. Six years later, analysis of all the findings in the evaluation study, loosely called The Testing and Measuring project, 'was guided by the conceptual framework established … between values education, school ambience and student academic diligence (Lovat & Dally, 2018, p. 7).' In time, the notion of 'surprise effect' was replaced with 'brain effect' or 'knowing effect' as neuroscientific and epistemological research helped to explain why the surprise was no surprise at all.

In somewhat similar ways, this 'project' by Minkang Kim and Derek Sankey will likely engender a surprise effect that should be no surprise at all. In introducing the 'brain effect' and 'knowing effect' in educational contexts, the authors show that concern for a scientific, evidence-based understanding of student academic achievement *necessarily* entails grappling with notions of value, emotion, empathy and the wellbeing of each and every individual student. Adopting a scientific approach to student learning and development in education *necessarily* involves interweaving the biological, sociocultural, relational and environmental factors involved.

In a sense, the wisdom that underpinned the AVEP effect, which is further evidenced in this book, is so deeply embedded in our knowledge tradition that the only real surprise was that we would have ignored or forgotten it. One of Socrates' pupils, Antisthenes (445–365 BCE), wrote that Pythagoras, a hundred years beforehand, had learned that the art of effective learning was in prioritising the wellbeing of the individual student and tailoring education to the individual student's needs. He wrote that Pythagoras regarded it as 'stupidity' itself that standardised expectations should be imposed on a diverse audience (Horky, 2013). Yet we live in times where educational systems inflict that very stupidity on generations of students. We also know through Pythagoras's biographers that he discovered – no doubt through trial and error – that his pupils excelled in mathematics (his prime concern) when immersed in multidisciplinary curricula (Stanley, 2010). Dance, theatre and music were especially important. Yet, we live in times where even kindergarten teachers are heard saying, 'we don't have time for play, for music.'

Similarly, the medieval Persian scholar, Abu al-Ghazali (1056–1111 CE), asserted that the most important element in any effective education is in instilling imaginativeness and eliciting wonder because these are the intellectual skills necessary to lifelong learning. The assertion is found in the context of his warning to teachers to avoid prescriptive and rote forms of learning because they dull the mind and kill off the desire to learn (Karim, 2003). Ghazali's warning fits well with updated findings from brain science (Narvaez, 2016; Gotlieb et al., 2016), which emphasise the indispensable interdependence of human emotion, sociality, morality and imagination in the business of learning. It also fits well with the ground-breaking work of Habermas (1972; see also Leydesdorff, 2000), whose 'ways of knowing' theory contains its own warning that the only true knowing comes when the cognitive interest to imagine is triggered. Yet we live in times that prioritise rote learning in schools and universities to the point where their measurable products have come to define the success of the institution and the key performance indicators of teachers, lecturers and institution heads. And we seem tin-eared to the educational damage this is causing.

In a word, this is why the work of Minkang Kim and Derek Sankey is so important and so timely. It focuses principally on the ramifications of the findings of cutting-edge brain science and complex systems theory for human learning and development. It offers a distilled version of the complexities therein in a logical, step-by-step and highly digestible form, such that the reader does not need a background in these sciences in order to understand the book's content. Nonetheless, one can be certain that the reader's understanding of the nuances of learning will be deepened by the reading, regardless of their background. However, its unique contribution is in offering insights into the wisdom of the past through sifting the findings of the most recent research in brain and systems sciences.

I proffer that the book lays bare why the surprise effect of AVEP was no surprise at all and why the only surprise left is that our systems of learning so persistently gravitate to that sign on the exit lane of the highway, 'Stop: You are Going the Wrong Way!' This excellent book gets us all back on the right road.

Terence J. Lovat
Emeritus Professor, University of Newcastle, Australia
Honorary Research Fellow, University of Oxford, UK
Honorary Professor, University of Glasgow, UK
Adjunct Professor, Royal Roads University, Canada

CONTENTS

PART 1 THE SCIENCE OF LEARNING AND DEVELOPMENT IN THE TWENTY-FIRST CENTURY   1

PART 3 LEARNING, WELLBEING AND THE ECOLOGY OF LEARNING ENVIRONMENTS   229

Minkang Kim (PhD) is Senior Lecturer in Human Development and Education at The University of Sydney. Born and educated in South Korea, she was Assistant Professor at Seoul National University before moving to Australia in 2010. Her academic focus is human development and neuroscience, particularly as these apply to educational research and practice. She has a special interest in the role of emotion, empathy and value in human learning, development and wellbeing Her teaching and research are underpinned by complexity (dynamic systems) theory and her research methods include the electroencephalogram (EEG), eye-tracking, micro-genetic observation and state space grids.

Derek Sankey (PhD), now retired, gained his PhD at the Institute of Education, London University, where he taught and researched from 1986 to 1995, playing a leading role in teacher education reform. He had previously directed a national project for the Farmington Institute, Oxford on the interface between science, religion and the humanities. He taught at the Hong Kong Institute of Education from 1995 to 2006. His academic background is in the history and philosophy of science, with a particular interest on the interplay of philosophy and neuroscience. His ongoing research is focused on the application of complexity theory and neuroscience in understanding the human self and its education.

PREFACE

When the American National Academy of Sciences published the second edition of *How People Learn* in 2018, they noted that, since the publication of the first edition in 2000, 'researchers have continued to make important discoveries about influences on learning, particularly sociocultural factors and the structure of learning environments. At the same time, technological developments have both offered new possibilities for fostering learning and created new learning challenges' (NASEM, 2018, p. 1). This new body of scientific research is revealing that, 'Learning is a remarkably dynamic process; from before birth and throughout life, learners adapt to experiences and their environment' (p. 3).

The aim of this new text is to introduce teachers to the *dynamics* of learning and development and the many changes that have taken place as a result of *scientific research* over the past quarter of a century, and their historical foundations. It is especially concerned with the interplay between the biological and sociocultural factors that influence learning, as well as the impact of learning environments. The focus of this book is on what scientific *research* and taking a *scientific approach* to understanding learning and development are able to offer education in deepening teachers' understandings of children and adolescents in their care – hence the emphasis in the book's title on 'the science' of learning and development, while recognising the contribution of many sciences in deepening our understandings as educators.

The past quarter of a century has witnessed a quiet revolution in two main areas of scientific research that are especially relevant to education, namely the changes that have occurred in our understanding of *brain science* and the *science of complex systems* as it applies to human development. These changes have challenged, and are still challenging, many previous beliefs about how children and adolescents learn and develop – and, indeed, about how we should view them as learning and developing, sentient, emotional, relational beings.

Although this book does not require any prerequisite knowledge of brain science or the science of complex systems, it nevertheless takes the reader, step by step, to the cutting edge of the dynamics of child and adolescent learning and development. In doing so, it encourages readers to take a critical approach to their study of learning and development and introduces them to some of the core philosophical issues that arise in regard to the science being discussed. Moreover, it situates the scientific study of learning and development firmly within the study of education, conceived as a discrete discipline in its own right, although always in cross-disciplinary dialogue with other relevant disciplines.

Past and present students may recognise some topics previously covered in lectures and seminars, but this book does not contain previous lecture or seminar notes. From the start, the aim was to provide an entirely new text, tailored to meet the varying needs of multiple readers, including preservice and in-service teachers, but also parents curious to know what is happening in the heads of their children, as they learn and develop.

It is our earnest hope that all readers will enjoy engaging with this book as much as we have enjoyed writing it. We wish you all the best for your continued learning and development, as you constantly adapt to the changing experiences afforded by the many dynamic, physical and sociocultural environments you will encounter on your continuing life's journey.

Minkang Kim and Derek Sankey

Sydney, Australia, 2022

ACKNOWLEDGEMENTS

This book could not have been written without the feedback received from literally thousands of students in different national settings, over very many years, when teaching about the science of learning and development in education. We are especially grateful to our wonderful doctoral students who contributed their critical and insightful thinking in designing and conducting the school-based EEG research reported in this book. In particular, we wish to mention Ling Wu, Chris Duncan, Soohyun Baek, Li Li and Kwan Yiu Yoyo Wu. We also acknowledge the immense contribution of school principals Bill Low, Gareth Leechman and Mohammed Riaaz Ali for inviting us into their schools to conduct this research, and their students, who so kindly participated. We are very grateful to Mary Helen Immordino-Yang for her friendly support over the years and her advice on a point of detail in the book. We acknowledge with gratitude our teachers and those colleagues who have critically encouraged our work. We recognise, with love, the supportive warmth of our families, Mum, Hujeong and Yoonyoung, Diana and Daniel, Roland and Anne, Judith and Ernie, and their families. Finally, and importantly, we are immensely grateful to Isabella Mead and Lauren Magee at Cambridge University Press for their insightful guidance and encouragement in preparing this text, Susan Jarvis for her editorial expertise and Lucy Russell for guiding the text through its initial conception.

The authors and Cambridge University Press would like to thank the following for permission to reproduce material in this book.

Figures 1.1, **2.5**, **8.8 (left)**, **9.9** and **10.9 (left)**: brain image © Getty Images/RapidEye. **Figure 1.2**: background image © Getty Images/Ariel Skelley. **Figure 1.3**: background image © Getty Images/FatCamera. **Figure 1.4** (left): © Getty Images/FRANK PERRY/Staff. **Figure 1.5**: © Getty Images/Bettmann/Contributor. **Figure 1.6**: © Getty Images/Heritage Images/Contributor. **Figure 1.7**: Linda B Smith. **Figure 1.10**: © Getty Images/Hulton Archive/Stringer. **Figure 2.2**: University of British Columbia Archives, photo by Unknown [UBC 5.1/3771]. **Figure 2.3**: Wikipedia Commons/United States National Library of Medicine, retrieved from https://commons.wikimedia.org/wiki/File:Wilder_Graves_Penfield.jpg. **Figure 2.4**: © Getty Images/GeoStock. **Figure 2.6**: brain image © Getty Images/PIXOLOGICSTUDIO/SCIENCE PHOTO LIBRARY. **Figures 2.8**, **8.8 (right)** and **10.9 (right)**: brain image © Getty Images/SCIEPRO/SCIENCE PHOTO LIBRARY. **Figure 2.9**: brain image © Getty Images/Dorling Kindersley. **Figure 2.10**: brain image © Getty Images/pukrufus. **Figure 2.11**: © Getty Images/koto_feja. **Figure 2.12**: © Getty Images/wetcake. **Figure 2.13**: © Getty Images/XH4D. **Figure 2.14**: nerve synapse image © Getty Images/CHRISTOPH BURGSTEDT/SCIENCE PHOTO LIBRARY. **Figure 2.15**: embryo image © Getty Images/SCIEPRO. **Figure 2.21**: courtesy of Barbara Arrowsmith-Young. **Figure 2.22**: © Getty Images/Imgorthand. **Figure 3.3**: © Getty Images/Mike Hill. **Figure 3.4**: © Getty Images/Kerri Wile. **Figure 3.5**: SCIENCE SOURCE/SCIENCE PHOTO LIBRARY. **Figure 3.6**: Wikimedia Commons/Stephen Morris, retrieved from https://commons.wikimedia.org/wiki/File:Belousov_Zhabotinsky_reaction_(4013035510).jpg, licensed under CC BY 2.0, https://creativecommons.org/licenses/by/2.0/deed.en. **Figure 3.7**: © Getty Images/ZedamNabil Photography. **Figure 3.9**: children playing soccer image © Getty Images/A-Digit. **Figure 3.10**: © Getty Images/PM Images. **Figure 3.11**: cover of Popper, K. & Eccles, J. (1977). *The self and its brain: An argument for interactionism* reproduced with permission of Routledge.

Figure 3.12: © Getty Images/Marc Piasecki/Stringer. **Figure 4.1**: © Getty Images/Klaus Vedfelt. **Figure 4.2**: © Getty Images/Fly View Productions. **Figures 4.3** and **7.3**: © Getty Images/FatCamera. **Figure 4.4**: temple image © Getty Images/jammydesign. **Figure 4.9**: © Getty Images/SCIENCE PHOTO LIBRARY. **Figure 5.3**: © Getty Images/Nastasic. **Figure 5.6 (left)**: © Getty Images/JGI/Tom Grill. **Figure 5.6 (right)**: © Getty Images/Sydney Bourne. **Figures 5.9** and **10.7**: © Getty Images/Ariel Skelley. **Figure 6.1**: © Getty Images/JGI/Jamie Grill. **Figure 6.4** (both images): © Getty Images/BSIP/Contributor. **Figure 6.5**: MARTIN SHIELDS/SCIENCE SOURCE/SCIENCE PHOTO LIBRARY. **Figure 6.6**: © Getty Images/Dimitri Otis. **Figures 6.7, 6.8** and **6.9**: facial images © Getty Images/mikkelwilliam (male); © Getty Image/ozgurdonmaz (females). **Figure 6.10**: brain image © Getty Images/BSIP. **Figure 6.11**: © Getty Images/Juanmonino. **Figure 7.1**: © Getty Images/Kean Collection/Staff. **Figure 7.2**: brain images © Getty Images/Jonathan Kitchen. **Figure 7.4**: © Getty Images/Carol Yepes. **Figure 7.5**: © Getty Images/adoc-photos/Contributor. **Figure 7.6**: © Getty Images/Wally McNamee/Contributor. **Figure 7.7**: map © Getty Images/FrankRamspott. **Figure 7.8**: © 2008 National Academy of Sciences, USA. **Figure 8.4**: images © *Journal of Moral Education* Ltd, reprinted by permission of Taylor & Francis Ltd, http://www.tandfonline.com on behalf of *Journal of Moral Education* Ltd. **Figure 8.6**: Wild Tribe Heroes. **Figure 8.11 (left)**: © Getty Images/SDI Productions. **Figure 8.11 (right)**: © Getty Images/Constantinis. **Figure 9.2 (left)**: © Getty Images/Stan Feldman/EyeEm. **Figure 9.2 (right)**: © Getty Images/lisegagne. **Figure 9.3**: Division of Rare and Manuscript Collections, Cornell University Library. **Figure 9.6**: © Getty Images/SDI Productions. **Figure 9.10**: Fielding International. **Figure 9.11**: images by kind permission of the Headmaster, Central Cost Grammar School, NSW, Australia. **Figure 10.1 (left)**: © Getty Images/Erik Isakson. **Figure 10.1 (right)**: University of Washington Institute of Science and Math Education, STEMteachingtools.org. **Figure 10.4**: flickr/LSE Library, 'Karl Popper c1980s', retrieved from https://www.flickr.com/photos/lselibrary/3833724834/in/set-72157623156680255/. **Figure 10.5**: © Getty Images/Hill Street Studios. **Figure 10.6**: © Getty Images/Photo and Co. **Figure 11.1**: © Getty Images/Peter Dazeley. **Figure 11.2**: Australian Institute of Health and Welfare (2020). *Australia's children*. Cat. no. CWS 69. Canberra: AIHW, https://www.aihw.gov.au/Zgetmedia/6af928d6-692e-4449-b915-cf2ca946982f/aihw-cws-69-print-report.pdf.aspx?inline=true, licensed under CC BY 3.0 AU, https://creativecommons.org/licenses/by/3.0/au/. **Figure 11.3**: © Commonwealth of Australia 2010. **Figure 11.5**: © Getty Images/Kohei Hara. **Figure 11.6**: © Kate Peters. **Figure 11.7**: Wikimedia Commons/Poliphilo, retrieved from https://commons.wikimedia.org/wiki/File:Statue_of_Dame_Julian.JPG, licensed under CC0 1.0, https://creativecommons.org/publicdomain/zero/1.0/deed.en.

HOW TO USE THIS BOOK

In using this book, you will quickly discover you don't require prior knowledge of the science. It is recommended that chapters be read in sequence, as they are progressive, deepening your understanding as you move through the book. You will also discover that the book does not simply provide the 'scientific facts' of what is currently known about how we learn and develop, but also offers historical background to current knowledge and understandings and raises important philosophical issues about how we understand ourselves in light of the science of human learning and development.

Various features are used throughout the book in order to enhance your own learning and development. You will find that many of these features help to deepen an understanding of the topics being presented.

CONCEPT MAPS

Each of the three parts of the book is introduced with a concept map that provides a visual explanation and connection between the concepts discussed in the chapters that follow.

REFLECTION

Reflection prompts appear throughout each chapter to encourage you to take a moment to pause and briefly reflect on important points raised in the text.

REFLECTION

Using the metaphor of the teacher's professional toolkit, in addition to a sound and accurate understanding of how children and adolescents learn and develop, what other professional 'tools' do teachers need, as an absolute priority, to do their job effectively?

RESEARCH LINKS

Links to research throughout each chapter provide extensive discussion on relevant and essential explorations of concepts and focus areas. They also provide important context for the discussion, as well as further understanding of scientific progress in the education field.

RESEARCH LINK 2.2

Curiosity enhances hippocampus-dependent learning and memory

Gruber, M., Gelman, B. D., & Ranganath, C. (2014). States of curiosity modulate hippocampus-dependent learning via the dopaminergic circuit. *Neuron*, 84(2), 486–96.

When do students best remember what they were taught? Many students find they are more likely to learn and remember the topics that they feel curious about (Figure 2.22), but until recently, little was known about the mechanisms by which students' curiosity affected learning and memory.

Matthias Gruber and colleagues (2014) studied brain mechanisms behind the relationship between curiosity and learning using typical memory tests and fMRI. In this study, participants underwent learning experiences by watching a set of questions and their answers on a screen. After watching each question, the participants were asked to rate how much they felt curious about the answer. While waiting for an answer to the question, participants were shown irrelevant face images.

APST AND ACECQA CURRICULUM SPECIFICATIONS

Each chapter ends by identifying the Australian Professional Standards for Teachers (APST) (AITSL, 2011) and Australian Children's Education and Care Quality Authority (ACECQA) standards (ACECQA, 2021) or focus areas most relevant to the material. The APST enumerate essential knowledge, professional skills and engagement for teachers, organisations and professional associations. The full standards are available online at: www.aitsl.edu.au/teach/standards. The ACECQA standards provide essential resources, guidance and support to governments to ensure quality outcomes across Australia in children's education. The full standards are available online at: www.acecqa.gov.au/nqf/national-quality-standard.

STANDARDS | **APST and ACECQA curriculum specifications**

APST Standard 1.2 Understand how students learn
Demonstrate knowledge and understanding of research into how students learn and the implications for teaching

APST Standard 6.2 Engage in professional learning and improve practice
Understand the relevant and appropriate sources of professional learning for teachers

ACECQA curriculum specifications
1.1 Learning, development and care
6.3 Professional identity and development

This chapter invites newly qualifying teachers to critically review what they currently know and understand from research about how students learn and the implications for teaching. It introduces the new science of learning and development as a touchstone to be used by teachers in deciding what are relevant and appropriate sources of professional learning for teachers, given the many

KEY POINTS FOR TEACHERS

Each chapter ends with a summary of important points for teachers to encourage reflection on the importance of interweaving research outcomes with the practical applications of each concept or focus area in the classroom.

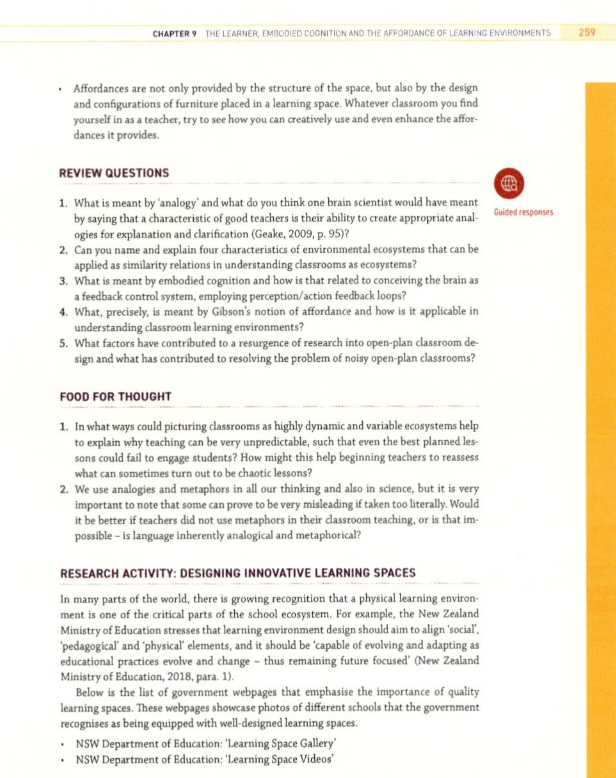

REVIEW QUESTIONS

The end-of-chapter review questions prompt you to reflect on the central learnings of each chapter. You can test your knowledge by responding to each of the questions after you have worked your way through the material. Guided responses to these questions are available in the book's online resources.

FOOD FOR THOUGHT

The end-of-chapter 'Food for thought' questions encourage further discussion on the central learnings of each chapter. They provide the opportunity to think critically about their application in real-life settings and to communicate your understanding of the material.

RESEARCH ACTIVITY

The end-of-chapter research activities provide an opportunity to further your knowledge on a particular concept or focus area mentioned in the chapter by engaging with material through traditional research. Prompts and links to recommended resources are provided.

ONLINE RESOURCES

The student online resources for *The Science of Learning and Development in Education* are freely available online at: www.cambridge.org/highereducation/isbn/9781108999786/resources. Visit the site to explore a variety of resources, including weblinks and guided responses to the review questions.

This margin icon is used throughout the book to indicate that a resource relating to the content under discussion is available online. The icon's descriptor can be used to help you easily identify each resource in the chapter's downloadable document.

PART 1

THE SCIENCE OF LEARNING AND DEVELOPMENT IN THE TWENTY-FIRST CENTURY

This book does not require any prerequisite knowledge of the science of learning and development, and it is very likely that teachers will find much in the book that is new. The main aim of Part 1 is therefore to start at the very beginning, assuming little more than a commitment to the professional practice of teachers. Teachers are fully aware they need to know how children and adolescents learn and develop. Traditionally within education, that process has been informed by a mix of speculative and scientific theory. However, the past quarter of a century has witnessed a massive growth in new scientific knowledge that can add considerably to teachers' professional knowledge regarding their students' learning and development and that also needs to be incorporated into future education research.

The centrality of education research and practice in this textbook, and the importance of this for teacher education, is explored at the start of Chapter 1 and also indicated in Concept map 1. All eight key concepts explored in the first five chapters are introduced In Chapter 1 using a light-touch approach, inviting readers to feel at home in this new field of study. A working definition of the science of learning and development as conceived in this textbook

Concept map 1 The science of learning and development

is stated early in Chapter 1, which also introduces the notion of the teacher's professional toolkit, as well as discussing the interplay between science and philosophical, critical analysis and the notion of critical realism that underpins this text.

The relationship between education and the science of learning is reinforced in Chapter 2 by focusing on the work of Donald Hebb (1949), a school teacher who became a distinguished brain scientist and laid down the foundations of synaptic plasticity that continues to underpin current research in the science of learning. Most of the learning occurring in schools today is Hebbian learning. Readers discover that those surrounding Hebb also made highly formative contributions to the science of learning, including Karl Lashley, Wilder Penfield (1958) and Hebb's doctoral student, Brenda Milner. Chapter 2 also provides an introduction to the basic anatomy of the brain, synaptic functioning, long-term memory and long-term potentiation (LTP).

Whereas Chapter 2 mainly refers to the science of *learning,* Chapter 3 focuses on the science of *development* and the major change that has occurred in developmental science over the past quarter of a century, with the introduction of dynamic systems theory (DST). This change has major implications for educational research and practice, not least because it challenges many of the theoretical assumptions that have guided education over the past 60 years, or so. The origins of DST in mathematics, chemistry, physics and meteorology are briefly explored as a means of introducing non-specialist readers to the core concepts of DST, including the idea that children are complex, emergent, self-organising beings.

Following the first three largely introductory chapters, Chapters 4 and 5 apply what has been learnt about the science of learning and development to two main educational concerns. Chapter 4 introduces the notions of working memory, executive functioning and metacognition, which are increasingly recognised as key to learning in schools. This chapter also provides readers with an introduction to research using the electroencephalogram (EEG). Chapter 5 focuses on the science of language, literacy and numeracy. The chapter details the origins of these fundamental areas of learning and locates where they occur in the brain and how they develop. The chapter ends with a discussion of how the science of learning and development is able to inform educational debate in the areas of literacy and numeracy. Picking up on the discussion of brain plasticity in Chapter 2, it also explores how plasticity is bringing new hope in remedying brain impairments such as dyslexia.

INTRODUCING THE SCIENCE OF LEARNING AND DEVELOPMENT IN EDUCATION

LEARNING OUTCOMES

By the end of this chapter, you will:

- Understand what is meant by the science of learning and development and start to appreciate its relevance for education research and practice
- Understand the relationship between theory and the interweaving of research and practice in education, and the importance of cross-disciplinary collaboration
- Be aware of quite recent paradigm changes in our understanding of learning and development, and know about some main conclusions of recent research
- Understand the importance of adopting a critical yet humanising approach to your studies in the context of education

Why the *science* of learning and development in education?

Everybody knows a lot about learning and development. We have all been learning and developing since before we were born, and we will continue until the day we die, so why do teachers (and parents) need to know about the *science* of learning and development? Because, strange as it might seem, it has all been happening in our brain, and we can only *know* about the brain by investigating it scientifically. All learning occurs in the brain and all development is related to changes occurring in the brain (Smith, 2009) – a brain that is massively complex, containing some 86 billion nerve cells (neurons) and just as many glia cells (Figure 1.1). Unlike neurons, glia cells do not produce electrical impulses; rather, they provide protection and support for neurons. Neurons are quite sparsely interconnected at birth, but they become massively interconnected in the early years, forming pathways that will be strengthened or weakened (or completely pruned) as the individual child learns and develops.

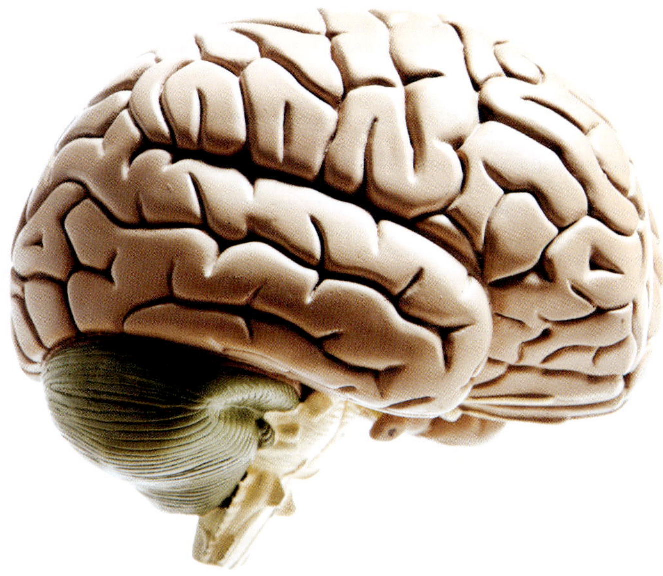

Figure 1.1 An adult human brain weighs around 1.5 kg with approximately 86 billion massively interconnected nerve cells (neurons) and just as many glia cells, which basically protect and support the neurons

You and your learning, developing brain

Who we are, as individual persons or 'selves', is largely a product of all that we have previously experienced on life's journey, starting when we were a foetus in the womb. But there is also a *constant interplay* between life's many and varied experiences and the genes that we inherited from our parents. Genes provide constraints on development; they ensured that when you came into the world and started to develop, you were on a human genetic trajectory and not the trajectory of a cat or dog. All of us are a product of the constant interplay between the physical impact on the brain of all that we experience, whether good or bad, and the influence of genes we inherited. Notice the emphasis on a *constant interplay* between our genes and experiences. This contrasts with the outdated claim that we are the product of a dichotomy that positions nature (genes) and nurture (experience) as opposites that are then said to somehow 'interact'. There really is no such dichotomy of opposites, though you will often find it repeated in textbooks.

That outdated dichotomy is partly based on a previous belief that genes are not impacted by environmental factors. Developments this century in the field of epigenetics show that idea to be false (Shenk, 2010). Genes are 'turned on or off' by environmental factors or 'markers' (Pembrey et al., 2006). The **nature/nurture dichotomy** can also lead to the false impression that somehow our genes are our destiny. If we have an unfavourable gene for some

nature/nurture dichotomy: a view of development that separates the biological and environmental influence and postulates how they are said to interact

dyslexia: an impairment with reading and spelling despite having the ability to learn. The 'central difficulty for a student with dyslexia is to convert letter symbols to their correct sound (decode) and convert sounds to their correct written symbol (spell)' (Australian Dyslexia Association, 2021).

homeostasis: 'the coordinated and largely automated physiological reactions required to maintain steady internal states in a living organism' (Damasio, 1999, p. 39)

interoception: the 'brain's representation of all sensations from your internal organs and tissues, the hormones in your blood, and your immune system' (Barrett, 2017, p. 56)

condition such as **dyslexia**, for example (explored further in Chapter 5), it has often been thought there is nothing that can be done to remedy the condition. But that is simply not the case (Geake, 2009), as we will discover in Chapter 2 when considering the therapeutic prospects that are emerging from current understandings of brain plasticity.

Another important point to note is that brains do not exist in isolation from everything else that makes you who you are, your body and your personal thoughts and feelings. Brains are intrinsically *embodied* (Immordino-Yang, 2016), biologically integrated within the body. Indeed, the primary role of the brain is to constantly monitor all that is happening in the body, keeping it all in metabolic balance, a process called **homeostasis** (Damasio, 2018), through a feedback sensing process called **interoception** (Barrett, 2017), which we examine in Chapter 11. Thus, the primary role of the human brain is not, as some philosophers seem to assume, to provide human persons with conscious self-awareness. The achievement of consciousness or conscious self-awareness is very much a Johnny-come-lately phenomenon in the evolutionary story of the brain, found not only in humans, but also in the higher apes, dolphins and elephants, and perhaps even capuchin monkeys (de Waal, 2009; Sankey & Kim, 2016).

So your brain is very much part of your body – it's an organ of the body, as are your heart and your lungs. Moreover, your embodied brain is also physically, socially and culturally *embedded* (Figure 1.2). For better or worse, as embodied brains we are inevitably situated in multiple physical and social environments that constantly impact what we do and who we become. As Linda Smith (2009, p. 82) notes in regard to learning and development, 'It's a brain in a body in a world that matters.' But please note that this is not a one-way process, because what we do and who we become impact the multiple environments in which we are situated.

Figure 1.2 The *embodied* and physically, socially culturally *embedded* brain. In understanding learning and development, it's 'a brain in a body in a world that matters'.
Source: Concept from Smith (2009, p. 82).

Finally, as you will learn throughout this textbook, brains are highly dynamic and complex, and they are also fragile – both physically and emotionally. A key message for teachers is that in working with children, we are working with embodied and embedded brains. It is a massive responsibility. And, because brains are physically and emotionally fragile, such work needs to be done with the greatest of care.

From conventional wisdom and craft knowledge to science

Teachers and parents are especially aware of learning and development. Although they can't see what is happening in children's brains, they certainly see it happening in the achievements, personalities and behaviours of the children in their care. Beginning with the explorations and discoveries in the early years of infancy, through the stabilities and growth of the childhood years and then the turbulences of puberty and adolescence, teachers and parents are encountering and constantly responding to the joys, challenges and worries of children's learning and development, on a daily basis (Figure 1.3).

However, although we may feel we know a lot about learning and development based on own experience and conventional wisdom, and even though teachers are especially aware of learning and development as a result of their professional practice and craft knowledge, there is a constant need to dig deeper by asking some difficult questions and trying to answer them. For example: *What are the processes involved in learning and development*, and *What is actually going on in the brain when students learn and develop?* Or, to make it up close and personal, *If all learning occurs in the brain and all development is a product of changes occurring in the brain, what's going on in your brain right now, and how is that contributing to your learning and development?*

Figure 1.3 What's happening in this child's head as she investigates the world, and what's going on in your brain as you read this book?

To answer these and similar challenging questions, we need to delve into the science of learning and development, but what does that mean, and shouldn't it be the *sciences* of learning and development? In using the singular term ('science') for the title of this book, we are emphasising *taking a scientific approach* to understanding learning and development in the context of education. In other words, we are placing an emphasis on scientific research. The two main areas of scientific research that are especially relevant to education, and which therefore form the focus of this text, are:

1. those aspects of brain science (especially the anatomy and functioning of the brain) that relate to learning and development, and
2. the science of complex, dynamic systems, or **dynamic systems theory (DST)** as it is known in developmental science.

dynamic systems theory (DST):
a theoretical framework in developmental science that focuses on the origins and self-organisation processes of novel functions

The teacher's professional toolkit

The purpose of this book is to provide teachers with an up-to-date and accurate account of some of what we now know about the science of learning and development, and its relevance to education. This book is written in the belief that the science of learning and development should be an *essential part* of every teacher's professional toolkit, along with and interwoven into subject expertise and pedagogical knowledge and skills.

The notion of a professional *toolkit* is a metaphor for the essentials needed to perform the professional role of the teacher. There may be many desirable things for teachers to learn in their professional education, but a sound and accurate understanding of how children and adolescents learn and develop should be professionally foundational. This book will be of considerable relevance to all teachers as they constantly strive to understand and teach students in their care, and also to the many parents who are interested in deepening their understanding of their children's learning and development.

REFLECTION

Using the metaphor of the teacher's professional toolkit, in addition to a sound and accurate understanding of how children and adolescents learn and develop, what other professional 'tools' do teachers need, as an absolute priority, to do their job effectively?

Theory and the interweaving of research and practice in education

Acquiring an understanding of how children and adolescents learn and develop might seem like a rather theory-driven enterprise; it has certainly tended to be so in the past. Books on the study of education seem to be inundated with theories of all kinds. However, ultimately education takes place in schools, in classrooms and other learning environments, and that should firmly link the study of *education* to *educational practice*. This book is not driven by theory, although it discusses some scientific theories; rather, it is motivated by a constant interplay between educational *practice* and scientific *research*, viewed as an *interwoven whole*.

Changing the emphasis from theory and practice to the interplay of *research and practice* contrast this text with many previous teacher education texts that embraced a '*theory into practice*' model. Student teachers, it was said, should first learn theory and then apply that *theory into practice* when planning lessons and addressing the pedagogical challenges they encounter in their classroom. If you think about it, this *theory into practice* model is a bit strange. It assumes that there exists a body of relevant theory that was generated independently of practice, but that somehow needs to be applied to educational practice and pedagogy. It seems to set up a dichotomy between theory and practice in order to knock it down. Surely, the most apposite educational theories are intimately related to practice and arise out of, and address, problems of practice. Or, to put it in a more straightforward way, aren't good teachers always *theorising* about their practice (Sankey, 1996)?

Another problem with the emphasis on 'theory' is that it deflects attention away from the key role played by research in the study of education. Theory often provides an important conceptual framework for conducting educational research but, arguably, what newly quali-fying and experienced teachers need most is a grasp of how evidence that is generated by ongoing research – including research into learning and development – relates to education. A holistic notion of the *interweaving of research and practice in education* is very compatible with a dynamic systems approach to learning and development – which, as you will learn shortly, isn't just one more theory to add to the list, but rather an overarching paradigm.

Education as a discrete science, informed by cross-disciplinary collaboration

A major problem with the **theory/practice dichotomy** in education is that it tends to position education as a discipline that is dependent on theories that arose when addressing problems in other disciplines, such as psychology or sociology. This carries the implicit assumption that education is not a science in its own right, but simply a derivative of other disciplines. If, on the other hand, the problems that arise in education are viewed as distinctively edu-cational, requiring educationally relevant research in order to address them, the idea that education should be viewed as a discrete science in its own right follows closely behind. Moreover, this view suggests that research conducted in relation to the science of education should be based in educational practice, solving educational problems. Ideally, it should be school based, using the natural setting of the school and classroom rather than a laboratory.

Nevertheless, disciplinary boundaries are a human construct and there is no need to create an additional dichotomy, separating the science of education from the remainder of science. Educational science should therefore work in cross-disciplinary collaboration with other sciences, particularly those related to the science of learning and development, incor-porating relevant insights and using the same technologies, even though the *research focus remains directed towards education* and resolving its most pressing problems. With regard to advancing knowledge and understanding about learning and development in education, it is therefore highly likely that research will increasingly use brain imaging technologies such **functional magnetic resonance imagining (fMRI)** and the **electroencephalogram (EEG)** (Figure 1.4). As you will discover in Chapters 4 and 10, EEG is very well suited to school-based research.

theory/practice dichotomy: a view in education that theory and practice are two distinct entities

functional magnetic resonance imagining (fMRI): a brain imaging technique that detects the changes in blood oxygenation and flow that occur in response to neural activity

electroencephalogram (EEG): a brain imaging technique that detects electrical activity in the brain using small metal discs (electrodes) attached to the scalp

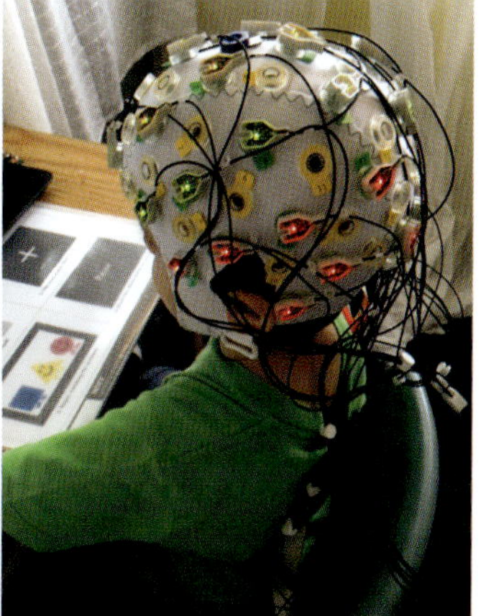

Figure 1.4 Two kinds of brain imaging – fMRI (left) and EEG (right)

The cognitive subconscious and other philosophical puzzles

In this text, we will also be interweaving the science of learning and development with philosophical issues, as they arise in context. Some of the scientific ideas we will encounter in this book raise very thought-provoking and educationally important philosophical questions that are relevant to teachers, especially as they relate to how we view children as learners. There are also important philosophical questions to be asked about the nature of science, the veracity (validity) of scientific evidence and what makes a scientific theory scientific. Moreover, Alva Noë (2004, p. vii) provided an important foundation for this text when he argued that science and philosophy are always closely interwoven: 'Philosophy flourishes in the midst of scientific research, not only because philosophical problems are in good measure empirical, but because scientific problems are in good measure philosophical.'

If science and philosophy are as closely interwoven as Noë suggests, then the ability to notice and articulate relevant philosophical questions related to the science of learning and development that arise in the context of education should also be part of each teacher's professional toolkit. As an example, consider the scientific research finding that much of our thinking, learning and development occurs below the level of conscious awareness, and therefore out of reach of conscious introspection. This is often referred to as the **cognitive unconscious** (Lakoff & Johnson, 1999), though this book uses the term **cognitive subconscious** (Sankey, 2006) because the word 'unconscious' could mistakenly imply a kind of coma, and that's completely the wrong idea.

cognitive unconscious: much of our thinking is operating below the level of conscious awareness and is therefore not available to conscious introspection

cognitive subconscious: an alternative term for the cognitive unconscious

To clarify the claim that much of our thinking is occurring below the level of conscious awareness, Lakoff and Johnson (1999, pp. 10–11) offer some examples of what is happening when holding a conversation:

- Accessing memories relevant to what is being said
- Comprehending a stream of sound as being language, dividing it into distinctive phonetic features and segments, identifying phonemes, and grouping them into morphemes
- Assigning a structure to the sentence in accord with the vast number of grammatical constructions in your native language
- Picking out words and giving them meanings appropriate to context
- Making semantic and pragmatic sense of the sentences as a whole
- Framing what is said in terms relevant to the discussion
- Performing inferences relevant to what is being discussed
- Constructing mental images where relevant and inspecting them
- Filling in gaps in the discourse
- Noticing and interpreting your interlocutor's body language
- Anticipating where the conversation is going
- Planning what to say in response.

The claim that much of our thinking is occurring below the level of conscious awareness has important implication for teachers in recognising that *the children we teach are both conscious and subconscious learners and decision-makers*. Teachers are therefore educating the conscious and subconscious understandings, beliefs and values of the children in their care

(Sankey, 2006). It also has implications for educational research because we *can't know about subconscious thought directly,* through reflection or introspection (Frith, 2012). Teachers should be wary of education research that uses methods such as self-reporting because it assumes that we do have direct access to our thoughts and feelings. The only way education researchers can access what is occurring in the brain beyond what children are consciously aware of is through brain imaging technologies, such as the EEG that you will encounter in this book, and fMRI.

Even more surprising, perhaps, some researchers have claimed that much of our decision-making is occurring in the brain *prior* to conscious awareness. We may think we are making a decision to act in a certain way, to say something or do something, but they argue that our brain has already decided! Could it be that our brain is quite literally faster than we think? The first experimental evidence that the brain might be deciding before conscious decision-making came from studies by Benjamin Libet, using EEG, which records the tiny electrical discharges occurring in the brain. More recent research supporting the results of Libet's experiment have used fMRI (Soon et al., 2008).

So imagine, as a teacher in school reprimanding a Year 9 student for breaking a school rule, what you would think if they replied, 'It wasn't me, it was my brain!' Or if you asked a student 'Why did you do that?' and they replied, 'I have no idea, it was below the level of my conscious awareness!' Would you agree and say, 'Yes, of course. I appreciate your brain is in control and much of your thinking and action occurs beyond the reach of your conscious awareness'? Or would you say, 'That's no excuse. Of course you know what you are doing and why you are doing it – please do as you are told'?

REFLECTION

If a child or adolescent said they didn't really know why they did something that was wrong, they could be speaking the truth. What do you think would be an appropriate response for a teacher to make in such a situation?

When hearing about the Libet experiment and its more recent versions, teachers need to be aware that, philosophically, this idea is highly controversial, not least because it seems to undermine the idea of free-will – the belief that as conscious agents we are free to choose what we do. To be fair to Libet, he did not agree that his result undermined free-will, but that is how it has usually been interpreted. Anyway, the point is that issues of philosophy certainly flourish in this area of research, to paraphrase Alva Noë (2004), cited above.

Whatever interpretation is given to the Libet experiment and its more recent versions, the subconscious nature of thinking and acting is now well established in science (e.g. Lakoff & Johnson, 1999), and it raises some very interesting philosophical questions about what we recognise as our human 'self' (*who* we are) and the learning, developing brain (*what* we are). For example, when we meet and greet other people, including children in our care, aren't we normally treating them as conscious 'selves', fully aware of their thinking and reasoning and in conscious control of their actions? But perhaps we are making a mistake: if much of human thinking is subconscious, we cannot be fully aware of it, which means *we often do not know why we do what we do* (Sankey, 2006). We will come back to this idea in Chapter 7, when we consider moral learning and development.

The cross-disciplinary relationship between brain science and education

brain plasticity: the ability of brains to be physically moulded and changed in response to experience

There's a historically long cross-disciplinary relationship between brain science and education. As you will discover in Chapter 2, one of the biggest ideas in brain science is **brain plasticity**, which basically refers to the way brains are physically moulded and changed in response to experience. *Without brain plasticity there would be no learning or development.* Brains have to physically change if they are going to learn and develop. Our current understanding of how brain plasticity works in the context of learning goes back to a Canadian schoolteacher in the 1930's, Donald Hebb, who was curious about what was going on in the heads of his students when they were learning. You will meet him in Chapter 2.

For now, we can simply note that the cross-disciplinary relationship between brain science and education was recognised a long time ago, so it's not a recent idea in education. Moreover, this cross-disciplinary relationship arises every time a teacher is curious about the *learning* brain and ponders what is going on in the brains of their students when learning. It also occurs every time an educational researcher uses brain imaging technology and draws on research findings across disciplines that are engaged with the science of learning and development when addressing relevant problems in the science of education.

RESEARCH LINK 1.1

Cross-disciplinary collaboration in the science of learning and development

Currently, there are a few organisations that pursue active collaboration between different disciplines, including biology, brain science, education and developmental sciences. Here are some examples. If you read their websites, you will see the ways in which cross-disciplinary collaboration is undertaken.

The International Mind, Brain and Education Society

The International Mind, Brain and Education Society (IMBES) aims 'to improve the state of knowledge in and dialogue between education, biology, and the developmental and cognitive sciences' (IMBES, 2018). It runs bi-annual conferences to foster dialogue between educational practitioners, scientists, public policy-makers and the public. Its official journal, *Mind, Brain and Education (MBE)*, publishes multidisciplinary original research that may touch upon important issues in the science of learning and development.

SIG 22 – Neuroscience and Education of the European Association for Research on Learning and Instruction

The European Association for Research on Learning and Instruction (EARLI) brings together researchers from the fields of education, cognitive and developmental science, and neuroscience. According to the SIG 22, 'taking interdisciplinarity as a basic principle, the SIG conceives the relation between educational research and neuroscience as a two-way street with rich bi-directional and reciprocal interactions between' them (EARLI, 2021). The SIG aims to 'provide forums to promote dialogue and collaboration between researchers in education and neuroscience' (EARLI, 2021).

Wellcome Trust's Education and Neuroscience Initiative

In 2014, the Wellcome Trust launched the Education and Neuroscience Initiative. This aims to 'develop and evaluate evidence-informed teaching and learning practices based on neuroscience' and to 'support teachers to better understand and access research on the science of learning, and use it to improve their practice' (Wellcome Trust, 2014).

Research link 1.1

Shaking the theoretical foundations

The relationship between education and child and adolescent *development* also goes back a long way. However, there is now growing awareness that we have entered a new, scientific era in regard to learning and development. Research has really progressed over the past 30 years or so and, as one might expect, recent research is challenging some of the old theories that have previous been influential in education.

The contribution of Piaget and Vygotsky

In the second half of the twentieth century, Jean Piaget (1896–1980) and Lev Vygotsky (1896–1934) were considered towering figures, producing theories of learning and development that were easily transferable to classroom teaching. Many of Piaget's ideas came from studying his daughter as she grew up. Piaget's stage development theory was dominant in many parts of the world, throughout the second half of the twentieth century, particularly in regard to early childhood and primary education, though it did not take hold in the United States, where behaviourism was the dominant theory. It remains influential in early childhood education in Australia.

Figure 1.5 Jean Piaget (1896–1980) – a towering figure in child development, in Europe in the mid- to late twentieth century

Piaget (Figure 1.5) emphasised children's self-initiated discovery in learning and development. Vygotsky (Figure 1.6), on the other hand, emphasised social and linguistic contributions. Pre-eminently, he believed children and adolescents learn and develop through social interactions and that their culture and language are the root of culture. From the 1970s onwards, as his ideas became better known in the West, Vygotsky became a major voice in the fields of psychology and education. Vygotsky's concept of the **zone of proximal development (ZPD)** proved particularly attractive to teachers. ZDP is the 'distance between the actual developmental level as determined by independent problem solving and the level of potential development as determined through problem solving under adult guidance or in collaboration with more capable peers' (Vygotsky, 1978a, p. 86).

zone of proximal development (ZPD): the putative 'distance between the actual developmental level as determined by independent problem solving and the level of potential development as determined through problem solving under adult guidance or in collaboration with more capable peers' (Vygotsky, 1978a, p. 86)

Figure 1.6 Lev Vygotsky (1896–1934) worked behind the Soviet 'Iron Curtain', so his ideas remained largely unknown in the West until long after his untimely death at 38 years of age

This provides a clear role for teachers in 'scaffolding' children's learning and development. However, from a developmentalist perspective, it doesn't explain the *process* of development. It is also associated with his now questionable view that animals 'cannot be taught (in the human sense of the word) through imitation, nor can their intellect be developed, because they have no zone of proximal development … animals are incapable of learning in the human sense of the term' (Vygotsky, 1978a, p. 88).

Historically, the work of Piaget and Vygotsky led the way in producing theories that seemed highly applicable to education, and both were convinced that the study of development should be a science. Vygotsky was able to draw on the work of Alexander Luria (1902–77), a leading Soviet brain scientist whose research investigated soldiers who had survived World War I with brain injuries. Luria's (1973) book, *The Man with a Shattered World*, is a classic that is still readily available in bookshops. Piaget, who trained as a biologist, said: 'Pedagogy is like medicine: an art, but one which is based – or should be based – on precise scientific knowledge' (cited in Dehaene, 2020, p. 237).

However, the theories of Piaget and Vygotsky had little or nothing to say about the brain processes involved in learning and development, which is hardly surprising considering when these writers were alive. Although important discoveries about the brain existed in their lifetime, much more has been learnt over the past 30 years or so, a period that has witnessed a massive increase in brain research worldwide. Largely as a result of using new imaging technologies, we now have a much more comprehensive understanding of the neurobiological processes associated with learning and development than these pioneers were able to draw upon for their research.

It is time to move on. Teachers are now able to re-evaluate and even push past much previous theory and embrace what we are finding out from the new science of learning and development. These are exciting times, and there are new things to learn and critical questions to ask about current theories of learning and development. Moreover, as this book will endeavour to show, it is entirely possible to engage with the new science of learning and development while retaining the important humanistic concerns that are a traditional hallmark of education as a caring profession (see especially Chapters 3 and 11).

Of course, it is not implied that we now have a complete scientific account of how human beings learn and develop – far from it. However, we are now on a scientific trajectory, which means we can apply the available technologies of science to collect empirical data that can test previous theory and develop new theories. If the history of science is anything to go by, that means we will never have a complete and settled scientific account of how human beings learn

falsification: the act of disproving a proposition, hypothesis or theory

and develop because scientific theory is always subject to **falsification** and change. (Popper, 1978).

Learning and development as dynamic systems

Changes in science often result from improved technology, and in brain science that includes the electron microscope and various measuring and imaging technologies, such as the EEG and fMRI. But there have also been theoretical shifts over the past quarter of a century, in particular the application of key insights drawn from dynamic systems theory (DST). DST, or complexity theory, has its origins in mathematics, physics and chemistry in the early to mid-twentieth century, before being systematically applied to development and learning in the mid-1990s.

In 1994, Esther Thelen and her colleague Linda Smith (Figure 1.7) in the United States produced a ground-breaking work, *A Dynamic Systems Approach to the Development of Cognition and Action* (Thelen & Smith, 1994). In the same year, in Europe, Paul Van Geert (1994) published his seminal *Dynamic Systems of Development: Change Between Order and Chaos*.

As you will discover, the application of DST to our understanding of learning and development brought some very big changes (Cantor et al., 2019; Kim & Sankey, 2010; Spencer et al., 2011). For much of the twentieth century, the main emphasis was on identifying and understanding the 'stages' and 'age phases' of development. We all know that children progress; four-year-old children can do a lot more than two-year-olds. So the big idea of twentieth-century child development theory was to categorise these putative 'stages' and 'age phases', then apply them to schooling – for example, making sure teachers don't introduce children to new ideas for which they are not *ready* because they haven't reached that stage or age phase yet (Brainerd, 1978) (Figure 1.8).

Figure 1.7 Linda B. Smith

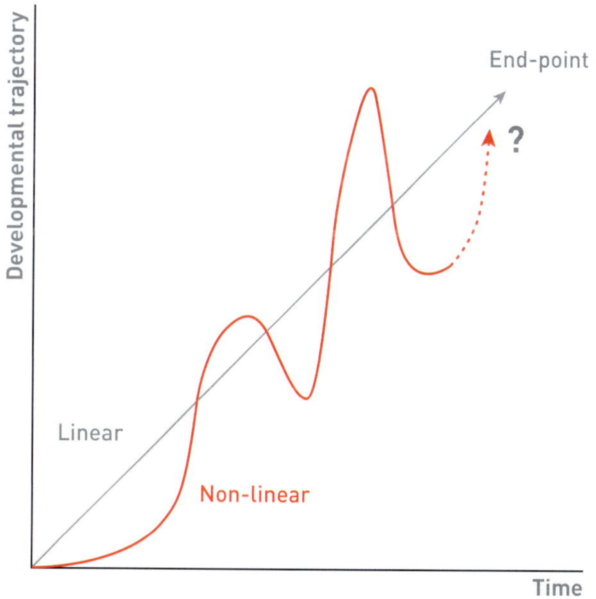

Figure 1.8 Does human development follow an onward and upward linear trajectory, or is it much more variable and non-linear?

It all appeared neat and predictable, one stage after another, onwards and upwards in a linear progression towards a known end-point we call adulthood. As noted, it also appeared highly relevant to classroom practice – these were ideas that teachers could use. However, a major problem with stage theories and age–phase theories is that they tend to provide over-simplified accounts of highly complex processes. They also over-generalise stability in development that is often highly variable and unstable in individuals. Moreover, at best there is a lot of overlap between putative stages.

The arrival of dynamic systems shook these neat, predictable and **linear** foundations, saying that child and adolescent development is **non-linear** and much more *'messy'*, more *variable and unstable*, than stage theory suggests. Moreover, there is no predetermined endpoint, no final stage of development. Partly, the limitations of stages theory come from the way research is conducted. Development may look neat and predictable at a low level of magnification, when using group-average, cross-sectional research methods, but at greater magnification and studying individuals over longitudinal time, these discrete stages and age-phases start to look very wobbly. This doesn't necessarily mean that stage theories are of no value in education, they can provide a *rough approximation* of how children develop, but from a dynamic systems perspective, that is all they provide, at best, a rough approximation.

linear: progressing from one stage to another in a single series of steps; or a straight line-like relationship between two variables

non-linear: a term referring to a relationship where the change of the outcome is not proportional to the change of the input

REFLECTION

What patterns do you think your own development has followed? Are you able to identify aspects of your development that followed linear stages? Which aspects of your development showed a non-linear pattern?

Logic and emotion, thinking and feeling

A second key foundational pillar of this textbook is the paradigm shift in our understanding of how rationality and emotion are interwoven in the brain (Figure 1.9). This is associated particularly with the seminal work of Antonio Damasio and his 1994 book *Descartes' Error* (see Chapter 6) (Damasio, 1994), although other researchers, including Esther Thelen and Linda Smith (1994), were also reaching similar conclusions. Since the onset of 'modern' science, in the seventeenth century, Western thought had drawn a hard and fast boundary between rational, logical thought and emotion. Emotion, we were told, was the antithesis of rationality – let emotion in and rationality would disappear. Damasio and others disagreed. His work with patients who had prefrontal lobe brain damage convinced him that if you eliminated emotion from human thought, it would cease to be rational; we need both, held together in balance.

Damasio was also fascinated by the way memory is laid down in the brain. There is no one place in the brain (no Cartesian theatre as it is sometimes called) where all

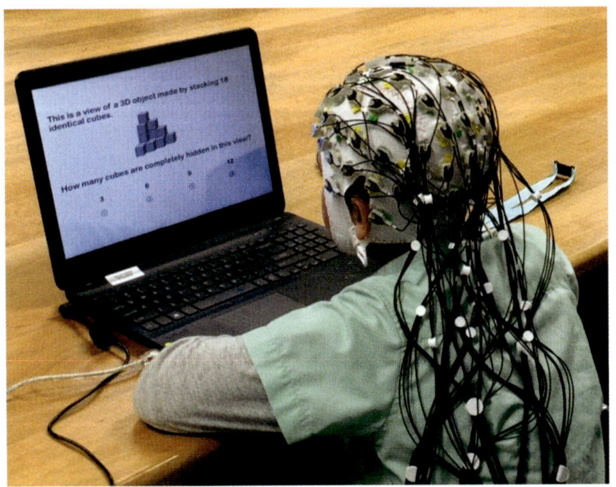

Figure 1.9 A nine-year-old student attempts a maths assessment question, but what emotions are coursing through his brain as he is thinking and do they impact his answers?

the memories that make us who we are come together. Memory is not stored as a whole; it is 'highly distributed' (Damasio, 1999, p. 220) across the brain, stored as parts in specialised areas of the brain. When we remember something, the various parts are drawn together and reconstituted as a whole in what Damasio (1994, p. 95) called 'a trick of timing'. This means that memories of a given event can vary over time and according to context; they are not as stable as many of us wish to believe (especially the weight lawyers place on *witness testimony*). And guess who was the first to suggest the idea of *distributed* memory? The Canadian schoolteacher Donald Hebb, who became a highly influential brain scientist.

Some main conclusions of recent research in the science of learning and development

Let's pause to list some of the main conclusions coming from research over the past 30 years into the science of learning and development, built on foundations laid by Donald Hebb and other pioneers you will meet in Chapter 2. An authoritative discussion of recent advances is provided in *How People Learn II*, published by the American National Academies of Sciences, Engineering, and Medicine (NASEM, 2018). The list provided below summarises what we have covered in this chapter with the sources – although, as you will discover in subsequent chapters, the relevant literature is very extensive. We now know the following information:

- Learning always occurs in the brain (Geake, 2009), including the learning of abstract concepts and bodily skills such as learning to grasp an object (e.g. Thelen & Smith, 1994) or how to pass the ball in sport.
- Brains are plastic (Doidge, 2007; Edelman, 1987; Hebb, 1949), which means they are constantly moulded and reshaped by the interplay of social, inter-social and cultural factors, including what children learn at school (Geake, 2009).
- Brains are embodied (Immordino-Yang, 2016; Lakoff & Johnson, 1999). They do not exist in isolation from the body, but are intimately involved in monitoring the workings of the body and its metabolism (Damasio, 2018).
- Embodied brains are also embedded in physical, social and cultural contexts, and are constantly responding to phenomena experienced in those contexts. Learning is shaped by both physical environments and social experience and, in turn, what is learnt often reshapes those environments (Dehaene, 2020).
- Much of our thinking and learning occurs below the level of conscious awareness and is not always available to conscious introspection and analysis. Action, resulting from past learning, often proceeds rational analysis and occurs subconsciously (Frith, 2012; Sankey, 2006).
- Perception, perceptual categorisation and memory are the main factors driving learning (Edelman, 1987, 1989; Thelen & Smith, 1994).
- Memory is not a retrieval of encoded mental representations; rather, it is always a reconstruction drawn from different areas of the brain, thus memories are not stable. Reconstruction varies over time and according to context (Damasio 1994; Hebb, 1949).
- Learning and brain development are intimately related. Both result from brain plasticity, where vast numbers of neural connections and networks are constantly shaped,

reshaped and pruned in response to stimuli from within the body and external physical and social environments (Edelman, 1987, 1989; Thelen & Smith, 1994).
- Each learner develops their own individually unique set of understandings and skills over the course of their life; these are shaped by a constant and dynamic interplay of multiple causes (seldom single causes) that are contextually situated (Smith, 2009).
- Each individual child's development is a dynamic progression, over time, and depends on a constant interplay of personal experience, biology, genetics, relationships and social, cultural contextual influences (Thelen & Smith, 1994; Van Geert, 1994).

The need for a critical and humanising approach in education

Perhaps we all know a lot about learning and development, but is it possible that what we know might be wrong? Could it be that some of what we know is based on folk wisdom or old theory that is no longer considered valid? Is it possible that what is stated as 'fact' in educational textbooks and/or government policy documents ought to be viewed with a degree of scepticism because they just might be wrong?

The key message is to please bring a critical attitude to your study. Be prepared to question what you already know and what you are learning. In other words, in studying the science of learning and development, adopt a scholarly approach, combining the thirst for knowledge and the joy of discovery with healthy critical thinking. And remember that it is not enough to say you think something is wrong. You need to clearly and thoughtfully say *why* it is wrong and, if possible, test your assumptions as rigorously as you can.

Taking a *sceptical* approach to established 'fact' is a very scientific position to adopt. It has its origins in the early years of modern science in the seventeenth century and the publication, in 1661, of a book called *The Sceptical Chemist* by Robert Boyle. Still remembered for Boyle's Law, Boyle was one of the founding members of the prestigious Royal Society of London for Improving Natural Knowledge, which was especially concerned with establishing the basic principles of science. Boyle urges scepticism and a dogged pursuit of reproducible results, moving beyond speculation to focus on the evidence of experimentation. The subtitle of his book is 'Doubts and Paradoxes'. What he was doubting was the many paradoxes in the ideas about chemistry in the seventeenth century, including the beliefs of the alchemists.

Earlier in this chapter, we referred to Antonio Damasio's (1994) ground-breaking book, *Descartes' Error*. The particular error to which Damasio was referring was Descartes' view of human rationality, which separated human rational thought (conceived as 'logical') from human emotion. Descartes believed, as so many others have, that if we let emotion into human rationality, it ceases to be rational. Damasio is not discounting logic; rather, he questioned the idea that human rationality can be modelled on logic, as Descartes supposed.

René Descartes (1596–1650) (Figure 1.10) was a French mathematician, philosopher and scientist who had a profound influence on much European thinking throughout the Enlightenment and until quite recently. In mathematics, he is known as the founder of analytic geometry. He is also often referred to as the father of modern philosophy. Descartes published his main philosophical ideas in two books, *Discourse on Method* (1637) and *Meditations on First Philosophy*. Philosophy, he believed, should be firmly grounded on the indisputable rational facts of mathematics. Postmodernism, which emerged in the late twentieth century, is in many ways a rejection of Descartes' science and philosophy and all that flowed from it during the Enlightenment.

Descartes was troubled by all the uncertainty in the seventeenth century, following the Reformation and the onset of modern science, including the **Copernican revolution**, which shattered the idea that the Earth is at the centre of the universe. Descartes wondered, in the face of such uncertainty, whether there was anything he could be absolutely certain about. He considered many options, including the evidence of his sensory experience – surely that was reliable. Evidence was driving the new science, but then he realised his senses could be tricking him. In his *Discourse on Method* (1637) and *Meditations on First Philosophy* (1641), he reasoned that the only thing he could be certain of is that he was *thinking*, and if he was thinking, then he must actually exist. So (in Latin) *Cogito ergo sum*, 'I think, therefore I am'. But is that correct? From the perspective of modern-day brain science, shouldn't it be: 'I am, therefore I think'?

Figure 1.10 René Descartes (1596–1650), outstanding French mathematician, scientist and philosopher, whose ideas were highly influential in the European Enlightenment and through to the twentieth century

Descartes is also renowned for his advocacy of mind/body dualism, separating the mind (or soul) from the brain. He viewed these as two utterly different substances. You will read more about this in Chapter 4, as this is an important philosophical issue that still arises within the science of learning and development. For now, we can note that Descartes firmly believed that the world and everything in it was like a machine, an intricate clock. Animals are machines and human beings are machines – except that seems to be wrong. Human beings aren't just machines, are they – aren't they pre-eminently minds or souls? Hence Descartes' advocacy of what is called **mind/body dualism**. The body, he believed, is *material and mechanical*, the mind or soul is not material and is *spiritual*. In many ways Descartes dilemma is still with us today, whenever mechanical analogies are used to describe animals and human beings.

Copernican revolution: a major shift in astronomy from an Earth-centred view of the universe to a Sun-centred view, as proposed by Polish astronomer Nicolaus Copernicus in the sixteenth century

mind/body dualism: a philosophical view that the mind is non-physical, and the mind and body are distinct substances or properties

REFLECTION

What do you think? Do our senses provide certainty and is the only thing you can be certain of is that you exist? And is the body mechanical and material, and the mind spiritual and non-material? Does this seem silly, or was Descartes saying something very important?

If we are to understand Descartes, we need to view him in the context of his own time and place. Very briefly, as historian of science Hugh Kearney (1971) pointed out many years ago, the scientific revolution, which included the Galileo episode, involved the relative fortunes of three competing Greek traditions, each having its own approach to observation and experiment: the Aristotelian (organic), the Hermetic (magical) and the Archimedean (mechanist) traditions.

In the battle of competing scientific traditions, victory would eventually go to the mechanist tradition – which, as we will see later, would dominate science until very recently. Descartes was not only a firm believer in the mechanical philosophy or tradition; he was arguably its leading advocate in the seventeenth century, along with Galileo (1564–1642), his slightly older contemporary. Many of the key assumptions of the mechanical philosophy of Descartes and others have been challenged and replaced by **complexity science** or dynamic systems theory.

complexity science: the study of dynamic, complex systems and how they give rise to emergent, dynamic, self-organising behaviours

Critical realism, naïve realism and constructivism in science

Is the world described by science *real* or is it purely a human *construct*? The contrast is between *scientific realism and constructivism*, but these two terms also need unpacking.

To come straight to the point, the position adopted in this text is often referred to as **critical realism** (Bhaskar, 1978). This contrasts strongly with what philosophers and social psychologists refer to as **naïve realism**, which is the belief that our senses are able to give us direct access to the world, as it objectively exists – that we perceive the world exactly as it is. Critical realists do not accept that, believing that what we perceive is mediated by our human senses and is interpreted using past experiences, present expectations and beliefs. Critical realism therefore *has a strong constructivist element*. It fully recognises that knowledge is a human and social construct, although it contrasts with strict constructionism or constructivism that has been a dominant force in education for the past half-century or more.

critical realism: accepts that science is a human construct, but insists scientific claims are informed and constrained by the nature of the world

naïve realism: the view that our senses are able to give us direct access to the world as it objectively exists; we perceive the world exactly as it is

According to both constructivism and critical realism, 'you construct the environment in which you live' (Barrett, 2017, p. 83). As an example, consider a tree falling in a forest. If no one is there to witness it, does the falling tree make a *sound*? The answer is that it *doesn't make a sound* because sound requires receptors (mammalian ears) and without ears what the tree emits is vibrations in the air (changes in air pressure). 'Without a perceiver there is no sound, only physical reality' (Barrett, 2017, p. 129). Barrett further explains that, 'Changes in air pressure and wavelengths of light exist in the world, but to us they are sounds and colours. We perceive them by going beyond the information given to us, making meaning from them using knowledge from past experience, that is concepts' (Barrett, 2017, p. 130). One way of saying this is that *the world exists as a reality, but is not given as a reality*. We have to perceive, categorise and interpret it, in order to know it.

This statement would be accepted by most practising scientists nowadays, and it is entirely consistent with both critical realism and constructivism. However, many constructivists go beyond this and adopt an *anti-realist* position that there is no real world, only a world of human constructions. This is not acceptable to critical realists (and the majority of scientists), who insist that scientific claims 'are informed, constrained and enabled in a non-trivial way by the world or reality' (Scott & Bhaskar, 2015, p. 63). You can construct knowledge as much as you like, but in the end reality (the way the world is) has its say. Jump from the top of a 20-storey building and you will likely die, for very sound scientific reasons. Reality inevitably constrains human constructions, which have to be *consistent* with the real world.

REFLECTION

Think about what you have just read and, if possible, discuss it with your classmates. Do you wish to affirm what is called naïve realism – that science provides a description of the world as it actually is? If not, what reasons would you give for preferring either critical realism or constructivism.

The problem of neuromyths in education

One major area of misunderstanding in education that has been widely researched is the prevalence of **neuromyths** (Howard-Jones, 2014). Neuromyths 'are mistaken beliefs about the human brain and its development, some based on a very loose extrapolation of sound scientific theory to unwarranted conclusions, others simply resulting from misinformation or misunderstanding' (Kim & Sankey, 2018). You may have heard, for example, that 90% of brain development occurs with the first three years (or five years, according to some versions) (Bruer, 1999). But it's wrong, it is based on a confusion between the rapid growth of neurons in the early years (neurogenesis) and the totality of brain development (Gogtay et al., 2004). See 'Synaptic Signalling and the Learning, Developing Brain' in Chapter 2.

neuromyth: mistaken beliefs about the human brain and its development (Kim & Sankey, 2018)

In another example, this time related to learning, some 97 per cent of student teachers at one university in Australia believed that, 'Individuals learn better when they receive information in their preferred learning style, e.g. visual, auditory, kinaesthetic' (Kim & Sankey, 2018). They are wrong. Certainly learners may have preferences, but there is no evidence that receiving information via that preference improves learning (Pashler et al., 2008; Rogowsky et al., 2015).

Are you surprised that this is a neuromyth – have you previously been told that individuals learn better when they receive information in their preferred learning style? The first thing to note is that it is not only student teachers in Australia who believe that this is correct; studies have shown that neuromyths are pervasive across different parts of the world and that this particular neuromyth is believed by teachers at levels quite close to those Australian students (Ferrero et al., 2016). Not surprisingly, when probing the origins of this neuromyth, almost 50 per cent of respondents said they learnt it from their teachers when at school, and 5 per cent said they learnt it in university lectures (Kim & Sankey, 2018)!

So how about a quick general knowledge brain quiz? Consider the statements in Table 1.1 and decide whether you believe they are true or false. There are no trick questions, but do read the wording carefully. A hint: 10 statements are true.

You can find the answers to this quiz in the online resources. You can find out more about the prevalence of neuromyth in education by doing the research activity for this chapter. In the readings for the research activity, you will find discussion of these neuromyths. One

Table 1.1 Brain quiz

1	There are sensitive periods in childhood when it's easier to learn things.
2	Skipping breakfast before going to school does not adversely affect students' learning.
3	Newborn babies prefer to listen to the language they heard while in the womb.
4	Normal development of the human brain involves the birth and death of brain cells.
5	Ninety per cent of brain development occurs in the first three years of life.
6	Production of new connections in the brain can continue into old age.
7	Vigorous exercise can improve mental function.
8	Short bouts of coordination exercises can improve integration of left and right hemispheric brain function.
9	When a brain region is damaged, other parts of the brain can take up its function.
10	When we sleep, the brain basically shuts down.
11	Information is stored in the brain in a network of cells that are distributed throughout the brain.
12	Differences in hemispheric dominance (left brain, right brain) can help explain individual differences among learners.
13	Brain development has finished by the time children reach the end of secondary school.
14	Learning occurs through modification of the brain's neural connections.
15	Mental capacity is essentially hereditary and cannot be changed by the environment or experience.
16	Environments that are super-rich in stimulus will improve the brain of pre-school children, when compared with normally rich environments
17	The visual input received by the right eye is sent only to the left hemisphere of the brain and vice versa.
18	Students taking lecture notes by hand will generally remember more and have deeper understanding of the contents of the lecture than those taking notes on a laptop.
19	The main areas for language acquisition are normally in the left hemisphere of the brain.
20	It has been scientifically proven that fatty acid supplements (omega-3 and omega-6) have a positive effect on academic achievement.

Brain quiz answers

of the studies cited for that activity (Kim & Sankey, 2018) investigated neuromyths believed by first-year undergraduate student teachers in an Australian context. The research reported in that article was based on these questions and provided their source.

Where's the human heart in all this?

In this chapter, you have been introduced to the science of learning and development, but is there something missing in all this? Are we missing something essentially human? At the end of a formal public lecture on the brain and learning, in the context of education, the session was opened up for questions. In addition to the presenter, other experts joined the stage.

There were a number of questions on the detail presented in the lecture, but then one member of the audience asked, 'Where's the human heart in all this?' One of the experts on the stage suggested that perhaps they needed to fetch a 3D printer. Presumably the aim of this reply was to trivialise the question: the human heart is not one of the structures needing to be considered in a presentation on the brain and learning.

Well, not literally of course, but the inappropriateness of that reply is that clearly the person asking the question was using the heart as a well-known metaphor. What she was actually pointing out was that the human dimensions of thinking, feeling and learning were missing in what the lecture had said about learning. It seemed to be 'throwing out the baby with the bathwater'. From experience, the idea that the science of learning and development has to be dehumanised in order to make it scientific helps to explain why some education students and teachers are sceptical about its relevance to education, along with the idea that it is just a recent fad that can't really be applied directly in the classroom, noted at the start of this chapter. In a study of teachers' attitudes to neuroscience, Debby Zambo (Zambo & Zambo, 2011) found three kinds of teacher responses. There are many who wholeheartedly embrace neuroscience; there are others who hold reservations (not sure of its value); and are still others who see no use for it at all – often for reasons similar to those just mentioned.

Having reached the end of this chapter, it is hoped that you have been able to embrace the science of learning while also fully appreciating the human dimensions. Underlying the concerns about throwing out the baby with the bathwater is a well-known philosophical issue called *reductionism,* also mentioned by Zambo (2013) as a stumbling block. We will take that issue up in Chapter 3, in the process of introducing you to dynamic systems theory. But for the moment, be assured that the approach taken in this book is very mindful that the science of learning and development *can* include the human heart and indeed *must* include it when the science is applied in the context of education.

The brain that changes itself

RESEARCH LINK 1.2

You can read these and other stories of restorative brain plasticity yourself, by going directly to Norman Doidge's (2007) book *The Brain That Changes Itself.*

These two stories are also retold in a documentary of the same name, available on YouTube, accessible through the online resources for this chapter.

If you watch the video of Barbara Arrowsmith-Young, you will see something of her work with children that is aimed at restoring what she calls their cognitive capacity. But see whether you can identify the neuromyth behind children wearing eye-patches. This was once really quite popular in remedial centres, though it was based on a neuromyth (statement 17 in the brain quiz, Table 1.1). However, pointing this out is not intended in any way to undermine her own remarkable story of recovery as a result of brain plasticity, or the exceptional work she is doing with children.

Research link 1.2

APST Standard 1.2 Understand how students learn

Demonstrate knowledge and understanding of research into how students learn and the implications for teaching

APST Standard 6.2 Engage in professional learning and improve practice

Understand the relevant and appropriate sources of professional learning for teachers

ACECQA curriculum specifications

1.1 Learning, development and care

6.3 Professional identity and development

This chapter invites newly qualifying teachers to critically review what they currently know and understand from research about how students learn and the implications for teaching. It introduces the new science of learning and development as a touchstone to be used by teachers in deciding what are relevant and appropriate sources of professional learning for teachers, given the many advances that have occurred over the past 30 years. Those who study this chapter should be able to re-evaluate and even push past much previous theory in light of research emanating from brain science related to learning and development, and insights drawn from dynamics systems theory.

This chapter is also concerned with what it means to be a professional teacher, working within the field of education. The chapter takes a high view of education as a science in its own right, concerned with distinctly education problems, yet able to work collaboratively across disciplinary boundaries. This chapter illustrates how cross-disciplinary engagement with other disciplines, such as brain science, can help teachers' own professional learning and development. Nevertheless, though positioning education as a discrete science and introducing the reader to some of the main conclusions coming from the science of learning and development that are especially relevant to teachers, it also emphasises that education is, by its very nature, a caring profession.

SUMMARY

- The brain is *embodied* (biologically integrated within the body) and *embedded* in multiple physical, environmental, social and cultural contexts. With regard to learning and development, it is the brain in the body in the world that matters.
- The view adopted in this text is that the science of learning and development should comprise an educationally relevant introduction to both those aspects of brain science that relate to learning and development, and the science of complex systems, or dynamic systems theory (DST) as it is known in developmental science.
- The chapter argues that the science of learning and development should be an essential part of every teacher's professional toolkit along with, and interwoven with, subject expertise and pedagogical knowledge and skills.

- Much previous teacher education embraced a *theory into practice* model, but this is problematic for a number of reasons. One important reason is that in emphasising theory, it deflects attention away from the key role that research plays in the study of education. Newly qualifying teachers and practising teachers need a grasp on research as it relates to education.
- Education science is arguably a discrete science in its own right, but it also needs to work in mutual cross-disciplinary collaboration with other sciences, particularly those that conduct research into the science of learning and development.
- Much of our thinking, learning and development occurs below the level of conscious awareness, and therefore out of reach of conscious introspection. This is known as the 'cognitive unconscious' or 'cognitive subconscious'. Teachers should be wary of education research that uses methods such as self-reporting because it assumes that we have direct access to (can directly introspect) our thoughts and feelings.
- Brain plasticity, which is one of the big ideas in brain science, refers to the way brains are physically moulded and changed in response to experience. Without brain plasticity, there would be no learning or development. Moreover, brain plasticity is now being used to remedy brain-based impairments, such as dyslexia.
- In the twentieth century, 'stage' and 'age-phase' theories of learning and development were built on the core idea that a child's progress is largely predictable, one stage after another, onwards and upwards in a linear progression towards a known endpoint: adulthood. However, such theories tend to provide an over-simplified account of highly complex processes. At best, they offer only a rough approximation of development.
- In developmental science, there have been a number of important theoretical shifts over the past quarter of a century. One major shift has been the application of complexity or dynamic systems thinking to learning and development. In particular, a dynamic systems account challenged the idea that development is linear and predictable by instead portraying child and adolescent development as mainly non-linear and much more unpredictable, unstable and messy than stage theory suggests.

KEY POINTS FOR TEACHERS

- The distinguished twentieth century educator and developmentalist, Jean Piaget, saw a close connection between the practice of teaching and medical practice. Pedagogy, he said, is like medicine: an art, but one that is based – or should be based – on precise scientific knowledge.
- Instead of viewing educational practice (pedagogy) as founded on very many seemingly unchanging (and often incompatible) theories and speculative beliefs regarding learning and development, try to inform your practice by ongoing research and sound scientific evidence.
- In the context of education and pedagogy, teachers should view each child and adolescent as an inherently embodied brain, nested in a world with multiple learning and developmental environments.

- In working with children, we are physically changing children's brains. It is a massive responsibility and, because brains are physically and emotionally fragile, it needs to be done with the greatest of care.
- When working with children and adolescents, try to keep in mind that much of their thinking and decision-making is occurring subconsciously. Teachers are therefore educating both the conscious and subconscious understandings, beliefs and values of children in their care.
- Stage theories of learning and development were very popular in the twentieth century, but although children do seem to go through 'stages' and 'age-phases', teachers should be careful not to use that in assessing children's learning and development, which is much more complex and non-linear than stage theories suggest.
- When applying scientific knowledge to educational practice, take care to be accurate. Make sure you are not misunderstanding the science or using it to come to unwarranted conclusions. Also, in studying learning and development, adopt a scientific, scholarly approach, combining the thirst for knowledge and the joy of discovery with healthy critical thinking.
- The brain is plastic, moulded and changed in response to experience, which is why teaching works. Plasticity also implies that children's genes are not their destiny. Genes are turned on or off by environmental factors, and this has important implications for the treatment of genetically based learning difficulties encountered in education, such as dyslexia.

REVIEW QUESTIONS

Guided responses

1. What did this chapter suggest are the two main components of the science of learning and development that are especially relevant to education?
2. What is meant by saying that your brain is intrinsically embodied and also physically, socially and culturally embedded – that so far as learning and development is concerned, it is a brain in a body in a world that matters?
3. What is meant by the term 'cognitive subconscious'? Why is this important for teachers?
4. What is meant by the term 'neuromyth'? Give two examples of neuromyths that are prevalent in education and say why they count as neuromyths.
5. Why is it necessary to adopt a critical but also humanising approach when studying the science of learning and development in the context of education?

FOOD FOR THOUGHT

1. Thinking back over your own education in school, do you think some of your teachers had a good working knowledge of brain science and the dynamic systems approach to development as part of their professional toolkit? What key items do you think it contained?
2. In this chapter, it was suggested that in teacher education, there should be a shift away from thinking about *theory and practice* to instead focusing on *research and practice*. Do you think focusing on research evidence rather than theory would make a difference to

how you view the relevance of an academic study of education to your practice as a school teacher? Give reasons for your response to these issues.

3. Based on what is said in this chapter, that much of our thinking is occurring below the level of conscious awareness and we are therefore not able to access it, do you think this changes the way you view children and their learning? For example, when teaching children, are you impacting their subconscious as well as conscious thinking? Does this make any difference to how you view the role of the teacher?

RESEARCH ACTIVITY: NEUROMYTHS IN EDUCATION

As you learnt above, neuromyths are 'mistaken beliefs about the human brain and its development, some based on a very loose extrapolation of sound scientific theory to unwarranted conclusions, others simply resulting from misinformation or misunderstanding' (Kim & Sankey, 2018; Sankey & Kim, 2016; Howard-Jones, 2014; OECD 2002, 2007).

One of the most prevalent neuromyths in education is known as 'VAK learning styles' – that is, individuals learn better when they receive information in their preferred learning style (Kim & Sankey, 2018). There are also other neuromyths that are common in education, including 'left/right brain learners' and what is often called 'brain gym'.

You can read about the most prevalent neuromyths and why they are wrong in the suggested readings below.

- Australian undergraduate students' beliefs in neuromyth: Kim, M. & Sankey, D. (2018). Philosophy, neuroscience and pre-service teachers' beliefs in neuromyths: A call for remedial action. *Educational Philosophy and Theory*, 50(13), 1214–27.
- Cross-national variation in neuromyths: Ferrero, M., Garaizar, P. & Vadillo, M. A. (2016). Neuromyths in education: Prevalence among Spanish teachers and an exploration of cross-cultural variation. *Frontiers in Human Neuroscience*, 10, 496.
- OECD Centre for Educational Research and Innovation (CERI)'s resource: OECD (2007). *Understanding the brain: The birth of a learning science*. Paris: OECD Publishing.

Research activity links

As a research activity, create your own survey instrument with 20 items, and a balance of true and false statements, making sure you include the most prevalent neuromyths. You can use the statements in the quiz (see Table 1.1) and in the four suggested readings above. Then try conduct a mini-research survey using the instrument with a group of at least 20 family members and friends who are not studying brain science.

In designing your survey, include a space after each question where the person answering can indicate the source of the answer they gave, by choosing one of eight possible options. This is similar to the survey conducted for the first suggested reading above (Kim & Sankey, 2018), though that survey only used six options (schoolteachers, TV programs, books or magazines, university lectures, peers, others). Select options that are likely sources, given the backgrounds of the participants you intend to use it on. After conducting the survey, use the Kim & Sankey reading as a guide when setting out your results.

In addition, for all the **false** statements, prepare a short debriefing information sheet that will be given to participants after the survey, explaining why those statements are false.

HEBB'S POSTULATE AND THE LEARNING BRAIN

LEARNING OUTCOMES

By the end of this chapter, you will:

- Know about the foundational neuroscientific work of Donald Hebb, how he transitioned from being a teacher and school principal to becoming a distinguished brain scientist and what is meant by Hebbian learning and Hebb's postulate
- Know some basic brain anatomy, including cortical, sub-cortical and limbic structures, and understand Santiago Ramón y Cajal's neuron doctrine
- Understand what is meant by an action potential, a synaptic potential and the synaptic plasticity hypothesis
- Know of Brenda Milner's discovery of two forms of memory (declarative and non-declarative), the plasticity of memory, how memory impacts who we are and how brain plasticity is offering hope for the future in overcoming dyslexia and other developmental deficits

Donald Hebb: The schoolteacher who made up his mind

You may sometimes hear critical voices saying that studying brain science in education is taking things too far – it's just a recent fad that can't really be applied directly in the classroom. Teachers don't need to know about the brains of the children and adolescents they teach and, furthermore, brain science in education is full of mistaken ideas (neuromyths), a problem you studied in Chapter 1. One well-known source, often cited in support of this critical attitude, is an academic article by American psychologist John Bruer (1997) titled 'Education and the Brain: A Bridge Too Far'. It was a catchy title and it has appealed to many who want nothing to do with brain dynamics in the context of education.

Bruer's argument was that the complexities of brain function are too far removed from children's actual behaviours to be of any practical use to teachers, but is that actually correct and why this emphasis on 'behaviour', what about children's learning and development? In fairness to Bruer, his paper was published a quarter of a century ago, and many advances in our understanding of the brain and the complexities of human brain development have occurred since then. Nevertheless, Bruer's comment and the rather negative attitude it supports is in stark contrast to Donald Hebb's belief, that we need to study the brain if we are to understand how children learn. That was half a century before Bruer and Hebb was a schoolteacher, so the application of brain science to education is not a recent fad.

In fact, Hebb was a very successful and innovative schoolteacher of English, and then distinguished principal of a school in Montreal, Quebec, Canada. He was fascinated to know what was happening in the brains of his students, when they were thinking, learning and memorising. He later became one of the most distinguished brain scientists of the twentieth century and **Hebb's postulate** (hypothesis) (also known as Hebbian learning) remains foundational in neuroscience to this day. You will find his name cropping up throughout this book, as most learning in school is 'Hebbian learning'. It explains 'what teachers have long known: that repetition is necessary for effective and reliable learning' (Geake, 2009, p. 54). So, in studying the science of learning and development in education, you are following in the footsteps of a very innovative and much-loved teacher and a distinguished brain scientist.

Donald Hebb was born in 1904, in the town of Chester in Nova Scotia, Canada. Both his parents were medical doctors. His mother was very interested in the progressive educational ideas of the Italian doctor Maria Montessori (1870–1952), so she home-schooled Donald until he was eight years old. This appears to have given him a very good start – when he started school, he was put a year ahead of his age group and he did well in elementary (primary) school. In high school, though, his rebellious attitude and disrespect for authority held him back and he failed Grade 11 the first time he completed it. When he subsequently passed, he entered Dalhousie University to study English, with the aim of becoming a novelist.

From an educational point of view, it is interesting to pause and think about Hebb's own learning experience. This is the story of a child and adolescent who is going to become a famous and distinguished brain scientist. He is going to lay down some very important foundations for understanding the processes involved in learning and memory, related to what is called **brain plasticity**. But you would hardly think so, given his high school and undergraduate track record. The point is that his story should be a reminder to all teachers that we can never be sure about how any child's life-trajectory might change, and how they might excel

Hebb's postulate (or Hebbian learning): Donald Hebb's theory of learning by association, when adjacent cells are excited simultaneously and repetitively, thus forging a connection that is strengthened through further repetition

brain plasticity: the ability of brains to be physically moulded and changed in response to experience

in future. And one of the main reasons is the plasticity of the brain – the way brains are physically moulded and changed in response to experience, which Hebb's postulate explains.

RESEARCH LINK 2.1

Neuroscience Education for Prekindergarten–12 Teachers

Dubinsky, J. M. (2010). Neuroscience education for prekindergarten–12 teachers. *Journal of Neuroscience*, 30(24), 8057–60.

Although, as you have read above, some academics believe that the findings of brain science studies are too far removed from the classroom context to be able to effectively inform practice (e.g. Bruer, 1997), other studies (e.g. Dubinsky, 2010) show that teachers believe they need to understand neuroscience, and that those who have received brain science training have found it enhances the quality of their classroom teaching.

A well-known example of an introductory brain science course, tailored to the needs of practising teachers, is the BrainU Program (brainu.org) directed by Dr Janet Dubinsky, a neuroscientist at the University of Minnesota. In this program, teachers in the United States are introduced to big ideas in brain science (such as those introduced in this text) and attend hands-on activities and experimentations.

A follow-up study of teachers who attended a two-week program discovered that external observers found the quality of learning and student engagement of a trained teacher's classroom is better than that of teachers who did not receive neuroscience training. Also, the teachers who participated in the training showed increased knowledge of brain science and confidence in applying it, compared with those who did not participate in the training (Figure 2.1).

Research link 2.1

Read Janet Dubinsky's (2010) article. Notice in particular what the research team observed in regard to teachers' performance in the classrooms. In what ways did the teachers improve as a result of neuroscience training?

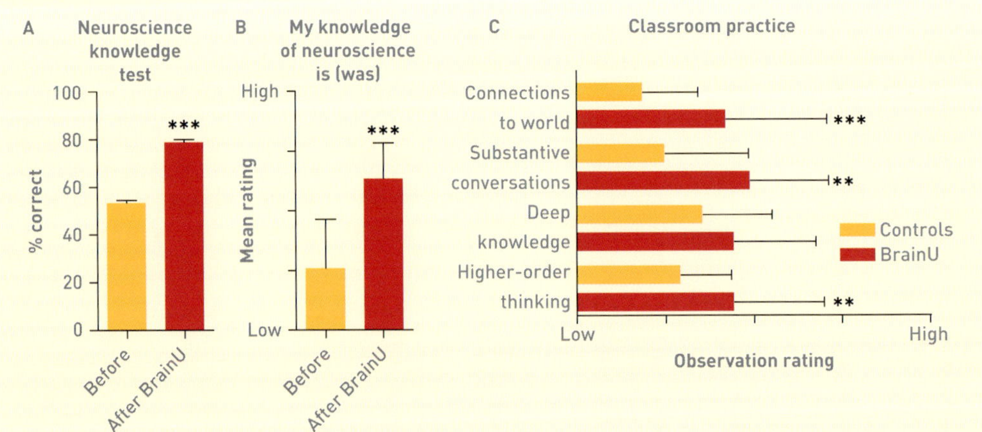

Figure 2.1 Changes in teacher neuroscience knowledge (A), confidence in that knowledge (B) and quality of pedagogical practices observed by trained assessors (C) as a result of attending a single two-week introductory BrainU workshop
Source: Dubinsky (2010, p. 8058).

From innovative teacher to leading brain scientist

So how did Donald Hebb (Figure 2.2) make the transition from teacher of English and school principal to be a pioneering brain scientist? Actually, that's a very good learning and development question. Part of the answer has to be hard work and determination, but we should also consider those who influenced him, those whose work and discoveries provided a *scaffold* for his own learning. It all began for Donald Hebb when working as a school principal. He enrolled for a part-time master's degree in psychology at McGill University, Montreal and was awarded the degree in 1932.

Two years later, disillusioned by the constraints of curriculum policy in Quebec, he left the teaching profession and Canada to undertake a PhD candidature under the supervision of a distinguished psychologist Karl Lashley (1890–1958). Lashley's influence greatly enhanced Hebb's learning and academic development. (If you ever consider extending your education by undertaking doctoral studies, choose a good supervisor to scaffold your learning.)

Figure 2.2 Donald Hebb (1904–85)

After gaining his PhD in 1936, Hebb returned to Montreal to work with the highly innovative neurosurgeon Wilder Penfield (1891–1976) (Figure 2.3) at the Montreal Neurological Institute. Penfield pioneered a clever surgical technique for mapping the brain, which helped to reduce the chance of unintentionally damaging functionally adjacent areas of the brain when removing tumours or areas of the brain that were causing epileptic seizures.

Before doing the operation, with the patient fully conscious on the operating table (using only local anaesthetic), he delicately touched an electrode on nearby areas of the brain and asked the patient what they felt. By stimulating different areas of the brain, he found that it produced sensations in various parts of the body, such as the arm, hand or thumb, and, importantly, the sites in the brain related to the 'ability to speak and comprehend language' (Squire & Kandel, 2009, p. 11). Sometimes he found patients would say they were having real experiences. For example, one patient stated, 'It sounded like a voice saying words.' Another said, 'I am seeing a picture of a dog and cat' (Squire & Kandel, 2009, p. 11). As a result of using this technique on more that

Figure 2.3 Wilder Penfield (1891–1976)

1000 patients, he was able to produce a map of which motor and sensory areas of the brain correspond to their associated body parts (Penfield, 1958).

When working with Penfield, Hebb researched how brain surgery and other brain injuries affect brain function. In 1939, Hebb left Montreal for another eight years, to again work with Lashley. But by 1947 he was back in Montreal, having been appointed professor of psychology at McGill University (Figure 2.4), where he had previously gained his master's degree.

Hebb had been a distinguished schoolteacher and when he taught in university, his teaching skills did not desert him. His students enjoyed his teaching and found it inspirational. He believed that the role of the teacher was to help facilitate and encourage student learning, but the passion to learn had to come from the students themselves. Moreover, he believed that university students should be evaluated on the basis of their ability to think creatively and come up with interesting, probing questions, rather than their ability to remember established knowledge.

Donald Hebb's neuropsychological theory of learning

So, what *is* happening when students are thinking, learning and memorising? Donald Hebb realised that if you are going to answer this kind of question successfully, you have to come to grips with

Figure 2.4 McGill University, Montreal, Quebec

what is known about the biology of the brain. You can't just pay attention to the overt, observable behaviour of students, as behaviourists believed; you have to try to understand what is happening in their brains. To be honest, he wasn't the first person to notice this – it goes back at least as far as Santiago Ramón y Cajal (1852–1934). Cajal (pronounced Ka-hal), who received a Nobel Prize in 1906, realised that his work on *the anatomy of the brain* had implications for understanding learning (Cajal, 1894, 1975; see also Kandel, 2006, pp. 61–8).

Hebb was coming at the issue from the other direction – from a quest to understand learning – but he realised the importance of brain anatomy and brain function. He built this cross-disciplinary idea into the title of his ground-breaking 1949 book, *The Organisation of Behaviour: A Neuropsychological Theory*. Notice the word 'neuropsychological' – combining neuroscience and psychology. In the Introduction to the book, he made his cross-disciplinary approach clear when he said psychologists needed to 'seek common ground with the anatomist, physiologist and neurologist' (Hebb, 1949, p. xii). And that's precisely what he did. In choosing to study with Lashley and in working with Penfield, he was activating his own neuropsychological approach. Indeed, he laid the foundation for the discipline of *neuropsychology* by focusing on how brain processes underpin psychological processes such as learning and memory.

behaviourists: those who accept and work withing the doctrine of behaviourism that 'psychology is the science of behaviour and is not the science of the inner mind – as something other or different from behaviour' (Graham, 2019)

Beyond the limitations of behaviourism

An interesting point is that Lashley was something of a neuropsychologist himself, while also being a distinguished behaviourist. **Behaviourists** tended to think of the brains as unknowable and they believed the way to make the study of learning scientific is to focus exclusively on what is directly observable. They therefore focused their research on observable behavioural responses to stimuli. The problem is that this leaves out what is happening between the stimulus and the response, including the unobservable processes within the brain that underlie perception, attention and thinking, as well as learning and memory.

Lashley, however, did not restrict himself to researching observable, behavioural responses: he was keen to find out where in the brain memory was stored (Squire & Kandel, 2009, p. 9). Lashley conducted numerous experiments with rats that had different areas of their brain cortex removed. His experiments led to the view that learning and memory could not be localised to a single brain region. This was consistent with the widely held assumption that learning through association, in classical conditioning, as discovered by Ivan Pavlov (1849–1936) at the start of the twentieth century, must be occurring within complex *circuits* in the brain.

However, one of the first scientists 'to challenge this assumption was Donald Hebb, who boldly suggested that the mechanism for making associations takes place inside single cells' (Squire & Kandel, 2009, p. 64). Hebb accepted that learning and memory involved *assemblies* of nerve cells (neurons) that were distributed over larger areas of the cortex, but he argued that what mattered was the *signalling* between the individual neurons. He believed it was synapses that were strengthened by *learned associations* and he postulated (hypothesised) that this would occur when two adjacent and interconnected cells were excited simultaneously and repetitively.

Hebb's postulate

Hebb was in search of an explanation for how neurons connected together to form assemblies or pathways in the brain and how these connections were strengthened. Thinking of just two cells, he proposed:

> When an axon of cell A is near enough to excite a cell B and repeatedly or persistently takes part in firing it, some growth process or metabolic change takes place in one or both cells such that A's efficiency, as one of the cells firing B, is increased. (Hebb, 1949, p. 62)

Hebb imagined two cells that were located close enough to each other such that when one of the cells fired and did so repeatedly, it excited the other cell and caused it to fire. As a result, the two cells became *associated*, causing a physical, metabolic change to occur in one or both cells. Thus, they became connected and the more these cells fired together, the stronger their association or connection would become, and the faster their synchronous firing would be. One easy way to remember the basic idea is: neurons that fire together, wire together (using 'wiring together' as a metaphor for 'connecting together').

But what was Hebb talking about? What are the 'axons' to which he refers and what does he mean by the cell 'firing' and by it firing 'repeatedly and persistently'? It is time to leave Donald Hebb for a while and get into some basic brain anatomy, before considering the structure of neurons and how they signal to form pathways in the brain, which will bring us back to Hebb.

Some basic brain anatomy and the neuron doctrine

We start with the **cerebral cortex** and the four **lobes of the brain**. The cerebral cortex is the outermost structure of the brain, what you see in photographs of the brain.

Cortical structures

The cerebral cortex (Figure 2.5) largely comprises up to six layers of cell bodies, and is traditionally referred to as 'grey matter', as it looks grey in a brain that is dead. When alive, however, it is a healthy pink colour. White matter, by contrast largely comprises myelinated axons (see Figure 2.11 below) that look whitish because of the white colour of myelin. The cerebral cortex contains around 20 per cent of the 86 billion neurons that comprise the brain. In size, the cortex is about 1800 square centimetres, but it is crimped up to fit into the skull (Donahue et al., 2018). The thickness of the cortex varies between 1 millimetre and 4.5 millimetres, with an average of around 2.5 millimetres. The bulging part of each fold is called a **gyrus** (plural gyri) and the grove that runs between two gyri is called a **sulcus**. You can see this in Figure 2.5.

cerebral cortex: the largest part of the human brain that largely comprises up to six layers of cell bodies and is traditionally referred to as 'grey matter'

lobes of the brain: the four large divisions of the cerebral cortex

gyrus (plural gyri): a bulging part of each fold on the cerebral surface of the brain

sulcus: a groove that runs between two gyri

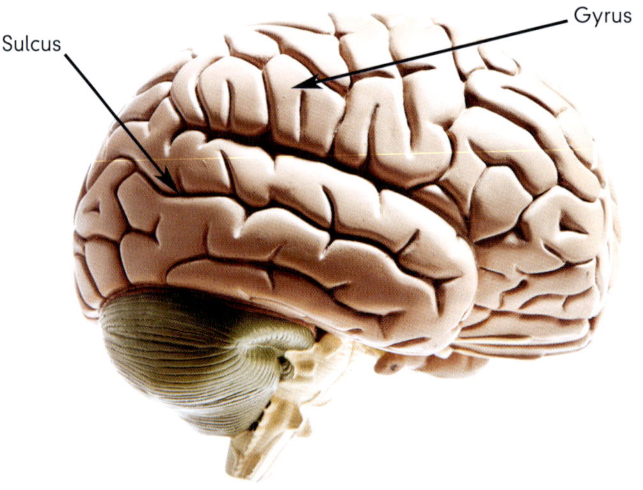

Figure 2.5 The cortex, crimped to fit into the skull

In Figure 2.6, you can see that the cortex is divided into four lobes: occipital (*pronounced like ox-sip-it-al*), parietal (*pronounced like per-eye-et-al*), temporal and frontal. Each of the four lobes is broadly associated with particular brain functions. The occipital lobe is associated with visual experience, the parietal lobe with spatial processing and the integration of input from different sensory receptors, and the temporal lobe with auditory input, speech and language, and aspects of emotion.

The frontal lobe is sometimes referred to as the brain's 'executive' (using a business metaphor), as it is involved in decision-making, planning, cognitive control, inhibitory control and memory. That said, it is important to appreciate that the majority of mental processes

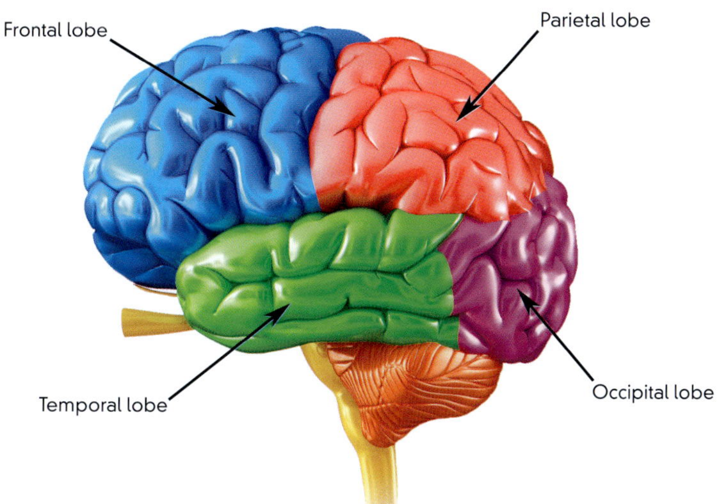

Figure 2.6 The four lobes of the brain

in the brain are distributed across multiple areas. There is an exuberance of *interconnectivity* both within and across brain regions. The brain truly is a highly dynamic and complex system, beyond anything else known to human beings.

Viewed from above, the brain is divided into two **hemispheres** (left and right) (Figure 2.7) and these two halves are joined together by a thick body of nerve fibres called the corpus callosum (Figure 2.8). In addition, there are many **sub-cortical structures**.

hemispheres: the two halves of the cerebral cortex that are joined together by a thick body of nerve fibres called the corpus callosum

sub-cortical structures: a set of structures located below the cerebral cortex; include limbic structures and the basal ganglia

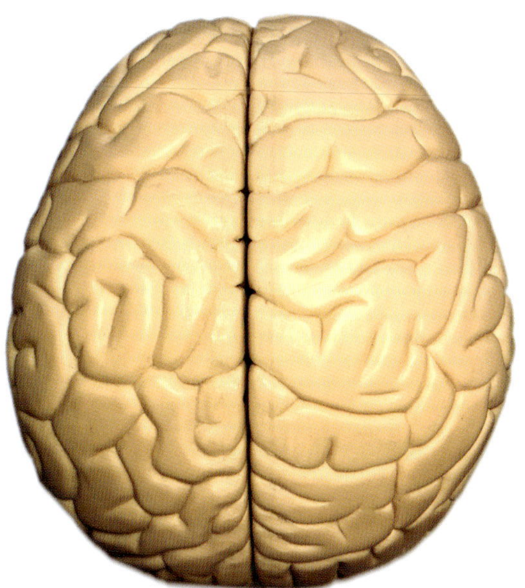

Figure 2.7 The two halves (hemispheres) of the brain, as seen from above

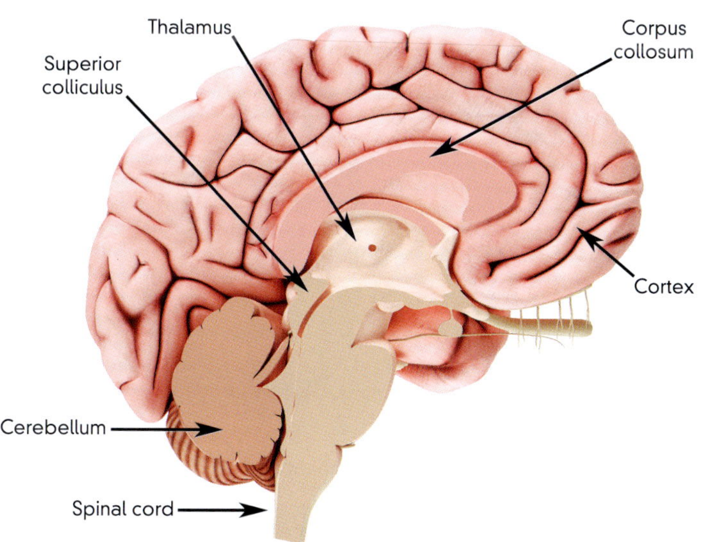

Figure 2.8 Section of a divided hemisphere through the corpus collosum

Sub-cortical structures of the brain and the limbic structure

The cortex can be viewed as a crimped-up blanket in the skull that covers other structures that lie beneath it, some of which make up what is called the limbic structure. Not surprisingly, these are known as subcortical structures. Figure 2.8 shows a section cut through the corpus collosum, between the two hemispheres. If performed as a surgical operation (as it has been in patients with severe epilepsy), it severs the nerve fibres and results in the two hemispheres no longer being able to communicate with each other. Figure 2.8 also shows the position of the **thalamus**, situated very near to the centre of the brain and at the top of the brain stem, which acts as a kind of 'relay station' in the brain, sending signals from **sensory receptors** (eyes, ears, skin, nose and tongue) to relevant areas of the cortex.

To the rear of the thalamus, the **superior colliculus** 'is a multisensory midbrain structure that integrates visual, auditory, and somatosensory spatial information to initiate orienting movements of the eyes and head toward salient objects in space' (Myoga, 2020, p. 601). When a teacher asks a child to pay attention, the teacher is seeking neuronal stimulation within the superior colliculus, though they may not realise it. The **cerebellum**, which literally means 'little brain', is tucked under the temporal and occipital lobes and is also listed as subcortical. It plays an important role in balance and motor coordination, but it is also implicated in attention and language tasks such as language fluency and correcting language errors.

A number of subcortical anatomical structures, pictured in Figure 2.9, are collectively known as the **limbic structure**, which is believed to have emerged with the first mammals. The

thalamus: a part of the brain situated very close to the centre of the brain and at the top of the brain stem; acts as a kind of 'relay station' in the brain

sensory receptors: endings of the sensory cells that are distributed across different parts of the body (e.g. touch and temperature receptors in the skin)

superior colliculus: 'multisensory midbrain structure that integrates visual, auditory, and somatosensory spatial information to initiate orienting movements of the eyes and head toward salient objects in space' (Myoga, 2020, p. 601)

cerebellum: a subcortical part of the brain that is tucked under the temporal and occipital lobes

limbic structure: a set of subcortical brain structures including the hippocampus, amygdala, thalamus, hypothalamus, fornix and cingulate gyrus

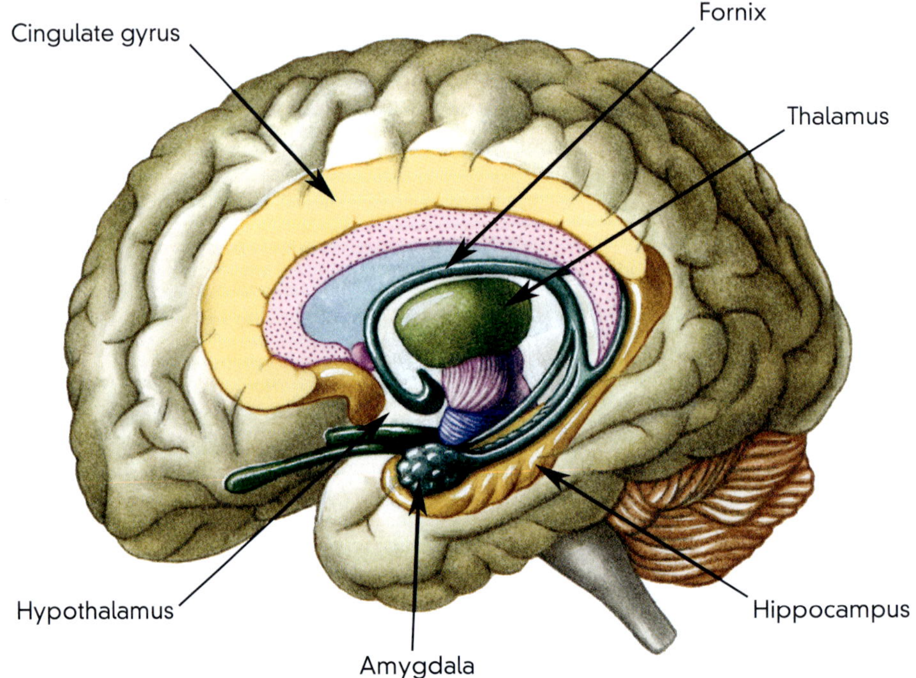

Figure 2.9 Diagram of the limbic structure

structures included (hippocampus, amygdala, thalamus, hypothalamus, fornix and cingulate gyrus) certainly exist in the brain, but whether they constitute a discrete structure, let alone a brain 'system' as often stated, has been a matter of contention (Barrett, 2017). One reason is that it is associated with the so-called 'triune brain theory' (putative reptilian, limbic and neocortex systems), devised by Paul MacLean (1913–2007) in the 1960s. As you will discover in Chapter 6, this is controversial not only anatomically, with 'its cognitive circuitry layered on top of so-called emotion circuitry' (Barrett, 2017, p. 82), but also in its claims about what humans have inherited from evolution.

Most of the components included in the limbic structure are situated just above the brain stem, with the exception of the **hippocampus** and the **amygdala**, which branch into the temporal lobes. There are two of each of the structures (hippocampus, amygdala, etc.), one on each side of the brain. We will be meeting the hippocampus and amygdala again, as they are particularly related to aspects of learning, memory and primary, lower order emotion, and so of particular interest to teachers.

In Figure 2.10, you can see that the cingulate gyrus is positioned above the corpus collosum. When viewing figures of brain anatomy, take care not to confuse the two, as they have quite similar shapes. The cingulate gyrus is involved in the regulation of emotion and pain. Together with the cingulate sulcus, they form what is known as the cingulate cortex. Functionally, it is situated at the interface of learning, emotion and action, and is implicated in disorders such as depression and schizophrenia. However, it is important to keep in mind that brain processes are highly distributed and involve many different parts. For example, Geake (2009, p. 103) notes that 'forms of imagination involving anticipation, mindedness and counterfactual thinking rely on the cingulate cortex, but also the lower parts of the of the frontal cortex, the cerebellum and the orbitofrontal cortex (the bottom part over the eyes).'

> **hippocampus:** 'the oldest part of the cerebral cortex responsible for spatial localisation, formation of declarative memory, and transfer of short-term to long-term memories' (BrainU, 2021)
>
> **amygdala:** almond-shaped part of the limbic system that is involved in emotional reactions, particularly fear

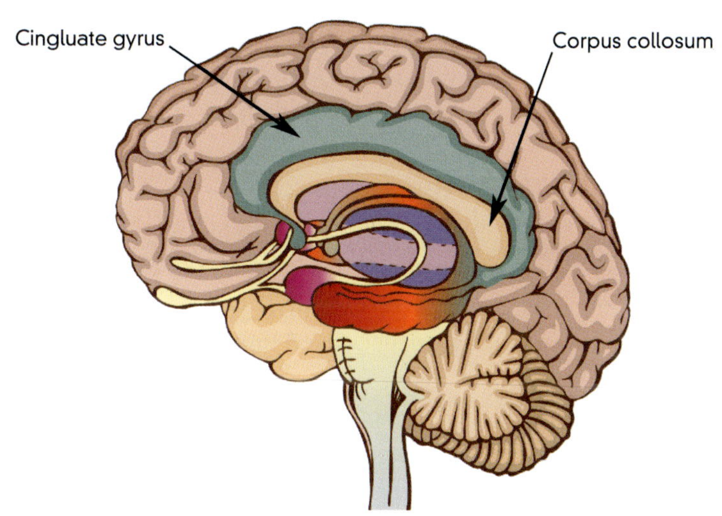

Figure 2.10 The corpus collosum and cingulate gyrus

The neuron doctrine, the structure of neurons and how they signal

Moving on from the macro structures that we have just been considering, it is time to go microscopic and consider the structure of **neurons** and how they fire, which will bring us back to Hebb's postulate. After that, we consider the formation of the brain in the womb.

Your brain weighs less than 1.5 kilograms (3.3 pounds). Within your brain, as you learnt in Chapter 1, there are around 86 billion microscopic neurons (Figure 2.11), along with glial cells, which provide protection and support for neurons. Previous textbooks have often

> **neuron:** a brain cell that is specialised for the transmission of information and characterised 'by long fibrous projections called axons, and shorter, branch-like projections called dendrites' (BrainU, 2021)

Figure 2.11 Close-up image of neuron showing cell body, axons and dendrites

glial cells: non-neuronal cells in the brain; radial glia cells in particular provide a fibre pathway that physically and chemically guides the migration process of new neurons

pyramidal cells or pyramidal neurons: excitatory neurons in the cerebral cortex, hippocampus and amygdala

cell body: also called the soma (using the Greek word for body); it is the neuron's powerhouse, providing energy to drive the cell. The cell body also houses the cell nucleus, which in turn houses the DNA that encodes the cell's genes.

dendrites: the main input component of the neuron that receives signals from other neurons close by

axon: a long nerve fibre that transmits action potentials away from the neuron's cell body towards the axon terminal

axon terminals: the 'very end part of an axon that makes a synaptic contact with another cell; the point where neurotransmitters are released' (BrainU, 2021)

stated that there are 10 times more **glial cells** than neurons, but science progresses and the general consensus now seems to be that there are around 85 billion glial cells, roughly the same number as neurons. There are different kinds of neurons, but the main excitatory neurons in the cerebral cortex, the hippocampus and amygdala (which are of most interest to us) are called **pyramidal cells or pyramidal neurons**.

Earlier in this chapter, we noted the very important foundational work of the Spanish brain anatomist Santiago Ramón y Cajal. He hypothesised that the neuron is the basic anatomical unit of the nervous system and that neurons are not physically joined together (as some thought), but rather each neuron is a separate entity that joins with others at highly specialised contact points (Cajal, 1888). Cajal's theory is now known as the *neuron doctrine*.

Cajal was also one of the first to realise that all neurons have a similar structure, with four main components. Each cell has a **cell body**, **dendrites**, an **axon** and a grouping of **axon terminals** (also called presynaptic terminals). In a lecture delivered at the Royal Society of London, Cajal (1894) speculated that each of these four components has a specialised role to play in receiving and transmitting signals. He suggested that signals flow in one direction, from the receptor dendrites, through the cell to the axon and then to the axon terminals. In the 1950s, the invention of the electron microscope showed that Cajal's speculation was correct: nerve cells are indeed connected at the synapses. Hebb was building on the neuron doctrine when formulating his theory of learning.

You can see these four components indicated in Figure 2.12. The cell body or soma (using the Greek word for body) is the neuron's powerhouse, providing energy to drive the cell. The cell body also houses the cell nucleus, which in turn houses the DNA that encodes the cell's genes. The dendrites (from the Greek *dendron* – tree) often have the appearance of intermingling tree branches. The dendrites are the main input component of the neuron that receives signals from other neurons close by (Figure 2.13). The axon is the output component. The axon can be very short – as small as 0.1 millimetre – to as long as a metre or more. Towards the end of the axon, it divides into many thin branches, each containing an axon (presynaptic) terminal.

REFLECTION

Having been introduced to quite a lot of (possibly new) information, concerned with naming the parts of the brain, what learning strategies might you apply, or recommend others employ in learning this information so that it is remembered in the long term?

Synaptic signalling and the learning, developing brain

The signalling that Cajal identified is now called an **action potential**. This is an all-or-nothing signal (it either fires or it doesn't fire) that is also one-way (from dendrites to axon terminal). Technically it is called a depolarising electrical signal, but basically it results from a change in the electrical potential across the external membrane (skin) of the cell, which has tiny pores that allow sodium ions (electrically charged atoms) into the cell and the subsequent movement of potassium ions out of the cell. It's as though the cell membrane has electrical shivers!

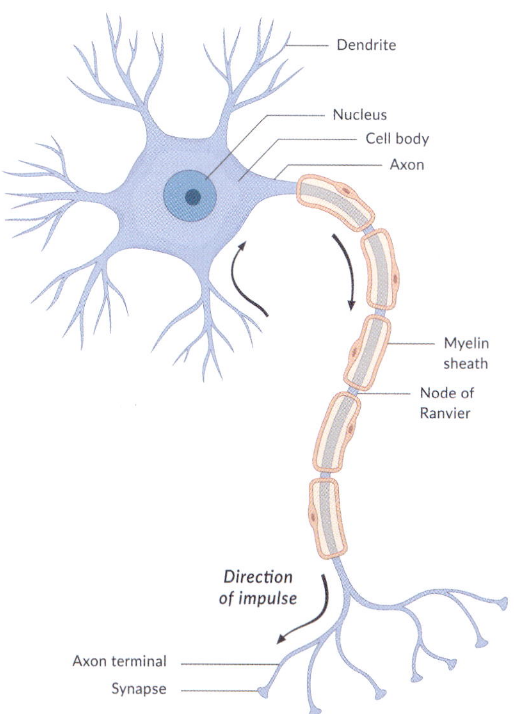

Figure 2.12 Diagrammatic representation of a pyramidal neuron

action potential: an electrical signal that travels from neuron's cell body to the axon terminal, where it triggers the release of neurotransmitters (BrainU, 2021)

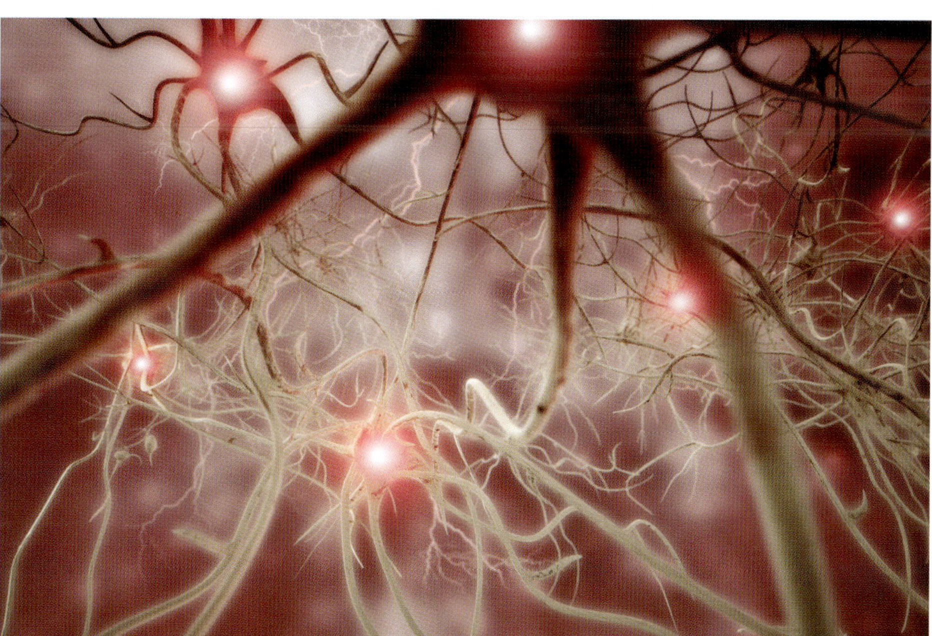

Figure 2.13 Profusely interconnected neurons signalling back and forth

synaptic potential:
the potential difference across the postsynaptic membrane that is triggered by neurotransmitters at a synaptic gap or cleft

neurotransmitters:
chemicals that are released from the tip of the axon or axon terminal, including glutamate, serotonin, dopamine, norepinephrine, acetylcholine and gamma-aminobutyric acid (GABA)

There is another kind of signalling called **synaptic potential**, which occurs as neurons signal to each other across what is called the synaptic gap or cleft. As noted above, Cajal (1894) believed that each neuron was a separate entity that joined with others at highly specialised contact points. That view was shared by the British neuroanatomist Charles Sherrington (1857–1952), who coined the term 'synapse' (from the Greek *synapsis* – conjoin together) to name the specialised gap (Sherrington, 1906).

This process is illustrated in Figure 2.14. The main point to notice is that the current produced by the action potential cannot jump directly across the synaptic gap to activate the very closely adjoining target cell. Instead, it does something rather amazing. The tip of the axon or axon terminal contains chemicals, which are called **neurotransmitters**; they include glutamate, serotonin, dopamine, norepinephrine, acetylcholine and gamma-aminobutyric

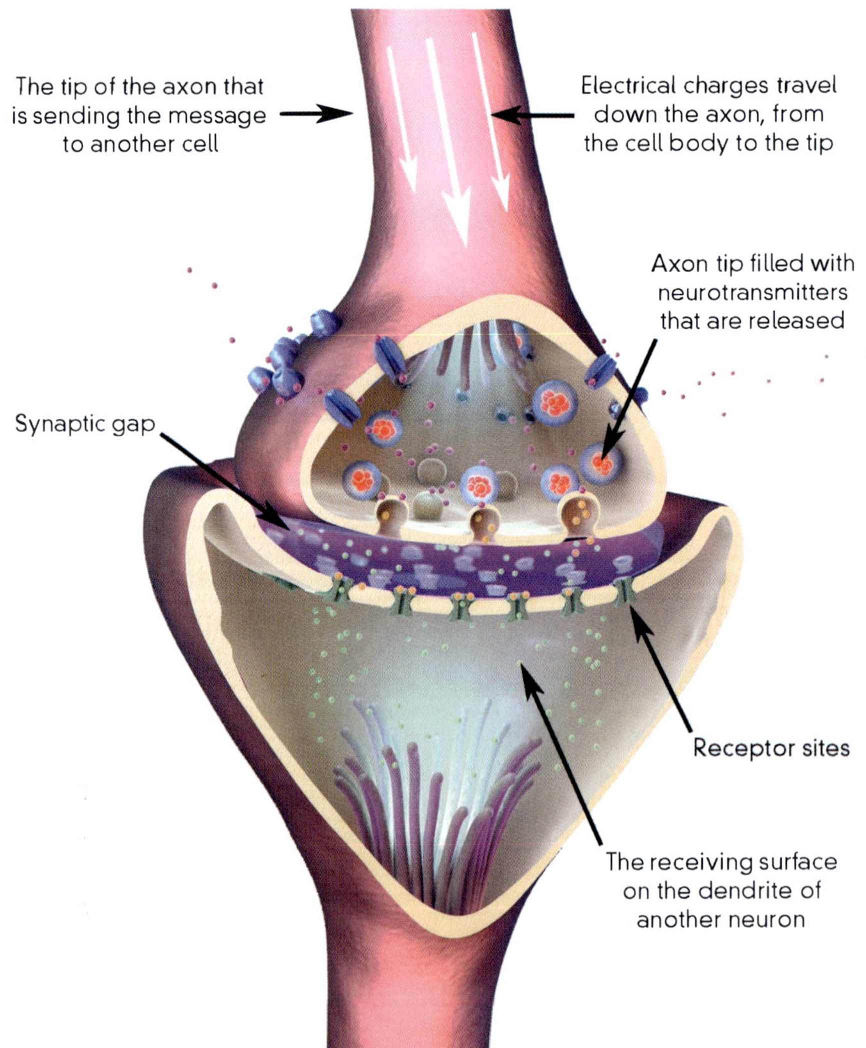

Figure 2.14 Synaptic potential

acid (GABA) – a wonderful set of strange-sounding names. These are contained in small packets called organelles (each packet is called a quantum) and, when the electrical signal arrives, some packets are released into the synaptic gap or cleft. These are then taken up at receptor sites on the neighbouring cells' dendrites, which in turn will or will not induce an electrical current in that cell that travels down through its axon, to repeat the same process in that cell.

Synaptic plasticity hypothesis

Neurons thus use two kinds of signals: action potentials and synaptic potentials. The important difference is that whereas an action potential from dendrites to axon terminal is a rapid, invariant, all-or-nothing signal, the synaptic potential is variable (graded) and modifiable, and that is precisely what is required for brain plasticity. To put it simply, if neurons were precisely 'wired' together, it is hard to see how they could produce change, and there has to be change if there is to be learning and memory – if no change occurs then nothing new happens – no learning or memory. Cajal (1894) hypothesised that what is happening in the processes of learning and memorising is that the patterns and intensities of the electrical signals are changing.

This ability to produce such a change is what we now call the **synaptic plasticity** *hypothesis*, which basically asserts that the extent to which an action potential in one cell excites (or inhibits) an action potential in a target cell is not fixed, but rather is plastic, which means it can be changed and modified. As we learnt in Chapter 1, brain plasticity refers to the way brains can be physically moulded and changed in response to experience. The bottom line in all of this is that without brain plasticity there would be no learning or development. Brains have to physically change if they are going to learn and develop. So how does that happen?

synaptic plasticity: activity-dependent modification of the strength or efficacy of synaptic transmission at pre-existing synapses (Citri & Malenka, 2008)

We are back to Donald Hebb. Building on Cajal's hypothesis of synaptic plasticity, Hebb (as we have previously seen) hypothesised that:

> When an axon of cell A is near enough to excite a cell B and repeatedly or persistently takes part in firing it, some growth process or metabolic change takes place in one or both cells such that A's efficiency, as one of the cells firing B, is increased. (Hebb, 1949, p. 62)

In short, Hebb is suggesting that change and plasticity (hence learning and memory) in the brain come from repetition – when cell A 'repeatedly and persistently' excites cell B, to use his own words.

The neurons in our brains are constantly signalling to one another, and in the process they are forming neuronal pathways that may be strengthened as a result of repetition or weakened if not used – 'use it or lose it'. Let's try an analogy. Imagine a field of corn, golden and ripe for harvest. Someone in a hurry decides to walk through the corn as a short-cut, rather than stick to the road. It makes a bit of a pathway, but it is not very visible. However, another person notices it and also takes the short-cut, and then another and then another. Through repetition, what was at first a slight trace in the cornfield becomes a clear pathway. In somewhat similar ways, neural pathways of thinking, learning and being are forged in the brain through repeated and persistent synaptic firings, connecting neurons and forming neuronal pathways.

However, equally if not more importantly, those connections and pathways that are not used are weakened or pruned. It has been estimated that by the age of two years, a child has produced around 40 per cent more neurons than an adult (Geake, 2009, p. 51), and consequently far more synaptic connections or possible connections than they will ever need. However, that doesn't mean the young child is more intelligent than an adult. Most of these connections are not salient (have no relevance or meaning) and do not relate to any behaviour or knowledge of the world. That will mainly come as a result of Hebbian repetition, resulting in how each of us comes to categorise the world. It is this dynamic process of neural 'sculpting' (creating neural connections and pathways and pruning others) that 'creates our unique set of memories and worldviews' (Geake, 2009, p. 52) and our own distinctive personalities.

Synaptic plasticity: The key to learning, memory and development

Notice that in stating his postulate, Hebb is also endorsing something more fundamental: learning and memory occur at the synapses as a result of synaptic plasticity. Back to our original question: What is actually happening when students are thinking, learning and memorising? The answer that is beginning to emerge is that it has a lot to do with what is happening in the brain, as neurons join with other neurons and form pathways in the brain through the strengthening or pruning of synaptic connections.

In short, thinking, learning and memorising are occurring at the synapses, as a result of experience. This means that at school, in providing rich learning experiences, teachers are physically changing the brains of their students – teachers are quite literally 'brain changers'. No wonder teaching is such a responsibility, having such a physical impact on the children and adolescents in our care – let's strive to do it with the greatest of care and love.

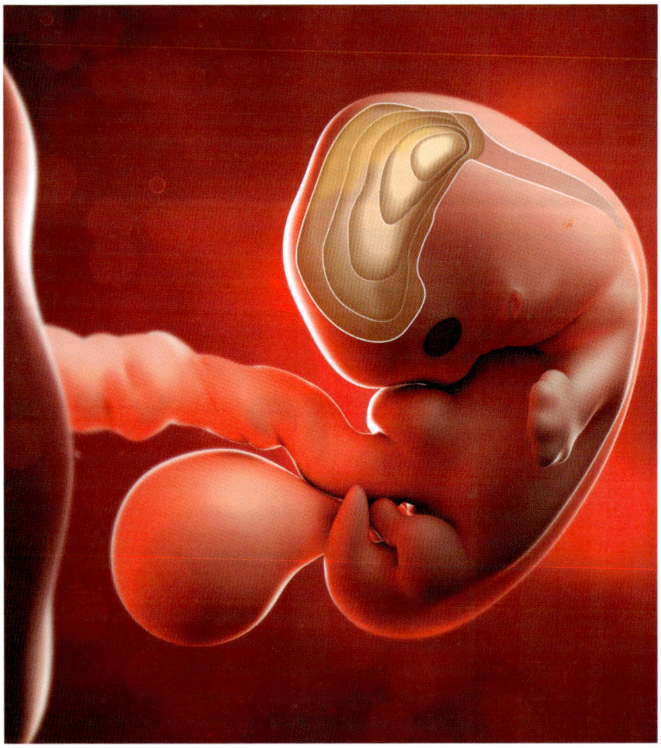

Plasticity and cell migration in the formation of the brain

Here is another question: Where did your brain come from and how did it originally form? These questions also involve brain plasticity, change and development.

By the time your mum realised she was pregnant, some four to five weeks after conception, your brain was already rapidly forming in her womb. As the early cells divided and then divided again and again, a neural tube started to form (Lagercrantz et al., 2011). The most forward part of the neural tube forms into the **telencephalon**, which expands outwards as a result of cell proliferation (see Figure 2.15). The subsequent formation of the

telencephalon: from the Greek words *telos* (end) and *enkephalos* (brain), telencephalons literally mean 'end-brain'; it includes all that we normally mean by the brain, the cerebral cortex and subcortical structures

Figure 2.15 The brain begins to form, expanding outwards from the neural tube to form the telencephalon

brain results from a process called **cell migration** (Leone et al., 2008), when newly generated neurons, formed from stem cells, migrate along radial glial fibres to their specialist location in the brain (see Figure 2.16). Cells then jointly self-organise to form a particular brain structure.

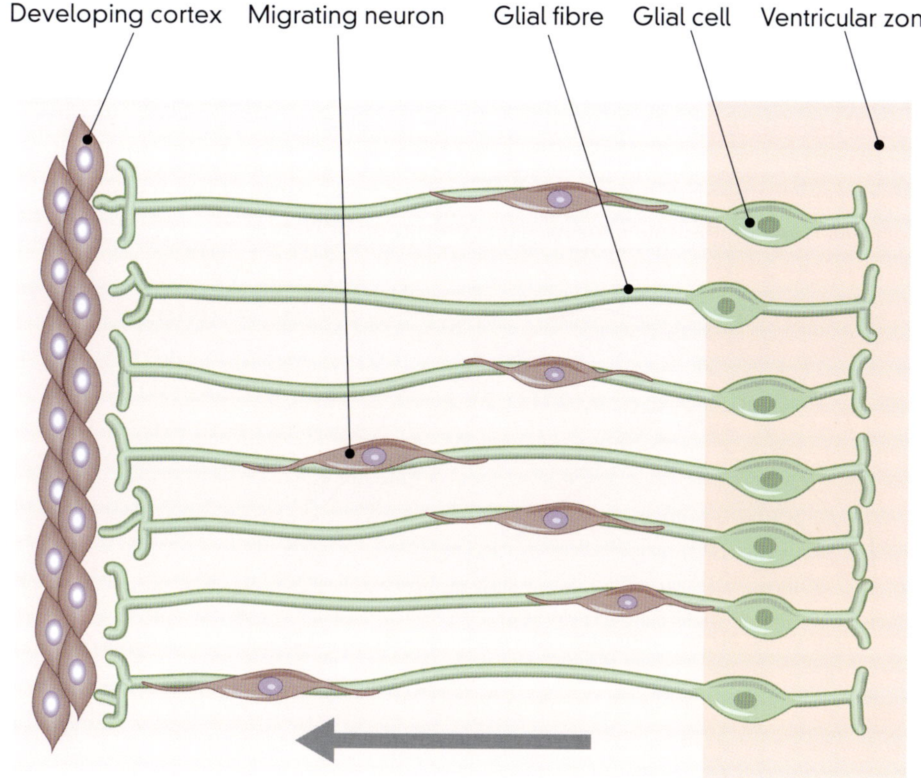

Figure 2.16 Newly formed neurons migrate along glial pathways to their specialist location in the brain

The generation of new neurons and cell migration in the foetal brain is a wonderful and amazing process, involving many different steps. You will not be surprised to learn that, in detail, there is much we still do not know about it and a lot of research is currently being undertaken, yielding new discoveries. For example, there are different kinds of glial cells with different functions, including what are called radial glia, which as just noted provide a fibre pathway that physically and chemically guides the migration process (Hansen et al., 2010). But we now know that radial glia also act as 'multipotent precursor stem cells', which means they are able to self-renew, as well as producing neurons and 'oligodendrocytes' (Huang et al., 2020), which are the cells that produce the insulating myelin sheath around the axons of neurons.

When migrating neurons reach their destination in the brain, the cell body produces two kinds of extensions: dendrites and axons. Dendrites 'extend from the cell body, often in the form of a tree, and form the input component or receptive area for incoming signals'. The axon provides the 'output component of the neuron' (Squire & Kandel,

2009, p. 31). Communication between neurons, via the synapses, leads to the formation of neural assemblies and neuronal pathways or maps.

Brain development is spread over a very extended period of time, from the moment of conception, when egg and sperm combine, to well into adulthood. In fact, to some degree brain development continues to the time of death. As summarised in Figure 2.17, we can consider two important timelines in development, one concerned with the neuronal processes of development and the other concerned with spatial areas of the brain. As the diagram shows, neurogenesis (formation of neurons) and cell migration begin soon after conception. By the time a full-term baby is born, migration has ceased, but neurogenesis, synaptogenesis (formation of synapses) and synaptic pruning continue into adulthood. Whereas synaptic pruning largely begins after birth, synaptogenesis and axon growth begin while the baby is in the womb, from about 16–24 weeks after conception onwards.

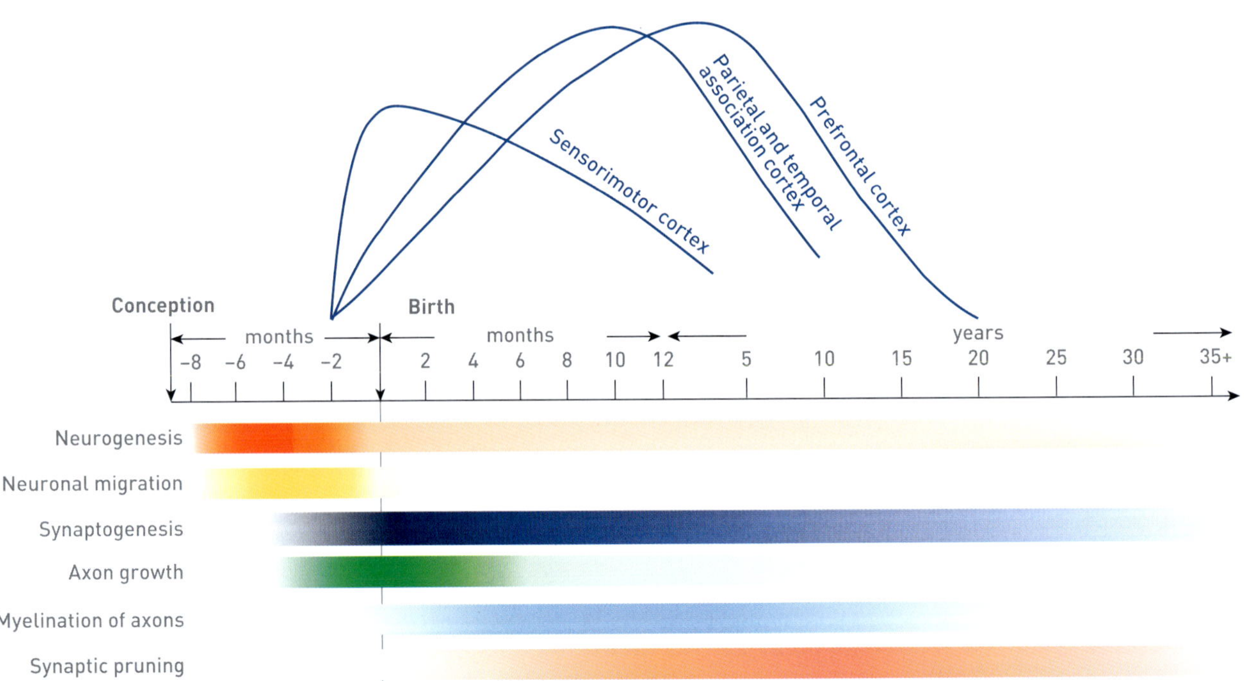

Figure 2.17 Timeline of human brain maturation processes
Source: Based on data presented in Casey et al. (2005) and Ouyang et al. (2019).

With regard to spatial development, as represented in the curvilinear lines above the timeline and taking synaptogenesis as the example, the sensorimotor cortex (located at the very front of the parietal lobe) develops rapidly during the last two months before a full-term birth, continuing for some two years after birth at a declining rate. The parietal and temporal lobes, along with the frontal cortex, also start to develop prior to birth, though not as rapidly as the sensorimotor cortex, and development continues for longer, into the teens.

The brain at birth is endowed with more neurons than it will retain. In the early years, myriad connections form, but in puberty many will be pruned and new connections will form. However, though the density of neuronal connections is much reduced, the strength of those connections is increased. At a biological level of description, learning and development result from the formation of neuronal connections and neuronal maps, which also underpin the development of our personality.

Figure 2.18 illustrates the growth of neuronal connections and their subsequent pruning and strengthening in puberty (Lagercrantz et al., 2011). At the time of full-term birth, a baby has almost all the neurons it will possess and also a primary repertoire (Edelman, 1987) of synaptic connections that mostly results from the constant and spontaneous firing of neurons (see 'Edelman's theory of neuronal group selection' in Chapter 3). Connections continue to be formed into childhood, but that is increasingly accompanied by synaptic pruning of unused connections and the strengthening of used connections, especially during puberty and into adulthood.

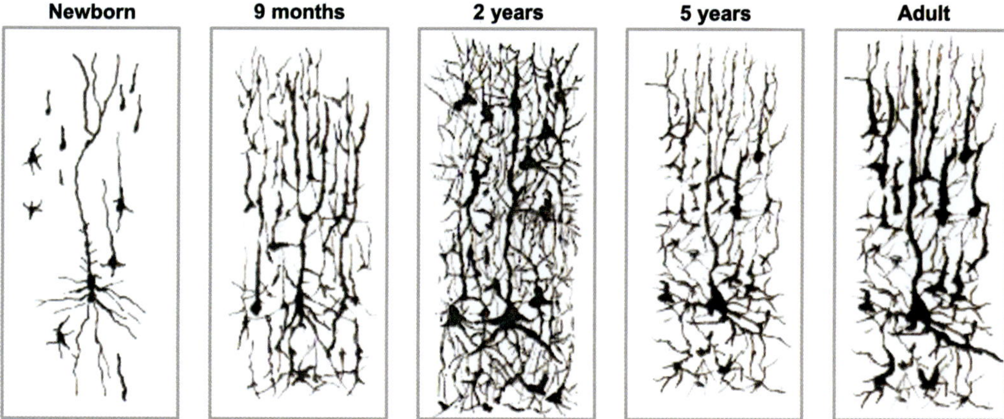

Figure 2.18 Diagrammatic representation of increasing density of neuronal synaptic connections in the early years and their subsequent pruning and strengthening

Plasticity, memory, how you learn and who you are

You may be surprised to learn that much of what we know about the biological processes involved in memory comes from the study of fruit flies and sea slugs (Squire & Kandel, 2009). It also comes from experiments conducted on mice, rats and monkeys, on the assumption that what we learn indirectly from the study of other animals can be applied in understanding how human brains work. For ethical reasons, most of what we know directly about the human brain comes from post-mortem studies, but there is another intriguing direct source of information that comes from studying people who suffer brain damage, especially in those cases when the damage is a result of brain surgery.

Brenda Milner and some early clues about memory and learning

We need to go back to Donald Hebb and his colleague Wilder Penfield in Montreal, Canada. In 1949, Hebb accepted Brenda Milner (1918–) from Manchester, England as one of his research students. Working in conjunction with Wilder Penfield, Milner made significant discoveries about learning and memory by studying a patient who went by the name of HM (to protect his identity). HM suffered from acute epilepsy. In 1953, it was decided the only way it could be relieved would be to remove parts of the medial temporal lobe on both sides of his brain, including the hippocampus, which as you now know forms part of the limbic structure.

The surgery relieved the epilepsy, but it left HM with a devastating memory loss. In particular, he was unable to retain his immediate memories as long-term memories. If you had met HM, he would have been very pleasant and would have spoken to you very sensibly, so that you would likely not have detected a problem. But leave the room and return in an hour, and he would claim he had never seen you before. From studying what HM could do and could not do over very many years, Milner was able to make four very important discoveries about the nature of memory (Milner, 1962; Milner et al., 1968; Squire, 2009):

1. The ability to acquire new memories is not related to other perceptual or cognitive abilities. It is distinct and is directly associated with the medial temporal lobe of the brain, including the hippocampus, the structure that was removed by surgery.
2. The medial temporal lobe is not required for immediate memory, because he retained that and could hold a perfectly ordinary conversation using it.
3. The medial temporal lobe, including the hippocampus, cannot be where long-term memories are stored, because she found HM could clearly and accurately remember events from his childhood.
4. His loss of memory was not total; there were things that HM could learn and remember perfectly well, suggesting there is more than one kind of memory.

Declarative and non-declarative memory

Based on these very important initial discoveries, it is now generally accepted that there are two main memory systems (Squire & Kandel, 2009, p. 15). They go by different names, but we will use the terms declarative and non-declarative. **Declarative memory** (also called, explicit memory, conscious memory, memory with record) is what we normally mean by memory: it is memory for facts, ideas, and events. **Non-declarative memory** (implicit memory, unconscious memory, memory without record) is related to performance and the subconscious use of skills, such as riding a bicycle or playing the piano without having to consciously think about it.

Of course, it could be argued that there is often a conscious component to any skill. Think of driving a car, for example. To begin with, there is a lot of information to learn about how to use the controls, how to steer and position the car on the road and so forth that has to be retained in declarative memory. However, once we learn to drive, we do it without having to consciously attend to that information. It has entered non-declarative memory. In the case of HM, it was found that he not only retained previously learnt non-declarative memories, but could actually improve new motor skills over time.

Teachers need to be aware of the difference between these two kinds of memory, especially when planning lessons that involve performative skills. The difference is not age or stage related; it is brain related – a result of how and where two biologically different kinds of memories are processed and stored.

declarative memory: memory for facts, ideas and events; it is also called explicit memory, conscious memory and memory with record

non-declarative memory: a type of memory related to performance and the subconscious use of skills, such as riding a bicycle or playing the piano

Think about your own learning and memory. What aspects of your learning are reliant on declarative memory? What aspects are reliant on non-declarative memory? Jot them down and see whether you can find 10 examples of each.

Long-term declarative memory and long-term potentiation (LTP)

Milner's work also showed that there is a biological difference between what we might call **short-term** and **long-term memory**. In the case of HM, he could hold a memory over a short period of time, allowing him to be entirely coherent in his normal conversations, but he could not retain it for long. That suggested that there is some kind of 'trigger event' in the brain required for long-term memory.

In 1973, picking up on Brenda Milner's insight about the role of the hippocampus in the medial temporal lobe, Tim Bliss and Terje Lømo, working in Norway, tried to stimulate a pathway in the hippocampus of a rat by using a high-frequency electrical stimulus (called a tetanus) (Bliss et al., 2003; Bliss & Lømo, 1973). They discovered that it worked, producing an increase in synaptic strength that lasted long-term – for hours and, if repeated, for days. This is what is now called **long-term potentiation (LTP)**. But what is going on and how does it work?

Some years later, experiments by Jeff Watkins and Graham Collingridge in Bristol, England discovered that, in a particular pathway in the hippocampus called the Schaffer Collateral Pathway, the neurotransmitter

> glutamate acts on two different types of ionotropic receptors in the hippocampus, the APMA receptor and the NMDA receptor. The AMPA receptor mediates normal synaptic transmission and responds to an individual action potential in the presynaptic neuron. The NMDA receptor, on the other hand, responds only to extraordinarily rapid trains of stimuli and is required for long-term potentiation. (Kandel, 2006, pp. 283–4)

The NMDA receptor doesn't function when receiving normal, low frequency firing because it has magnesium (Mg^{2+}) ions that act as a plug (a bit like the cork on a wine bottle) that blocks the receptor, preventing calcium that is in the synaptic gap to enter. However, at high frequency the magnesium cork is displaced, which allows calcium into the receptor and allows a chemical reaction that induces LTP, which Bliss and Lomo discovered is associated with long-term declarative memory. You can see this represented in Figure 2.19.

This was done artificially, but it is believed that this is precisely what happens in a salient or meaningful learning experience (Squire & Kandel, 2009). A key message for all teachers is to make learning meaningful – we remember best what is salient and meaningful, and for very good neurobiological reasons. A salient or meaningful learning experience is remembered because it induces LTP, and that requires the stimulus of high-frequency firing in the synapses, allowing the magnesium plug to be removed. So perhaps teachers, in providing students with meaningful learning experiences, should think of themselves as cork-pullers!

short-term memory: memory over a short period of time that cannot be retained for more than a day or so

long-term memory: memories that are stored in a variety of places in the brain and are retained over long periods of time

long-term potentiation (LTP): a long-lasting synaptic strength that has been induced by high frequency electrical stimulus

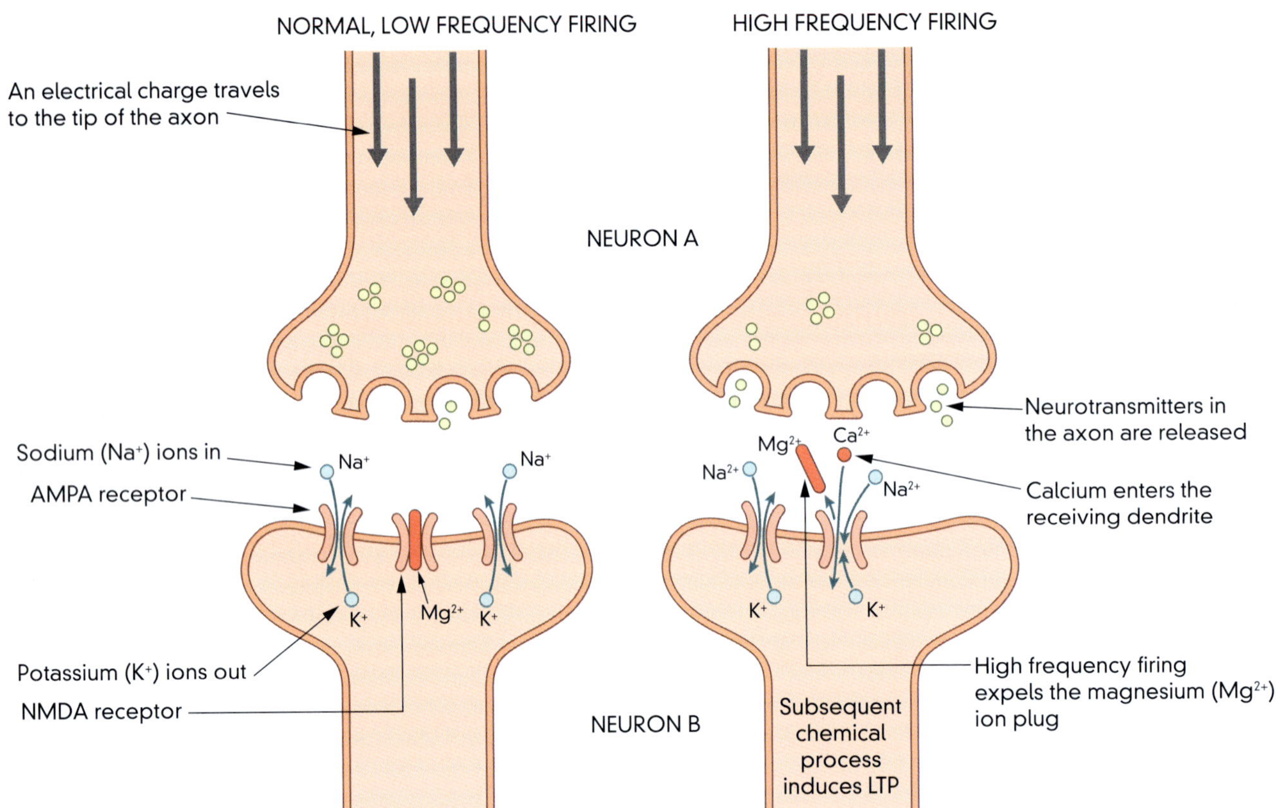

Figure 2.19 The NMDA receptor in the Schaffer Collateral Pathway, and the induction of LTP
Source: Based on Squire & Kandel (2009, p. 126).

But there is just one more piece to the jigsaw. Squire and Kandel (2009) recall that Holder Wigström and Bengt Gustafsson in Sweden discovered that the signal not only has to fire at high frequency, but also has to 'fire *repetitively* so as to substantially depolarise the postsynaptic neurons and thereby remove the (Mg²⁺) plug from the mouth of the NMDA receptor channel' (Squire & Kandel, 2009, p. 127). Now who have you met in your reading who emphasised the need for *repetition*? Yes, this is Hebb's postulate. Indeed, this discovery in Sweden 'provided the first direct evidence for Hebb's proposal made in 1949' (Squire & Kandel, 2009, p. 127). Synapses that have this property are now called *Hebb synapses*.

Donald Hebb, school teacher and headmaster, who became a pioneering brain scientist, had wondered what was happening in the heads of his students when they were learning, and he had speculated that 'some growth process or metabolic change takes place' (Hebb, 1949, p. 62). Without knowing the detail, Hebb was pointing the way towards the discovery of LTP.

Memory makes you who you are, and the children you teach who they are

Memory is essential to learning and development, but memory is even more fundamental than that; our memories make us who we are. Memory also enables us to become educated and, because humans have the ability to communicate what we have learnt and memorised, we are able to create cultures that can be remembered and transmitted through time, from one generation to the next.

It is often claimed that each and every one of us is a product of the interweaving of what we have inherited from our parents and the multiple and continuous experiences we have of the world (physical and social). Actually, most of what we know about the world is not built into our brains at birth, it is acquired through experience (Dehaene, 2020). However, it is not simply our experiences of the world that shapes the 'self' or person, it is what is remembered of those experiences – the names and faces of family and friends and the multitude of things we have enjoyed or not enjoyed: what we learnt at school, the teachers we encountered, those who inspired us and those who did not. Your memory is a record of all of your personal experiences; they make you who you are.

A main implication of this is that, although all brains employ the same neuronal processes (described in this chapter), differences in experience and the memories those experiences evoke mean that 'no two human brains are, have ever been or ever will be, identical' (Geake, 2009, p. 46) and 'This applies to identical twins' (2009, p. 46), whose slightly different experiences in the womb and the memories they form mean that their brains are already differently connected by the time they are born. For teachers, the key message is that every child is unique – their uniqueness is not a proposition of social theory, rather it is inscribed in their brain.

So, when you are working with children in school, try to think of them as functioning brains that hold memories of personal experiences – some good, some not good – that make them who they are. The boy who always misbehaves is a functioning brain that is not producing behaviour that others (teachers, parents, peers) approve of. Try to appreciate this child as a functioning brain, moulded by experiences that have been forged into memories, physically encoded on their brain. Step back, if you can, from what you see – the clothing, the face, the body language – and what the child did, to appreciate that you are dealing with an embodied brain that has memories. Moreover, try to see a child as an embodied brain that has emotions (including fear and joy), which are also memorised in their brain.

If we lose our memory, we lose something of our 'self'. Memory 'is the mental glue that binds together and interconnects our life's experiences'. A life without memory 'is a life in dissolution' (Squire & Kandel, 2009, p. 209). In the worst cases of dementia, we may lose ourselves entirely, stripped of the memories that made us the person we once were, no longer able to maintain enduring, meaningful interactions with other human selves. Moreover, it is not only in later life that disorders of learning and memory impact persons: they can also impact infants. Furthermore, many psychological and emotional problems are, at least in part, related to memory, post-traumatic stress disorder (PTSD) being one such devastating condition.

Thankfully, for the most part memory is central to all that is worthwhile and meaningful in life. In fact, memory and meaning are closely related: we remember best those things that are meaningful (Immordino-Yang, 2016), whether meaningfully positive or negative. There are those, of course, who prefer to think of the human self and its sense of identity (Who am I?) as a human social construct. Indeed, there is 'a long tradition in philosophy for claiming that selfhood is socially constructed' (Zahavi, 2009, p. 551) and that is bound to be correct, because social and cultural experiences are involved in shaping the brain and are laid down in memory. But without memory there would be no recollection of those experiences, and thus no sense of identity. Quite literally, we are what we remember.

Think carefully about what has been said in the three paragraphs above about viewing children as embodied brains and the importance of memory in making us who we are. Do you agree, and how might these ideas influence your practice as a teacher?

Brain plasticity offering hope for the future

At the close of Chapter 1, it was emphasised that it is possible to engage with the science of learning and development and also retain the important humanistic concerns that belong to education as a caring profession. This is an important point because, as those who teach the science of learning and development know, some students come to lectures and seminars with very genuine worries about the word 'science'. This may partly result from not enjoying or not being successful in science in their own school experience, but it may also be a conscious or subconscious feeling that science will undermine the hope, joy and faith they bring to becoming or being a professional teacher. It may also undermine their belief that, as a teacher, they can really make a qualitative difference to the life of a child. This concern is often related to what they perceive to be the **determinism** of science, although they may not use that word. For example, the idea that genes determine who we are and our destiny.

determinism: a philosophical view that every event is determined completely by pre-existing causes. One example is the idea that genes determine who we are and our destiny (genetic determinism).

Stanislas Dehaene (2009), who you will meet throughout this text, picks up on this point in the context of discussing dyslexia. He says when giving lectures on reading and its impairments, he can generally spot parents who have a dyslexic child. Many of them, he says, 'look on each scientific advance as a stab in the back bringing them nothing but bad news' (Dehaene, 2009, p. 256). He also says he can usually also spot teachers, who respond with a similar mixture of discouragement but also relief – 'it's all bad news but luckily it's brain science and so not my worry'. As a key researcher in brain science, with a particular interest in its applications to education, Dehaene (2009, p. 256) says that although he 'can certainly empathise with these feelings of despondency, they are totally misguided. They betray two frequent misconceptions about brain development.'

The first misconception is that our genes are our destiny, that they 'dictate inalterable, ironclad laws that will govern the rest of our lives' (Dehaene, 2009, p. 256). The second misconception that is frequently found in education betrays an underlying belief in Descartes' mind/body dualism. The idea is that education operates at the level of the mind, not the brain, so all the various forms of therapy that might address a prenatal anomaly, such as dyslexia, might help to ease the mind, but they can't fix the brain. In response to the despondency and misconception Dehaene finds in some parents and teachers, he says:

> I would like to emphasize to families with dyslexic children that genetics is not a life sentence. The brain is a 'plastic' organ, which constantly changes and rebuilds itself

and for which genes and experience share equal importance. Neuronal migration anomalies, when they are present, affect only very small parts of the cortex. The child's brain contains millions of redundant circuits that can compensate for each other's deficiencies. Each new learning episode modifies our gene expression patterns and alters our neuronal circuits, thus providing the opportunity to overcome dyslexia and other developmental deficits. (Dehaene, 2009, p. 257)

This is a powerful message from one of the world's leading brain researchers, who has a particular appreciation of the implications of his research for education. It is not wishful thinking, it is science.

The brain 'is a plastic organ', he says, which basically means that brains are physically moulded and changed in response to experience. Brain plasticity will be taken up again in Chapter 3, and it underpins this whole text because it is foundational to brain science. As Dehaene notes, it is increasingly recognised as having strong therapeutic implications, so the plasticity of the brain is being used to fix brains that are damaged.

Compelling case studies of how plasticity can bring healing to the brain are presented by Norman Doidge (2007) in his book *The Brain That Changes Itself*. One example he gives concerns a patient called Cheryl Schiltz, who had her vestibular apparatus (sensory organ for balance) almost destroyed by drugs she was given after a routine operation. It caused her to lose the ability to stand or sit upright without tumbling forward. She was told by her doctor that it was a *permanent* condition, but she would get used to it over time (a terrible prognosis). Now, however, as a result of treatment using brain plasticity, employing redundant pathways in her brain, her balance has been completely restored to the point where she can confidently ride a bicycle on the road.

During treatment, Cheryl wore a specially designed helmet that looks a bit like a construction-worker's hat (Figure 2.20). Inside the helmet was a sensor, called an accelerometer, that detected the movement of her head as it moved up and down and also sideways. The accelerometer was linked to a computer screen and, via 144 electrodes, to a plastic strip placed on the tongue. As Cheryl tilted forward, the monitor showed her what was happening to her head, and she also felt electrical tingles on the front of her tongue. As she tilted backwards, she felt tingles on the back of her tongue – similarly sideways, left and right.

The apparatus was alerting her to what was happening to her posture and she was then able to correct it. She was able to feel release from tumbling as soon as she put the strip on her tongue and, over time, a residual effect became apparent – the time she could stand without tumbling after taking the strip out of her mouth increased. The brain was

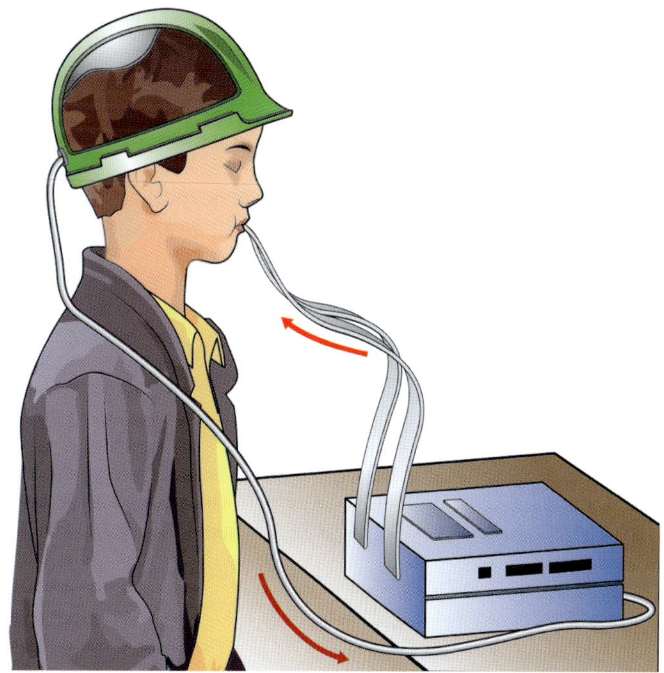

Figure 2.20 The apparatus using brain plasticity therapy to restore the sense of balance

Figure 2.21 Barbara Arrowsmith-Young

gradually bypassing the damaged area of her brain using alternative pathways, so she no longer needs the apparatus because the debilitating tumbling has been remedied.

A second example in the book concerns Barbara Arrowsmith-Young (Figure 2.21), who was born with a devastating array of learning impairments, including major problems with literacy and numeracy. For example, Barbara couldn't read a clock because she couldn't understand the relationship between the big hand and the little hand (Doidge, 2007). Her teachers told her mother that she would never learn like ordinary children, but she now runs a centre in her home city of Toronto, Canada, assisting children with learning difficulties. How could she go from being a child with such learning impairment to become a teacher using brain plasticity to assist other children with impairments? Her road to recovery began when she was told about Alexander Luria's (1973) book *The Man With a Shattered World*. She realised that her brain damage was rather similar to that soldier, which told her the location of the lesion in her brain. She then heard about brain plasticity, which gave her hope that her teachers might have been wrong, and maybe she could overcome her difficulties.

RESEARCH LINK 2.2

Curiosity enhances hippocampus-dependent learning and memory

Gruber, M., Gelman, B. D. & Ranganath, C. (2014). States of curiosity modulate hippocampus-dependent learning via the dopaminergic circuit. *Neuron*, 84(2), 486–96.

When do students best remember what they were taught? Many students find they are more likely to learn and remember the topics that they feel curious about (Figure 2.22), but until recently, little was known about the mechanisms by which students' curiosity affected learning and memory.

Matthias Gruber and colleagues (2014) studied brain mechanisms behind the relationship between curiosity and learning using typical memory tests and fMRI. In this study, participants underwent learning experiences by watching a set of questions and their answers on a screen. After watching each question, the participants were asked to rate how much they felt curious about the answer. While waiting for an answer to the question, participants were shown irrelevant face images.

As shown in Figure 2.23, in both immediate and one-day delayed memory tests, participants showed better memory for information that they were curious about. Moreover, incidental learning (i.e. remembering irrelevant people's faces) is more likely to happen when the person is in a state of high curiosity (Gruber et al., 2014). A critical question is what is happening in the brain when people feel curious? fMRI results showed that

Figure 2.22 Curious children (aged five and eight) examining a snail in the grass

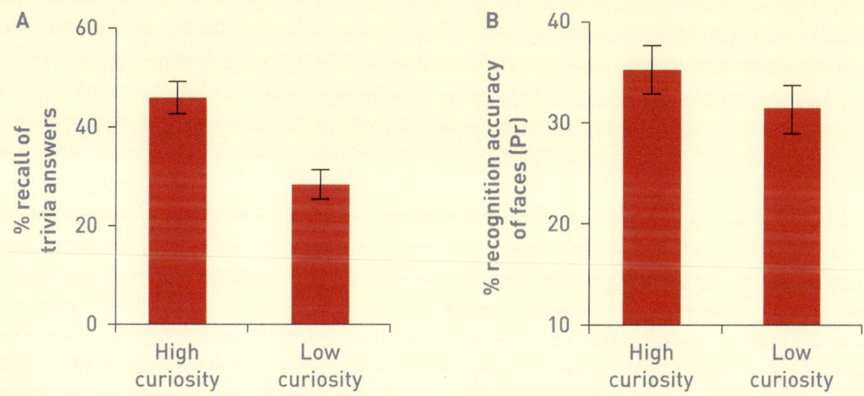

Figure 2.23 Curiosity enhances learning of information (A) and the recognition of faces (B)
Source: Gruber et al. (2014, p. 491).

activity in the midbrain and the nucleus accumbens (dopamine circuit) was enhanced when participants were at states of high curiosity. Furthermore, anticipatory activity in the midbrain and hippocampus, and the functional connectivity between these regions, showed positive correlation with retaining memory of images that one saw incidentally during the curious state. Gruber and colleagues (2014) concluded that the 'curious state' enhances dopamine release, which facilitates long-term potentiation (LTP) in the hippocampus (Figure 2.23).

STANDARDS   **APST and ACECQA curriculum specifications**

APST Standard 1.1 Physical, social and intellectual development and characteristics of students

Demonstrate knowledge and understanding of physical, social and intellectual development and characteristics of students and how these may affect learning

APST Standard 1.2 Understand how students learn

Demonstrate knowledge and understanding of research into how students learn and the implications for teaching

APST Standard 6.2 Engage in professional learning and improve practice

Understand the relevant and appropriate sources of professional learning for teachers

ACECQA curriculum specifications

1.1 Learning, development and care

1.6 Diversity, difference and inclusivity

6.5 Research

This chapter has provided newly qualifying teachers with introductory knowledge and understanding of the neural processes (synaptic plasticity) occurring in the brains of children and adolescents that underpin their overt physical, social and intellectual development. It does this by introducing readers to a distinguished teacher and school principal in the 1930s who realised, against the then prevailing view of behaviourism, that if educators are going to understand how children learn and develop, they have to come to grips with what is known about the biology of the brain. He became a highly distinguished brain scientist and laid important foundations for contemporary brain science by postulating what is called Hebbian learning. Most learning that occurs in schools is in fact Hebbian learning.

The chapter has also provided newly qualifying teachers with knowledge and understanding of research into how students learn and the implications for teaching. This includes providing an introduction to basic brain anatomy, mechanisms of synaptic plasticity, development of neuronal connections and the neural mechanisms of memory formation. By understanding neural mechanisms of learning and development, teachers will gain a new appreciation of the neurobiological reasons for the diversity and individual differences in learning and development found across students in classrooms. Knowledge and understanding of the biological basis of difference should help engender professional attitudes of acceptance, inclusivity and care.

SUMMARY

- The relationship between brain science and education can be traced back to Donald Hebb, who was a distinguished Canadian schoolteacher and school principal. He became convinced that if we are to understand how children and adolescents learn, we have to engage fully with what is known about the biology of the brain. He therefore left teaching and became a highly distinguished brain scientist.

- Hebbian learning remains a main foundation of contemporary brain science. Most of the learning that occurs in school is Hebbian learning. Hebb postulated that learning occurs when neurons and clusters of neurons become connected or *associated* as a result of repeatedly firing together in response to a stimulus, thus causing a change to occur in one or both cells. The more these connections are used, the stronger and faster the connection becomes.
- The cortex (outer layer of the brain) is divided into four lobes: frontal, temporal, parietal and occipital. The cortex forms about 80 per cent of the brain. The cerebral cortex comprises the cell bodies of the 80–100 billion neurons in the human brain.
- Subcortical structures in the brain include the corpus collosum, thalamus, superior colliculus, cerebellum and limbic structure, which basically includes the brain stem, thalamus, hippocampus and amygdala.
- The brain is formed in the womb as a result of cell migration, as newly generated neurons, formed from stem cells, migrate to different parts of the developing brain and self-organise into different brain structures.
- A neuron in the brain is a cell that is specialised for the transmission of information. It is characterised by long fibrous projections called axons and shorter, branch-like projections called dendrites. With each neuron, tiny electrical signals flow in one direction, from the receptor dendrites, through the cell to the axon and then to the axon terminals.
- When the electrical signal arrives at the axon terminal of a neuron, neurotransmitters (chemicals) are released into the synaptic gap or cleft. These are then available to be taken up at receptor sites on the neighbouring cells' dendrites, which in turn induce an electrical current in the receiving neuron that travels down through its axon, to repeat the same process in that cell.
- The neurons in our brains are constantly signalling to one another; in the process, they are forming neuronal pathways that may be strengthened as a result of repetition or weakened if not used. It is this dynamic process of neural 'sculpting' (creating neural connections and pathways and pruning others) that creates our unique set of memories and our own distinctive personalities.
- Memory is essential for learning. Researchers have discovered that a salient or meaningful learning stimulus, within a pathway in a part of the brain called the hippocampus, produces a high-frequency signal in the *presynaptic neuron*. This, in turn, allows calcium to flow through a specialised receptor on *the post-synaptic neuron* (called a NMDA receptor), which induces a process called long-term potentiation (LTP). LTP appears to be a necessary process in the formation of long-term declarative memory – the kind of memory involved in learning facts, events and ideas taught in school.

KEY POINTS FOR TEACHERS

- If you want to know what is happening when students are thinking, learning and memorising, you need to come to grips with what is known about the biology of the brain, and that requires a willingness to engage in cross-disciplinary conversation with brain scientists.

- In studying the science of learning and development in education, you are following in the footsteps of a very innovative and much-loved teacher, and distinguished brain scientist, Donald Hebb.
- Hebb's story should be a reminder to all teachers that we can never be sure how any child's life-trajectory might change, and how they might excel in future. One of the main reasons for this is the plasticity of the brain.
- Hebbian repetition results in how each of us comes to categorise the world. It is this dynamic process of neural 'sculpting' (creating neural connections and pathways, and pruning others) that creates our unique set of memories and world-views, and our own distinctive personalities.
- At school, in providing rich learning experiences, teachers are physically changing the brains of their students – teachers are quite literally 'brain changers'. No wonder teaching is such a responsibility, given that it has such a physical impact on the children and adolescents in our care. We should strive to do it with the greatest degree of care and love.
- Teachers need to be aware of the difference between declarative and non-declarative memory, especially when planning lessons that involve performative skills. The difference is not age or stage related; it is brain related – a result of how and where two biologically different kinds of memories are processed and stored.
- Teachers should always strive to make learning as meaningful as possible. We remember best what is salient and meaningful, and for very good neurobiological reasons. A salient or meaningful learning experience is one that induces long-term potentiation (LTP), a trigger for long-term memory.
- Memory is essential to learning and development, but it also makes us who we are. If we lose our memory, we lose something of our 'self'. Squire and Kandel (2009, p. 209) said that memory 'is the mental glue that binds together and interconnects our life's experiences'. A life without memory 'is a life in dissolution'.
- Memory also enables us to become educated and, because humans have the ability to communicate what we have learnt and memorised, we are able to create cultures that can be remembered and transmitted through time, from one generation to the next.
- When working with children in school, try to think of them as functioning brains that hold memories of personal experiences – some good, some not good – that make them who they are. Moreover, these brains have emotions (including fear and joy) that are also memorised.

REVIEW QUESTIONS

Guided responses

1. How does Hebb's theory of learning differ from behaviourism?
2. How do neurons communicate with one another via synapses?
3. What does synaptic plasticity mean?
4. Does the number of neuronal connections continuously increase as we become older?
5. Where in the brain is students' long-term memory of what is learnt in school stored?

FOOD FOR THOUGHT

1. As noted in this chapter, the resistance to studying the brain partly derives from behaviourism, which strongly influenced education, especially in the United States, before the onset of cognitive psychology. Based on what you have read in this chapter, what would you say to a classmate who argues that teachers don't need to understand the brain; all they need to concentrate on is what the children in their class actually do, and respond to that?

2. How does it make you feel when reading in this chapter that our memories make us who we are and if we lose our memory, we lose something of our self? Within the social sciences, there is often a strong emphasis on personal identity. Do you see any conflict between these two sets of ideas, or are they entirely compatible?

3. Why do you think some people view science as cold and detached, as a body of dry facts that has no place for human feelings? Given what you have experienced in this book so far, in studying the science of learning and development, do you think this is an accurate portrayal of science? How would you bridge a perceived gap between science and the human emotions?

RESEARCH ACTIVITY: COULD YOU TEACH YOUR OWN STUDENTS WHAT IS HAPPENING IN THE BRAIN WHEN WE ARE LEARNING?

Teachers not only teach their own subject areas, but also need to help students understand what is happening when they are learning. Students in your classroom (e.g. pre-school aged children, primary or secondary students) could therefore benefit by learning basic brain anatomy and how learning and memory occurs in their brain. Do you think you could teach what you learned in this chapter to your own students in a developmentally appropriate manner? Or do you think that, even though it is possible with adolescent students, it is impossible to teach how the brain works to young children?

Actually, even young children are able to learn how the brain works when given engaging learning opportunities. Below are teaching toolkits designed by professional neuroscientists to teach how brain works to young children. Please look at their materials and think about the listed questions.

- British Neuroscience Association: 'Resources for teaching, learning and sharing neuroscience' and 'Webinars for kids'
- Washington University: 'Neuroscience for kids'
- Frontiers for Young Minds: 'Everything you and your teachers need to know about the learning brain'
- Two fun activities for primary school children (recommended by the British Neuroscience Association, cited above; see Figures 2.24 and 2.25)

Research activity links

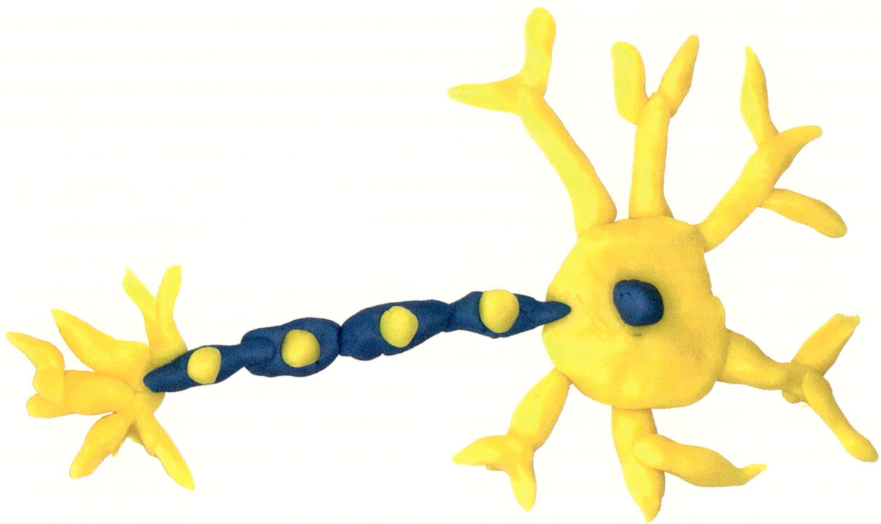

Figure 2.24 Play-Doh neuron made by a five-year-old child

Figure 2.25 Brain hemisphere hat

Questions

1. In these resources, which topics and concepts from this chapter are being taught to children?
2. What is the age-level of the children or young people engaged in the learning? How is the same topic or concept introduced differently across different groups of learners?
3. In what ways do the suggested activities arouse children's or young people's curiosity?

THE DYNAMICS OF LEARNING AND DEVELOPMENT

LEARNING OUTCOMES

By the end of this chapter, you will:

- Know why dynamic systems theory is often referred to as an overarching research paradigm and appreciate how its emphasis on the *processes* of change makes it different from much previous developmental theory as applied in education
- Know what is meant by saying that development is an emergent, self-organising process, with feedback, and appreciate where this account of development originated
- Have a basic understanding of how Edelman's theory of neuronal group selection contributed to Thelen and Smith's dynamic systems account of development, especially the role played by value and salience in the neuronal selective process
- Know what is meant by the philosophical problems of mind/body dualism and reductionism, know how reductionism contrasts with the core dynamic systems notion of emergent self-organisation and be able to discuss these issues critically

Dynamic systems as a coherent, overarching research paradigm

Towards the end of the twentieth century, there was a widespread feeling – especially within education – that the main parameters of human development were pretty well known and established. There were multiple theories and different approaches to development on offer as well as plenty of learning theories that, at least in part, overlapped with theories of development.

As noted in Chapter 1, during the second half of the twentieth century and well into the twenty-first century, the developmental theories of Piaget provided a clear framework of *developmental change*, based on the notion that learning and development progressed through discernible and discrete developmental *stages*. For example, in a paper published in 2008, we read that Piaget's 'work on children's quantitative development has provided mathematics educators with crucial insights into how children learn mathematical concepts and ideas' (Ojose, 2008, p. 26). Moreover, it was widely believed that these 'crucial insights' could be applied directly in the classroom, as 'knowledge of Piaget's stages helps the teacher understand the cognitive development of the child as the teacher plans stage-appropriate activities to keep students active' (Ojose, 2008, p. 29). For this author at least, Piaget's stages were not only efficacious for lesson planning but also classroom management, 'keeping students active'.

Piaget's stages also provided the notion that if education and schooling are to be successful, students need to be developmentally *ready* for what they are asked to learn – for example, there is no point in trying to get young children who can only think concretely (Stage 3) to learn abstract ideas (Stage 4) (Inhelder & Piaget, 1958). The concept of *readiness for learning* was applied generally across the curriculum, and was viewed as especially relevant with regard to 'a child's cognitive developmental readiness for reading and mathematics' (Arlin, 1981).

Details of Piaget's four stages (Figure 3.1) are readily available, so it is not necessary to repeat them here. For our purposes, however, it is important to note that, although it was widely recognised that any given child's progression through the stages would be highly individualised, it was nevertheless believed that all children would systematically and predictably progress on a linear, onward and upward trajectory from a known starting point to a known end-point. Moreover, it was possible to determine the stage that each child had reached and predict what would come next. Variability existed between children, but there was little recognition of variability within each child's development, or that development could actually regress. However, this is a problem 'because, as teachers often discover, each child's trajectory in every aspect of their maturation is both progressive and

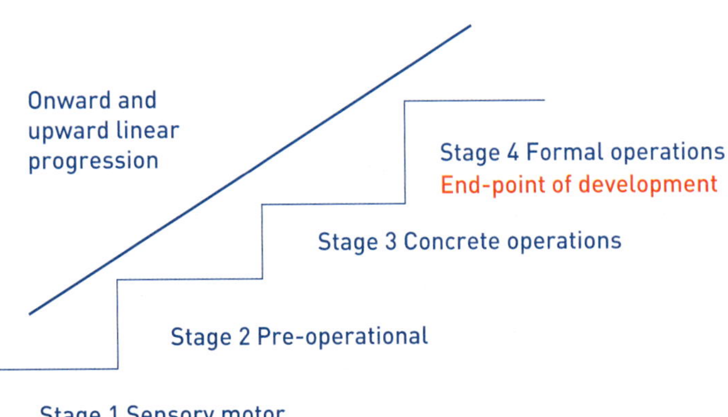

Figure 3.1 Summary of Piaget's four linear stages of cognitive development, onwards and upwards from a known starting point to a known end-point

regressive' (Kim & Sankey, 2010) – a point to which we will return shortly, as it is a key idea in a dynamic systems account of development.

From the 1970s, Vygotsky's emphasis on the social/cultural aspects of learning and the importance of language development was added into the mix. Indeed, by the end of the twentieth century and well into the twenty-first century, there were plenty of theories and approaches on offer – more than enough to keep student teachers busy in their studies and plenty to aid the teacher in meeting the learning and development needs of children and adolescents in their care. However, the winds of change were starting to blow. Problems were starting to emerge.

A proliferation of theories but no overarching coherence

One major problem experienced by students when attempting to learn these very many various theories and approaches was to see how they might all fit together. It was as though we had many bits of a jigsaw puzzle, but not the picture – and there were also some key bits missing. More fundamentally, it was difficult to grasp what the discipline of human development (child, adolescent, adult development) actually entailed. Paul Van Geert (2008), one of the pioneers of dynamics theory, suggests that one way to get a good idea of what an academic discipline such as human development entails is to look at introductory student textbooks. He then adds that the main picture revealed through such textbooks, in the mainstream discipline of human development or developmental psychology, is that development

> is basically a collection of perspectives and approaches (theories), of influences on development (e.g., genes, environment), of aspects or dimensions (e.g., physical, cognitive), of phenomena' (e.g., attachment, conservation), spread out across the life span or part of it, in phases or ages … Developmental psychology is apparently not a first-principles-based science. There seems to be no fundamental developmental mechanism, the understanding of which forms the key to a thorough understanding of the emergence of developmental phenomena. (Van Geert, 2008, pp. 242–3)

There are lots of bits, he says, but 'no fundamental developmental mechanism', nothing to pull all the bits together, a discipline with no overall coherence and 'no key to a thorough understanding of the emergence of developmental phenomena' (Van Geert, 2008, p. 243). Another way of making the same point is to say that the discipline of human development appeared to contain multiple theories and approaches but no *overarching paradigm* (Kim & Sankey, 2010) that unifies them into a whole, such as quantum theory in physics or the process of natural selection in biology. It also had nothing that fundamentally addressed the basic questions of what is actually *developing* in human development and how is it developing. There is a picture of change, but what is the *process?*

That changed when dynamic systems theory (DST) started to be applied to human development in the mid-1990s. DST provides developmental science with the 'fundamental developmental mechanism' or process that Van Geert (2008) maintained the discipline was missing. It is not just one more theory to add to the ever-growing list, but a scientific research *paradigm* (Kuhn, 1970) or scientific *research program* (Lakatos, 1970) that is able to incorporate some aspects of previous theory while also providing a critical perspective on it and going beyond it. Hence DST may also be referred to as a coherent metatheory (Overton, 2007), using the Greek word *meta* as a prefix, meaning beyond (in the sense of being more comprehensive or transcending). As a rough guide, one can picture it in terms of a Venn diagram (Figure 3.2).

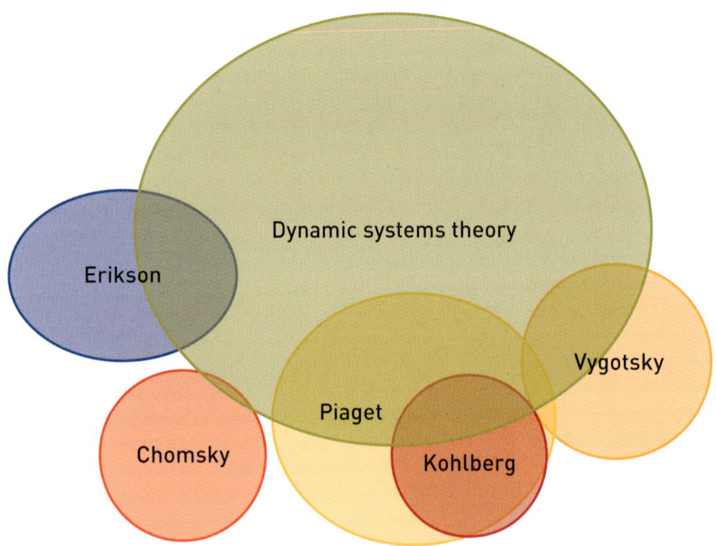

Figure 3.2 DST as an overarching research paradigm or research tradition that is able to accommodate many aspects of previous theories (though rejecting others) but also going beyond them

Some fundamental problems with 'stage' and 'age-phase' theories

From the perspective of DST, there are a number of problems with conventional stage and age-phase theories. One major problem is that they tend to provide an account of development that is unrelated to the multiple contexts and real-life situations in which development occurs. Another problem is that most have little to say about the role of the brain as the locus of learning and development. Moreover, what they do say seems to be very reliant on common sense – based on what we observe on a regular basis as babies become infants, infants become children, then adolescents, onward and upward towards the inevitability of adulthood. One can see evidence of 'stages' and 'age-phases' whenever one observes children over a period of time; as relatives sometimes exclaim, 'She's really grown up since I last saw her!' But common sense and casual observation can be misleading.

Certainly there is a time on the trajectory through life when a child can't do something, then they can. And it seems to follow a familiar pattern – all children seem to go through the same developmental 'stages', one after the other. It seems predictable, just like we expect that the Sun will *rise* in the eastern sky tomorrow morning. However, science often challenges common sense. The ancient theory of an Earth-centred universe was based on commonsense observation that the sun rises in the morning and sets before the onset of night. But the Sun doesn't rise or set. There is a Sun and there is an Earth, but the Sun doesn't go around the Earth and therefore doesn't rise in the sky, although we see it doing so every day of our lives. The Polish priest Nicolaus Copernicus (1473–1543) challenged that idea in 1543 with his book *De revolutionibus orbium coelestium* (On the Revolutions of the Celestial Spheres) (Copernicus 1543).

Development occurs: infants really do develop towards childhood and then go through puberty and adolescence towards adulthood; however, there are no pre-determined stages or age-phases through which they pass, even though that is what we seem to observe. There is no designated staring point or end-point, and development doesn't follow a linear, onward-and-upward path as stage theories claims. The core problem with stage and age-phase theories is actually a manifestation of a more fundamental issue, a view of development as *orderly* and *systematic*. Consider, for example, how the following authors defined human 'development'.

> Development refers to the systematic continuities and changes in the individual that occurs between conception … and death. By describing changes as 'systematic' we imply that they are orderly, patterned and relatively enduring, so that temporary mood swings and other transitory changes in our appearances, thoughts and behaviours are therefore excluded. (Shaffer & Kipp, 2007, p. 2)

Notice the emphasis on *systematic continuities* in development that are said to be 'orderly, patterned and relatively enduring'. Even more telling, notice that in emphasising *continuities*, this definition excludes discontinuities. It also brackets out (intentionally ignores) all that is contrary to systematic continuity in development – 'all that is messy and highly variable, including mood swings and transitory changes of thought and behaviour. Indeed, it seems to discount much of what makes us the persons we are. But why? And is the exclusion of discontinuity and the messy, variable side of child development helpful for teachers' (Kim & Sankey, 2010, p. 82) when trying to understand the complexity of students they encounter in school and classroom? Isn't that what they most want to understand when dealing with 'difficult' children?

One major reason for bracketing out **variability** and the messy aspects of development is the way child and adolescent development has traditionally been researched. Briefly, it has been common practice in developmental studies to conduct *group-average* and **cross-sectional sampling** that compares groups of subjects at a number of different ages at a specific moment in time (Kim & Sankey, 2010). Development is said to have occurred if statistically reliable differences are found in the mean levels of performance at the different age levels (Kim & Sankey, 2010).

This approach carries the added implication that if variability among individuals is high, it is difficult to detect group differences (Kim & Sankey, 2010). So the variability and diversity that we see across individuals as they develop is a problem for this kind of research; it is viewed as *noise* in the system and noise is bad because it doesn't produce statistically reliable results. As a result, the instabilities and variabilities, and other messy aspects of development that would be of particular interest to teachers, are usually systematically excluded in cross-sectional, group-average research.

But this is an example of the tail wagging the dog. A view of development as orderly and systematic that has been adopted for a methodological reason (eliminating 'noise' in the data in order to detect group differences) has become an ontological claim about the very nature of child development – that it is, in fact, orderly and systematic, when actually it is not like that at all. We will come upon a similar philosophical problem later in this chapter when discussing 'reductionism' in science.

variability: differences or inconsistencies observable from both *within* and *across* individuals. During the processes of learning and development, both the amount and the quality of variability can change. Often, increased variability provides flexibilities in performance.

cross-sectional sampling: a data-collection approach that gathers data from individuals of various ages or developmental levels to study the differences among them

REFLECTION

In your learning so far, have you encountered two or more different theories of learning and/or development that seem to be incompatible – for example, those that focus on group averages and those that focus on individuals? Were you able to resolve the apparent inconsistency, or did you just ignore it?

Development as messy, exploratory, opportunistic and context sensitive

As we noted in Chapter 1, the neat and predictable view of development, with its systematic continuities and orderly, progressive, incremental trajectory towards a known end-point, was simultaneously challenged in 1994 by Esther Thelen and Linda Smith (1994), working in the United States, and by Paul Van Geert (1994), working in Europe. These scholars applied the conceptual framework of complexity theory, or 'dynamic systems', to understand the *process of change* in human development. One important change involved re-setting the research focus. Instead of focusing on group averages, DST is interested in the development of the individual. Instead of providing a cross-sectional snapshot, like a photograph capturing a moment in time, DST is interested in what happens over an extended, longitudinal time-span, like a video. As a result of changing the focus, Thelen and Smith state that the picture that emerges is of development as 'messy'. It is also 'exploratory, opportunistic' and 'context sensitive' (Thelen & Smith, 1994, p. xvi).

Figure 3.3 When coming into land, the seascape that had appeared orderly and placid from high above, from afar, starts to look dynamic and chaotic from below

Now, that's a very different picture, which brings all the individual variables and differences into play. Thelen and Smith referred to the change of perspective as the difference between 'the view from above and the view from below' (Thelen & Smith, 1994). When flying in a plane very high above the sea, it looks blue and placid, but as the plane descends to land the same seascape can look stormy, dynamic and chaotic. It is not the sea-state that has changed, it is the perspective – from above and then below, from afar then magnified when near. It is as if DST places development under a microscope. Previous theory is not necessarily wrong; it is just a view from above, at a low level of magnification, a general approximation, a group average, a cross-sectional view (Figure 3.3). What is required to understand the dynamics and complexity of the developmental change process is a view from below, with high magnification, looking at the detail of what is happening to individual children as they develop.

That provides a very different perspective on the developmental *change process*. To re-emphasise the descriptive terms used by Thelen and Smith (1994), 'development is messy' and what 'looks like a cohesive, orchestrated process from afar takes on the flavour of a more exploratory, opportunist … process' when viewed with higher magnification from below. It also appears to be a highly 'context sensitive' process. (Thelen & Smith, 1994, p. xvi). Let us look at these terms in a little more detail.

Development is not a linear, onward and upward progression as Piaget portrayed it, progressing from one clearly defined stage to the next. That is just a rough approximation; rather, if we look at the developmental trajectory of each and every individual child, the picture that emerges is one of considerable *individual difference and variability*. Although sometimes linear and quantitative (increased functioning and ability), child development is often non-linear and qualitative (Kim & Sankey, 2010, p. 83). We cannot determine which 'stage' a child has reached because any given child may vacillate between the putative stages, on one day appearing to have reached one stage and on another operating at a different stage.

Moreover, it is entirely possible for a child to regress, to go backwards on their developmental trajectory, before moving rapidly ahead again. Although the idea of development going backwards might seem contradictory, from a dynamic systems perspective behavioural instability is 'particularly important for understanding development, because it is frequently associated with transitional events' (Howe & Lewis, 2005, p. 249). In other words, the *discontinuities* and *transitory changes* that we saw were ruled out by Shaffer and Kipp (2007), above, are a core component of a dynamic systems account. They are significant because they provide a springboard for transition and advancement in the overall developmental trajectory of the individual.

Development is exploratory, opportunistic and context dependent. In the processes of learning and development, a lot depends on what the child is actually exploring, what catches their attention and what they find interesting at any given time, and what the child actually discovers in their exploration. Development taps into novelty opportunistically. It is therefore a very active process, as Piaget and other stage theorists would also insist, but it is much more tentative and fluid than the neatly rule-driven image portrayed by conventional development theory. As teachers know well, learning and development really does depend on what the child experiences and the physical, social and emotional context in which those experiences occur – *learning and development are inherently context dependent*. That implies a child's learning and behavioural performance will be situationally and contextually dependent – a claim to which we will return to in Chapter 9, when considering the affordance of learning environment, and in Chapter 10, when considering learning and assessment.

However, as noted above, a big problem with stage and age-phase theories is that they tend to provide an account of development that is unrelated to the multiple contexts and real-life situations in which development occurs. As teachers, remember that the *context* in which learning and development is occurring always matters, as does *history* – all that has happened before in a child's learning and development. With regard to learning, the idea that prior learning impacts present learning has often been emphasised, especially by constructivists, but from the perspective of a dynamic systems account it is not just prior learning that impacts a child's learning at any given moment, it is *prior everything* – including, for example, whether the child had breakfast and the emotional state of their home before coming to school.

REFLECTION

Reflecting on your own development and that of others you have observed, can you think of some examples of child development that clearly show the process of development as 'messy', 'exploratory', 'opportunistic' and 'context sensitive'?

Dynamic systems as a radically different account of learning and development

The differences between a dynamic systems approach and conventional cognitive theory, were clearly delineated when Thelen and Smith (1994, p. xiv) proposed a 'a radical departure from current cognitive theory. Although behavior and development appear structured, there are no structures. Although behavior and development appear rule-driven, there are no rules. There is complexity' (Thelen & Smith, 1994, p. xiv). In such a system, there is a 'multiple, parallel, and continuously dynamic interplay of perception and action, and a system that, by its thermodynamic nature, seeks certain stable solutions. These solutions emerge from relations, not from design' (1994, p. xix).

Moreover, the adoption of a dynamic systems perspective brings a radical change of vocabulary. Thelen and Smith (1994, p. xix) make it clear that in their dynamic systems account, they 'deliberately eschew [intentionally avoid] the machine vocabulary of processing devices, programs, storage units, schemata, modules, or wiring diagrams', instead substituting a 'vocabulary suited to a fluid, organic system, with certain thermodynamic properties'. So the child, with their embodied brain, is not a 'machine' – that's the wrong kind of vocabulary – but a 'fluid organic system'. They are not just replacing the language of Piaget, of 'schemas' and 'stage' theory, but also much information processing language, with its mechanistic analogies.

In addition to the problems of a 'machine vocabulary' in conventional accounts of learning and development, another problem is that, even when using a biological and organic vocabulary, it has often been analogous to the rather simplistic notion 'that "from small acorns mature oak trees grow", with predetermined starting points and programmed teleological ends; the result of some overarching design built into the organism, following stages, responding to schema of one kind or another'. However, 'there is no design, no stages, no schemas, just a process of *emergent self-organisation*' (Kim & Sankey, 2009, p. 287, emphasis added). The notion of emergent **self-organisation** is central to a dynamic systems account of developmental change.

self-organisation: 'describes characteristic relationships in emergent systems created by feedback mechanisms that either amplify an effect (positive feedback) or dampen an effect (negative feedback)' (University of Southampton, n.d.)

Figure 3.4 Is it possible that the flapping of a butterfly's wings could produce a tornado?

Development as an emergent, self-organising process with feedback

Could the flapping of a butterfly's wings in Brazil lead to a tornado in Texas? Although this possibility appears contrary to common sense and also Newtonian science, the answer is yes, at least in principle. This is known as the *butterfly effect* (Figure 3.4). Given the slightest perturbation (change) in the initial conditions of a weather system, over time it could *self-organise* through *feedback* (each event feeding into and changing the next) and *develop* into an unlikely and unpredictable major weather event. 'Self-organisation' and 'feedback' are closely interrelated, leading to the emergence of new and largely unpredictable developmental forms. Here is development on a grand scale in the natural world. Could it possibly apply to other forms of development – even human development?

Can you think of a real-life example of the butterfly effect in learning and/or development, particularly in school and classroom contexts, where what seems like a very minor change leads to a much more significant development?

The origins of dynamic systems theory

This question about a butterfly in Brazil was posed in 1972 by meteorologist Edward Lorenz, in a ground-breaking presentation at the Annual Meeting of the American Association for the Advancement of Science. His presentation laid down an important marker in the new science of dynamic, complex systems, and of course it introduced us all to the notion of the butterfly effect. There were, however, two earlier and very important pioneers of the theory. One was the British mathematician Alan Turing (1912–54), a mathematical genius (dare one say the greatest mathematician of the twentieth century) (Figure 3.5). He is especially known for his leading role in cracking the Nazi Enigma Code in 1942, which changed the course of World War II, and his pioneering work on the logic of computation. He is often referred to as the 'father of modern computation'.

Turing was also 'intrigued by biological patterns such as leaf arrangement and the spots on butterfly wings and dalmatian dogs and proceeded to show how biochemical reactions of sufficient complexity

Figure 3.5 Alan Turing (1912–1954)

(nonlinearity)' (Goodwin, 2001, p. 106) could produce those kinds of biological, stationary wave patterns. This focus on pattern formation in nature led to him writing an astonishing paper, 'The Chemical Basis of Morphogenesis' (Turing, 1952), which provided a mathematical formula for the biological process by which neurons become differentiated in the prenatal brain. Nobody had ever done that before.

At much the same time, Boris Belousov was working at the Soviet Ministry of Health in the Soviet Union. Already a distinguished scientist, he was researching how human bodies extract energy from glucose. In the process of mixing chemicals, something very unexpected happened. As he added in the last chemical, the clear mixture changed to being coloured but then it changed back again. If you add blue ink to clear water it will change to being blue, but it won't change back to being clear again. Even more strangely, the chemical mix started to oscillate from being clear to coloured and then back to being clear again. Unfortunately, working behind the Iron Curtain meant Belousov was not aware of Turing's work, because if he had been, he would have realised that what he had actually discovered was that, under certain conditions, chemicals may exhibit *emergent self-organisation*.

Belousov wrote up his discovery as a journal article, but the response he received from the reviewers was that his findings were impossible; they were incompatible with the laws of physics (as formulated at that time). The reviewers concluded that his results must be the result of careless work. Belousov was devastated and soon afterwards he abandoned science altogether, a victim of bias towards established knowledge in the process of peer review, with its presumed 'objectivity' and 'reliability'. One wonders why the reviewers didn't try to repeat the experiment, but clearly they didn't. In fact, in a version of the experiment known as the Belousov-Zhabotinski reaction it is highly repeatable. In what appears to be a mix of quiescent (inactive and dormant) chemicals, novel patterns emerge as though from nowhere, as a result of self-organisation, with feedback (Figure 3.6). This discovery was especially important because, as Nobel Laureate Ilya Prigogine (1997, p. 67) notes, although, 'Today many other oscillatory reactions are known … the Belousov-Zhabotinski reaction remains historic because it proved that matter far from equilibrium *acquires new properties*.'

Figure 3.6 The Belousov-Zhabotinski reaction. Patterns *emerge* as though from nowhere, as a result of self-organisation, with feedback.

dynamic (or complex) system: a system of many components whose interactions give rise to the collective emergent patterns of the entire system

The science of **dynamic systems** has its roots in mathematics, chemistry and physics, but it is widely applied, in particular to phenomena that are both *complex* and are said to exist *far from thermal equilibrium* as *non-linear, self-organising*, 'dissipative structures' (Prigogine, 1997, p. 66). These are structures that draw on a high energy source (in humans, the source is oxygen and nutrients) to do work before giving some of the energy back to the environment. These are also 'open systems' in which, with a sufficient injection of energy, new and ordered structures and patterns of behaviour may *spontaneously emerge*. All *living structures are emergent, dissipative, self-organising systems*; from the dynamic patterns of gene activities in developing organisms (Kauffman, 1995) to the behaviour of ant colonies and the functioning of human brains (Solé & Goodwin, 2000). The children you meet in school are *emergent, dissipative, self-organising systems*.

Children as complex, emergent, self-organising beings

What does all this mean for understanding child and adolescent development in education? The main thing to notice is that child and adolescent development is not a stand-alone phenomenon, with its own unique processes. Rather, it is deeply embedded in processes that are shared by all living creatures and many non-living systems, as emergent, self-organising, complex structures – or, to put it more simply, *development is a result of complexity, and that entails the process of emergent self-organisation.*

As teachers, try to view children and adolescents as *complex beings* and try to view their development as a product of *emergent* self-organisation. It is not predetermined, fixed by their genes or by putative developmental stages or age-phases that are somehow imprinted in their make-up, providing a ready-made pathway for them to follow. Rather, patterns of development *emerge* through self-organisation, like the patterns in a Belousov-Zhabotinski reaction. That is why children are all different and can vary so much in their own personal development. But exactly the same self-organising processes that allow for variability also produce the stabilities and similarities we observe in development. Development is not so random that anything can happen; rather, it is *constrained by process, context and history.*

As Thelen and Smith (1994, p. 73) note, the 'cornerstone of a dynamic theory of [human] development is this emergent nature of behaviour assembled in real time'. Thus:

> even behaviors that look wired in or program-driven can be seen as dynamically emergent: behaviour is assembled by the nature of the task, and opportunistically recruits the necessary and available organic components (which themselves have dynamic histories) and environmental support. (Thelen & Smith, 1994, p. 73)

However, notice that *there is no 'self' in self-organisation.* The word 'self-organisation' within a dynamics account of development is not referring to the human 'self' and its autonomy – how a child might consciously decide to organise their own self. Rather, we are saying that development self-organises *spontaneously,* without the need for some inner mechanism or external agent. Think, for example, of the patterns that emerge in desert sand dunes.

There is no 'self' involved in the self-organisation of sand dunes. Patterns of similarity and difference *emerge* as a result of the desert wind acting on the particles of sand (Figure 3.7). There is no 'hidden hand', no predestined master plan or stages lurking within the system, just *process, context and history.* (*Process* – sand dunes self-organise through constant feedback,

Figure 3.7 Similarity and difference in the development of sand dunes as a result of self-organisation

there is no master plan; *context* – the desert, the abundance of sand, heat, force of wind, wind direction, etc.; and *history* – all that has happened before combines to produce that particular sand dune, at that given moment in time.) The patterns produced through self-organisation involve what are called *feedback loops*, where what has happened previously feeds into and changes what happens next, which in turn feeds into and changes what happens next, constantly and dynamically shaping development.

Figure 3.8 The formation and self-organising development of a mountain stream, a metaphor for human development

For another example, drawn from the natural world, consider the formation and subsequent development of a mountain stream (Figure 3.8), used as an analogy for development by Thelen and Smith (2006). The formation of a *new* stream is not pre-determined; it results from rainfall or the melting of snow and, at the point of formation, a newly formed stream doesn't have a ready-made course to follow, as its course is not pre-designed. Rather, it moves in and out of small rocks and bigger boulders, gathering other streams on its way down the mountain side. Then:

> At some places, the water flows smoothly in small ripples. Nearby may be a small whirlpool or a large turbulent eddy. Still other places may show waves or spray. These patterns persist hour after hour and even day after day, but after a storm or a long dry spell, new patterns may appear. Where do they come from? Why do they persist and why do they change? No one would assign any geological plan or grand hydraulic design to the patterns in a mountain stream. Rather, the regularities patently emerge from multiple factors: The rate of flow of the water downstream, the configurations of the stream bed, the current weather conditions that determine evaporation rate and rainfall, and the important quality of water molecules under particular constraints to self-organize into different patterns of flow. (Thelen & Smith, 2006, p. 263)

Children are dynamic systems. If, as teachers, we are going to understand children and how they learn and develop, we constantly need to bear in mind that they are dynamic systems. As Howe and Lewis (2005, p. 273) note:

> If our minds were not inscribed in flesh, we would not have to worry about the properties of complex systems. But our minds are greatly dependent on our brains, and brains are designed by evolution to self-organise rapidly under the sway of experiences and the emotions that colour them. Therefore, to understand developing minds we need to understand developing brains, and the principles of self-organisation provide a foundation for doing so.

So, as a teacher, whenever you engage with children in your care, try to be aware that your students are developing brains and that their learning and development is a highly dynamic, self-organising process, occurring moment by moment in real time. It doesn't come in chunks, stage by stage; children are constantly learning and developing as they engage with all they experience, including the experience of talking to you, at any given moment in time.

Try to remember too that *development is always multi-causal* and that, given any particular task, children will opportunistically draw on multiple available cues, moment by moment. The smallest, most seemingly insignificant cause can, through ongoing feedback, self-organise to produce a significant developmental effect.

Moreover, the causal components contributing to each child's development and attendant behaviours are said to be *softly assembled*, which means the multiple interacting and interweaving causal components 'can be freely combined from moment to moment on the basis of the context, task and developmental history of the organism' (Gershkoff-Stowe & Thelen, 2004), which gives behaviour and development 'an intrinsic sense of exploration and flexibility' (Spencer et al., 2011, pp. 261–2) and the fluidity of an organic system, as mentioned above.

Emergent self-organisation in motor skill development

RESEARCH LINK 3.1

A dynamic systems account of motor skill development is well established in developmental science, and it is also applied in physical education. Nowadays, motor movement is viewed as an emergent property that self-organises from interactions of organismic constraints, task constraints and environmental constraints (Newell, 1986, Newell & Jordan, 2007; see Figure 3.9).

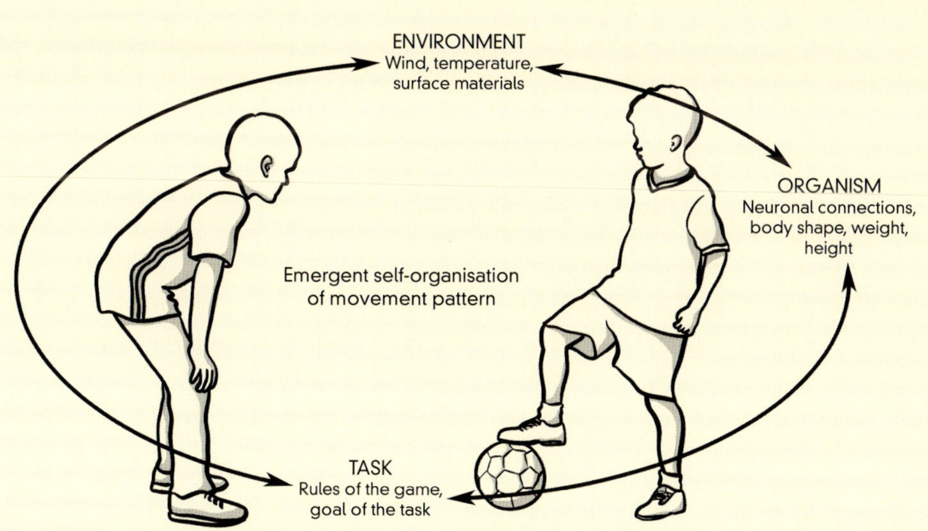

Figure 3.9 Emergent self-organisation of motor skills

Constraints place limits on degrees of freedom (e.g. possible motor skill repertoires in a given context). For example, during infancy, production of walking will not emerge until the strength and balance of an infant's legs are sufficient to allow for independent walking.

▶▶

Moreover, when an infant is just about to transit from crawling to walking, the quality of floor surfaces can improve or hinder the development of motor patterns (Ulrich et al., 1998).

A dynamic systems account of motor skill development rejects traditional views that motor skill development follows a simple progression from one skill to the next. From a dynamic systems perspective, changes in motor ability are considered to be non-linear, meaning that as one constraint changes, it in turn affects the other constraints, producing a new movement pattern. This means that no single component of constraints holds a privileged position over others, determining the movement patterns.

Moreover, the notion of variability – one of the key concepts from DST – provides very important insights for the teaching of motor skills. Hodges and Franks (2002, p. 809) strongly recommend that sport coaches should allow 'error and variability early on in learning' in order to facilitate each individual athlete's discovery of new movements. Meanwhile, instructions that build upon individual athlete's 'stable, pre-existing movement behaviours do not facilitate the acquisition of skills' (2002, p. 809). This type of instruction may prohibit early variability that is necessary to break away from 'undesirable movements' (2002, p. 809).

Edelman's theory of neuronal group selection

epigenetic: biological processes that do not depend directly upon gene expression. Development emerges in response to what the organism is currently experiencing and what has happened previously.

selection: a process in which environmental or biological influences decide which type of constituent parts succeed better than others

value: within Edelman's TNGS, the word 'value' refers to evolutionarily derived constraints favouring behaviour that fulfil homeostatic requirements or increases fitness in an individual species (Edelman, 1989, pp. 287–8). These constraints are provided by a selectional system in the brain, comprising diffuse ascending systems such as the dopaminergic system (Edelman, 2004, p. 180).

We now need to bring together what was learnt in Chapter 2 about the learning brain and brain plasticity, and what this chapter has so far presented about the dynamics of human development. Previously, we noted that a major problem with previous development theories is that most had little to say about the role of the brain as the locus of learning and development. From a dynamic systems perspective, brain development is at the core of human development. In their ground-breaking book, *A Dynamic Systems Approach to the Development of Cognition and Action*, Thelen and Smith (1994) drew heavily on Gerald Edelman's (1987, 1989, 2004, 2006) theory of neuronal group selection.

Edelman's theory seemed to be just the kind of theory that was required, because it provides a coherent, process-oriented **epigenetic** account of brain development from the womb onwards. The word 'epigenetic' (epigenesis) means that development emerges in response to what the organism is currently experiencing and what has happened previously (history). At the level of brain cells and also what we observe as overt behaviour, events that lead to increased complexity depend on the events that preceded them.

Experience is driving development, but how? What is the basic process? Edelman's theory posits that learning and development involve the key process of **selection**, applying the notion of natural selection to the strengthening or weakening of neuronal connection and neuronal pathways. As always in science, Edelman was building on what was already known, particularly the foundational work of Donald Hebb on synaptic plasticity. However, one big difference between Edelman and Hebb is the emphasis Edelman places on *selection,* which he argues is guided by *value and salience* – what the organism values and finds meaningful. **Value** is imposed in the brain by the brain – it's not just experience that is strengthening the connections between neurons; it also involves the *value and salience* attributed to experiences by the organism (Edelman, 1987, 1989) – the learning, developing child.

Learning to grasp a rattle

To help clarify these ideas and bring them together, let's consider what is happening when a baby, at about six months of age, learns to grasp a simple rattle that is presented to them (Figure 3.10). This is actually one of many experiments studied by Esther Thelen and her team (Thelen & Smith, 1994, 2006). The baby is seated upright in a chair with a comfortable but firm support. The baby sees the rattle and hears it, maybe smells it, but doesn't know how to grasp it. Conventionally, learning to reach out and grasp the rattle had been explained as resulting from it being genetically pre-programmed into the brain. Thelen and Smith's research, however, found that each baby has to work out how to do it by trying all manner of movements and then selecting those that help them locate the rattle in space, *selecting* those that are of *value* and are *salient* (have meaning), not least because they work, and grasping feels good.

Figure 3.10 A six-month-old infant attempting to grasp a rattle

What one sees when observing the baby is a large variety of random bodily movement – arms and legs moving in all directions. To appreciate what is happening, it is important to understand what Edelman (1987) is referring to when he talks of primary and secondary repertoires, which you met in 'Synaptic signalling and the learning, developing brain' in Chapter 2. The **secondary repertoire** of synaptogenesis occurs after a child is born, but at the time of birth the baby already has more neurons than it will retain and it also has a **primary repertoire** of synaptic connections that mostly result from the constant and spontaneous firing of neurons in the absence of environmental stimulation: 'Neurons are excitable cells: they can fire off spontaneously, and their action potentials self-organise into massive waves that travel through brain tissue' (Dehaene, 2020, p. 102). This spontaneous process is in addition to the process of synaptogenesis described by Hebb (1949), where synaptic connections result from environmental stimulation.

The point is that, by the age of six months, the baby already has a rich primary repertoire of connections and a swiftly expanding secondary repertoire, and these provide a vast

secondary repertoire: neuronal groups (or local circuits) that have been selected among populations of synapses by strengthening some synapses and weakening others. Certain circuits and neuronal groups in such repertoires are more likely to be favoured over others in future encounters with signals or similar types (Edelman, 1989, p. 46).

primary repertoire: large numbers of variant neuronal groups (or local circuits) within a given anatomical region that are formed via the epigenetic generation of diversity (Edelman, 1989, p. 44)

number of potential firing patters. Some of these connections and firing patterns will prove important for learning and development, and will be strengthened by repeated experience. Others will not be used and will weaken or die (be pruned). The random body movements made by the baby result from random firings in the cerebral cortex of the brain. Once in a while, a firing pattern will result in the baby touching the rattle. This is recognised by the brain as emotionally good, valuable and salient, causing a chemical reaction that reinforces that firing pattern. When the move is successfully repeated, that particular firing pattern is strengthened and when repeated again, it is further strengthened. That is how selection works in the brain, not only for babies and infants, but for everyone.

theory of neuronal group selection (TNGS): an account of brain function proposed by Gerald Edelman (1987) that emphasises the role of value and salience in the formation of, and subsequent modifications to, the strengthening of synaptic connections and synaptic maps (or pathways)

Edelman's **theory of neuronal group selection (TNGS)** provided Thelen and Smith with a neurobiological theory that seemed entirely compatible with their observations, explaining how we are able to learn complex skills of all kinds, from learning how to grasp a rattle to the development of abstract language without being pre-programmed like a serial computer. However, it was doing much more than that: it was also providing a model of learning and development that is thoroughly human, in integrating notions of value and meaning (salience), into a biological theory of how we learn and how, over time, we develop into highly *individual persons or selves*. Although the neurobiological processes involved in learning and development are universal, the same for everybody, what they produce, as a result of neuronal selection (selecting for value and salience), is personal individuality, personal 'selves'.

Two philosophical problems

To bring this chapter to a close, we need to address two major philosophical issues that often arise in response to claims about neurobiology and human persons or selves. One is the problem of mind/body dualism and the other is reductionism. These are two separate issues, though there are overlaps, as you will see.

Both arise in the context dynamic systems theory and its account of learning and development. Basically, DST is non-dualist or what is also called monist, which asserts that *mind and brain are one and the same thing*. DST is also non-reductionist, which means it provides an emergentist account of learning and development, as you discovered previously in this chapter (emergent self-organisation). It is appropriate to discuss these two issues now because experience shows that by this point in studying the science of learning and development, some students and their tutors begin to feel a bit uneasy about the claims being made. For example, in emphasising the brain, aren't we overlooking the role of the human *mind* and its learning? And aren't we in danger of reducing the human 'self' to nothing more than its brain, and the molecules and atoms that compose its brain?

Mind/body dualism

What is the relationship between you and your brain? Or, within an educational context, what is the relationship between your students and their brains? In 1977, two very famous twentieth-century scholars, Sir Karl Popper (1902–1994) and Sir John Eccles (1903–1997) wrote a book called *The Self and Its Brain* (Popper & Eccles, 1977) (Figure 3.11). This was a very deliberate and highly controversial title. Can you imagine why? Karl Popper is often regarded as the greatest philosopher of science in the twentieth century and John Eccles, an Australian born in Melbourne, won the Nobel Prize for Physiology or Medicine in 1963 for his work on the synapses. So these were very high-profile scholars. Why, then, was the title of their book so controversial?

Bringing the issue up to date and relating it directly to education, recently when reviewing an academic paper on brain science and education, one anonymous reviewer said: 'The idea that learning occurs in a brain and is necessarily the product of a brain is reductionist, pseudo-scientific and false … learning is not an activity of a brain, but of a person, mediated by a teacher and often in a classroom in the company of other persons.' In this comment, which is not uncommon in education – especially the philosophy of education – you may notice the same distinction between the brain and the person (self) that is implied in the title of the book by Popper and Eccles. So, back to our original question: What's the relationship between you and your brain? Are you and your brain two different entities – two different things – or are they one and the same? According to Popper and Eccles (1977), the self *has* a brain – but is that right? Should it not be that the self *is* a brain? When applied to education, the reviewer is claiming that learning is not an activity of the brain (as this book claims it is); rather, it is an activity of a person and, in classrooms, this is mediated by another person (a teacher), often in the company of other persons. So what is this 'person' that exists separately from its brain and what does it mean to say that 'learning is not an activity of a brain'? For example, isn't it often the case that damage to the brain impacts a child's ability to learn?

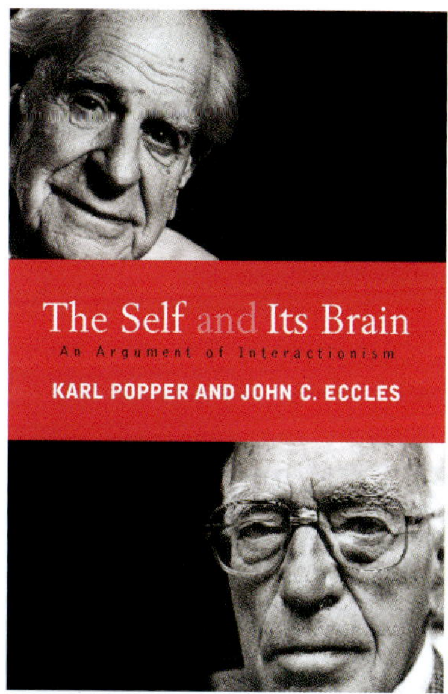

Figure 3.11 Karl Popper and John Eccles' book *The Self and Its Brain*

REFLECTION

What is the relationship between you and your brain? Or, in an educational context, what is the relationship between your students and their brains? Do you view students as persons (perhaps socially constructed persons) who are distinct from their brains? If possible, discuss this perplexing issue with classmates.

The key issue has been a longstanding problem in Western philosophy called mind/body dualism or mind/brain dualism (see also Chapter 1). It is especially associated with the seventeenth-century French philosopher René Descartes (1596–1650), who you met in Chapter 1, although it goes back further than that – at least as far as the ancient Greek philosopher Plato (c. 427–347 BCE). At the end of Chapter 2, we noted that there are a number of teachers who are resistant to neuroscience. Part of the reason seems to be the emphasis they place on role of *persons* (not brains) in the learning process, as seen in the comment by the above reviewer.

Persons, it is often said, are *social beings* with *autonomy* and *agency*. As noted in Chapter 2, they are also *emotional beings*, with feeling and a 'heart' (used metaphorically to mean feelings and emotions). This, it is asserted, is precisely how students in our schools should be viewed, as human, social persons, not as biological selves. Education is not a biological science, it is claimed; it is a social science or perhaps a human science, belonging within the humanities. However, it is not at all clear why we have to create this dichotomy – why can't it be both social and biological? Moreover, it is difficult to make such claims without them appearing to separate the person (mind) from its biological body, and nowadays that is generally considered very problematic, philosophically, biologically and even educationally. Instead, what is usually maintained is some kind of **mind/brain identity theory** or monism.

mind/brain identity theory: the philosophical view that processes of the mind are identical to states and processes of the brain

The many strands feeding into the philosophical debate are beyond the scope of this book, but if you are interested, they are very clearly stated in K.T. Maslin's (2001) book, *An Introduction to the Philosophy of Mind*. However, as Maslin (2001, p. 71) notes, the basic point is that 'dualism affirms that mental events can never be identical with physical events'. By contrast, 'the mind/body identity theory denies what dualism affirms … The mind is identical with the living brain and mental events are just brain events. There are no ghostly immaterial substances or events to constitute our minds' (2001, p. 71).

But this immediately raises a worrying problem: it suggests that what we call our mind, or our personal self, simply *is our brain*. To come back to our original two questions, it means there is *no relationship between you and your brain*, or between *your students and their brains*. You *are* your brain. So the question asked at the start of the section is actually flawed: it is a dualist question (dualism is built into the question), suggesting that you are different from your brain when, according to mind/brain identity theory, they are one and the same thing. Now, how does the claim that you *are* your brain make you feel?

The worrying problem of reductionism

reductionism: reducing a phenomenon to its basic constituent parts. Science is therefore methodologically reductionist when analysing a phenomenon in terms of its constituent molecules, atoms and sub-atomic 'particles', although that does not necessarily entail ontological reductionism – the claim that a phenomenon is *nothing but* its parts.

You are your brain and perhaps we should go further and say you are *nothing but* your brain. Does that feel right? The students you meet as a teacher are brains and *nothing but* their brains – is that correct? This *nothing but*, or *no more than* claim is what is called **reductionism,** the word used by the education reviewer in the quotation above. For very many people in education, it is totally unacceptable, and one likely reason why they feel very uneasy about the *science* of learning and development.

That is echoed in what the reviewer cited above said: 'learning is not an activity of a brain, but of a person'. Whether intentionally or not, the reviewer seemed to be embracing dualism separating person/brain in order to reject reductionism – the idea that learning is nothing but the activity of the brain. However, dualism and reductionism are not the same and need to be kept apart. The statement that *you are your brain* is opposed to dualism and asserts mind/brain identity, but that does not necessarily imply reductionism. The statement that you are *nothing but,* or *no more than,* your brain is reductionist (Figure 3.12).

Figure 3.12 Are these people studying atoms and molecules in motion? Is the *Mona Lisa* reducible to its constituent parts or is it more (very much more) than that?

To be absolutely clear, although what you read in this textbook accepts mind/brain identity (you *are* your embodied and embedded brain), it doesn't embrace reductionism (you are *nothing but* or *no more than* your brain). The rejection of reductionism in this textbook is entirely in line with complexity theory (Kauffman, 2008) and hence the dynamic systems approach to learning and development that underpins the textbook as a whole and that has been introduced in this chapter. That's why it is appropriate to discuss it now. Basically, a dynamic systems approach is *emergentist*, not reductionist.

Earlier in this chapter we said, 'As teachers, try to view children and adolescents as complex beings and try to view their development as a product of *emergent* self-organisation'. There's the key word: *emergent*. **Emergence** is the opposite of the reductionist notion that we are *nothing but*, can be reduced to what constitutes our brain. To see how this works, let's examine a classic reductionist statement, as advanced by Francis Crick (famous for his work in discovering the double-helix structure of genes).

In 1994, Crick published a book in which he claimed, 'You, your joys and your sorrows, your memories and your ambitions, your sense of personal identity and free will, are in fact no more than the behaviour of a vast assembly of nerve cells and associated molecules' (Crick, 1994, p. 3). He called this his 'astonishing hypothesis', but to be honest it is pretty much what most neuroscientists believe – so perhaps not very astonishing after all. But is he correct?

There's an important distinction to be made. Science has traditionally been reductionist in the way it conducts its research, taking things apart, understanding the parts (how they are composed and function) and then examining the parts that constitute the parts, all the way down to the laws of physics. That is *methodological* reductionism, and it is what scientists do. But the reductionist methodology scientists use to investigate the physical world (methodological reductionism) does not necessarily imply *ontological* reductionism, the claim that the world and everything in it (including human beings) is *nothing but* and is reducible to its parts. Indeed, it can be argued that 'reductionism alone is not adequate, either as a way of doing science or as a way of understanding reality' (Kauffman, 2008, p. 3).

Consider, too, distinguished neuroscientist Michael Gazzaniga's (2011) claim that, 'Reductionism in the physical sciences has been challenged by the principle of emergence. The whole system acquires qualitatively new properties that cannot be predicted from the simple addition of those of its individual components' (Gazzaniga 2011, p. 134). The point Gazzaniga is making is that you can reduce anything to its parts, downwards to the laws of physics, but if you reverse the process, from the bottom upwards, *you will not get the macro story from the micro story*. Something happens when going from one organisational level to the one above, called a phase-shift. So, 'while the substrate of the human self *is* the brain and its neurobiology, that is not the whole story, at all. Included in the new story, going from the microbiological level to levels above, is the emergence of social and interpersonal interaction, including our sense of responsibility and freedom to choose in our dealings one with another' (Sankey & Kim, 2016, p. 123).

In short, that means you are very much more than the sum of your biological and chemical parts. Physically, of course, you are the sum of all your parts, as is everything else in the natural world, but what *emerges* from those parts is all that we refer to as the thinking,

emergence: the ability of individual components within a large system to work together to give rise to novel patterns of behaviour that are often not predicted by an understanding of the behaviour of each constituent part

feeling, emotional person, or self. Nevertheless, the self doesn't *have* a brain, as Popper and Eccles (1977) believed; rather, the brain *is* the embodied and physically, socially, culturally embedded self. That seems to be a very important point for all teachers to hang on to when engaging with the science of learning and development. But what do you think?

STANDARDS   APST and ACECQA curriculum specifications

APST Standard 1.1 Physical, social and intellectual development and characteristics of students

Demonstrate knowledge and understanding of physical, social and intellectual development and characteristics of students and how these may affect learning

APST Standard 1.2 Understand how students learn

Demonstrate knowledge and understanding of research into how students learn and the implications for teaching

APST Standard 6.2 Engage in professional learning and improve practice

Understand the relevant and appropriate sources of professional learning for teachers

ACECQA curriculum specifications

1.1 Learning, development and care

1.6 Diversity, difference and inclusivity

This chapter has opened up a genuine problem that newly qualifying teachers often encounter when engaging with professional learning in teacher education. They are often presented with a proliferation of theories of learning and development that can seem unrelated and inconsistent and that, taken together, offer a bewildering collection of perspectives and approaches that they are told should be applied to practice. Moreover, they are often introduced to these multiple theories in a relativistic or eclectic manner; all are given equal merit and together they are said to constitute the body of educational theory. There is often little or no opportunity for newly qualifying students to consider that some may actually be problematic by present-day standards.

This chapter introduces students to the science of complex systems, or dynamic systems theory (DST) as it is known in contemporary developmental science, which doesn't add to the stack of previous theories but instead provides an overarching metatheory or paradigm, able to accommodate aspects of previous theory but also go way beyond it. In this chapter, you have also been introduced to Gerald Edelman's theory of neuronal group selection, as a component of DST, concerned with the dynamics of the processes occurring in the brain that drive learning and development.

By understanding DST and neuronal group selection, which currently underpin much contemporary research relevant to education, newly qualifying teachers will be equipped with a comprehensive appreciation of how children learn and develop, and the biological reasons for diversity and individual differences among the students in their classroom. In the process, they will also encounter and critique the pitfalls of reductionism, which can so easily undermine viewing students as emergent human persons, and also untangle the problem of mind/body dualism – both problems found in research into the science of learning and development.

SUMMARY

- In the late twentieth century, the study of human (child, adolescent and adult) development appeared to contain multiple theories and approaches, but no overarching theory or paradigm. That changed when dynamic systems theory (DST) started to be applied to human development in the mid-1990s. DST is not just one more theory to add to the pile; rather, it provides developmental science with a fundamental developmental mechanism (metatheory or paradigm) that the discipline was missing.

- Development is a result of complexity, and that entails the process of emergent self-organisation. Although genes provide constraints on development, the developmental process is not pre-programmed. Rather, patterns of development *emerge* through self-organisation, like the patterns in the Belousov-Zhabotinski reaction.

- The patterns produced as a result of self-organisation involve what are called *feedback loops*, where what has happened previously feeds into and changes what happens next, which in turn feeds into and changes what happens next and so on, constantly and dynamically shaping development.

- A dynamic systems account of human development necessarily incorporates brain development and needs to be consistent with what is known about the way the brain functions. Gerald Edelman's theory of neuronal group selection (TNGS) provided Esther Thelen and Linda Smith (pioneers of DST) with a coherent, process-oriented epigenetic account of brain development. According to the TNGS, learning and development involve the key process of selection where connections that are used are strengthened, while others are weakened or pruned. Neuronal selection in the brain is guided by the organism's perception of value and salience.

- Conventionally, the process whereby an infant learns to reach out and grasp a rattle had been explained as genetically pre-programmed in the brain. A dynamic systems account says that each baby has to work out how to do it by trying all manner of movements and then selecting those that help to locate the rattle in space, *selecting* those grasping movement that are *salient* (have meaning) and of *value*, because they work and grasping feels good.

- Mind/body dualism, the idea that mind and body are totally different, is a philosophically problematic view with a long history and is especially associated with René Descartes. It is still quite prevalent in educational thinking, whenever education researchers, policy-makers or teachers believe it is enough to know about the minds of children without being worried about their brains. Most brain scientists embrace what is called mind/brain identity: mind and brain are the same thing. In other words, you *are* your embodied and embedded brain.

- However, that does not imply another philosophical problem called reductionism, this time mainly coming from those involved in brain science. Reductionists believe that persons – you, your friends, the children you will teach – are *nothing but* their brain, *no more than* the biological components (neurons, molecules, atoms) of their brain. This view is generally not acceptable in education. It is also opposed by dynamic systems theorists who, in place of reductionism, espouse the notion of emergence. In short, although you *are* your embodied and embedded brain, you are also manifestly *more than the sum of your biological parts*, with a unique set of memories and your own distinctive personality. Persons are what brains produce; they result from (are an emergent property of) a highly complex brain.

KEY POINTS FOR TEACHERS

- Child and adolescent development does not have its own unique processes; rather, it is deeply embedded in processes shared by all living creatures and many non-living systems that are emergent, self-organising, complex structures. Teachers should try to view the children in their care as *complex beings* and their development as a product of *emergent self-organisation*.
- Teachers should be aware there is no 'self' in self-organisation. The word 'self-organisation' within a dynamic systems account of development is not referring to how a child might consciously decide to organise their own behaviour; rather, it means that development self-organises spontaneously, without the need for some inner mechanism or external agent.
- Teachers should try to remember that development is nearly always *multi-causal* (seldom the result of a single cause) and that given any particular task children will opportunistically draw on multiple available cues, moment by moment. Development and attendant behaviours are *softly assembled*; they can be combined freely on the basis of the context, task and developmental history of the organism.
- Although it has generally been recognised that there are considerable differences and much variability *between* children in their learning and developmental trajectories, dynamic systems research finds variability (stabilities and instabilities) within each child's trajectory, including developmental regression. Teachers should be aware that instabilities are often associated with transitional events, providing a springboard for transition and advancement.
- When referring to children's learning and development, teachers should be careful with the vocabulary they use, especially avoiding the use of machine analogies (children are not machines, they are learning and developing organisms) and also the mechanical language of processing devices, programs, storage units, schemata, modules or wiring diagrams.
- A major problem with previous development theories is that most have little to say about the role of the brain as the locus of learning and development. Teachers should be aware that, from a dynamic systems perspective, brain development is at the core of human development.
- Gerald Edelman provides an account of brain development that emphasises neuronal group *selection,* guided by *value and salience*. It is not just experience that is strengthening the connections between neurons; it is also what the organism values and finds meaningful. Teachers should always try to ensure that the learning and developmental

experiences they provide are viewed by their students as meaningful and have value – that they are worthwhile.

- When reading papers and articles coming from the discipline of neuroscience, teachers should be aware that most – although not all – neuroscientists are monists: mind is identical with the living brain and mental events are just brain events, the view advocated in this text. However, most neuroscientist are also methodological and ontological reductionists, who believe that persons (the children in your care) are *nothing but,* or *no more than* their brain.
- Teachers should be aware that within complexity science, and hence from a dynamic systems approach to learning and development, reductionism has been challenged. The children you teach are very much more than the sum of their biological and chemical parts; what *emerges* from those parts is all that we refer to as the thinking, feeling, emotional and socially and culturally embedded person or child.

REVIEW QUESTIONS

Guided responses

1. From the perspective of dynamic systems theory, what are the main limitations of conventional stage theories such as Piaget's theory?
2. Why are cross-sectional studies in human development limited?
3. What is meant by saying development entails the process of emergent self-organisation?
4. What roles do value and salience play in the brain's processes of learning and development?
5. What is reductionism? How is it different from an emergentist approach to human development and learning?

FOOD FOR THOUGHT

1. Before reading this chapter, did you tend to view child development as a collection of theories, approaches and influences? Where did you gain your ideas about development before reading this chapter – did they come from your own observations of children, from the media or did you not really think about it? Does it help teachers in their role if they appreciate that the children they teach will not follow a clearly defined developmental pathway, because it is all much more messy and variable than that?
2. If one of your classmates said that learning was not an activity of a brain, but of a person, mediated by a teacher and often in a classroom in the company of other persons, what might you say in response? What would you say about the origins of that comment and would you be sympathetic to what your classmate was trying to say? Would you agree with them, or would you say they were fundamentally wrong?
3. Do you feel that you are *nothing more than* the sum of your biological parts, or do you feel that you are more than the sum of those parts? Do you think it matters whether teachers think that the children they teach are nothing more than nerve fibres and atoms and molecules in motion? Might that influence how they view their students?

RESEARCH ACTIVITY: EMERGENT SELF-ORGANISATION OF TEACHER–STUDENT INTERACTION IN A CLASSROOM

An example of dynamic systems processes within education can be seen in the interactions between teachers and students within the classroom. In any given classroom, whether intended or unintended, teacher and student behaviours constantly feed back to each other via various channels (e.g. verbal communication, non-verbal gestures). Over time, specific patterns of classroom interaction develop and become stable. On occasions, these stable patterns become weakened or are replaced by other patterns of interactions. Thus, 'studying moment-to-moment interactions is important to fully understand the development of teacher–student relationships' (Mainhard et al., 2012, p. 1028).

The space state grids (SSG) method, underpinned by dynamic systems thinking, offers a useful tool by enabling educational researchers to visualise real-time teacher–student interaction trajectories on a two-dimensional grid. In Figure 3.13, the horizontal axis shows the different types of teacher behaviour and the vertical axis shows student behaviour. The arrowed line represents the change in interaction patterns over the course of a few minutes, and the thickness of the dots indicates the duration of each interaction state (Mainhard et al., 2012, p. 1029). In this case, we can notice that students in this class were largely attentive and active, spending a lot of time asking the teacher questions. For a short time, the teacher was not paying attention to students' learning, and this teacher does not use 'questioning' during her teaching.

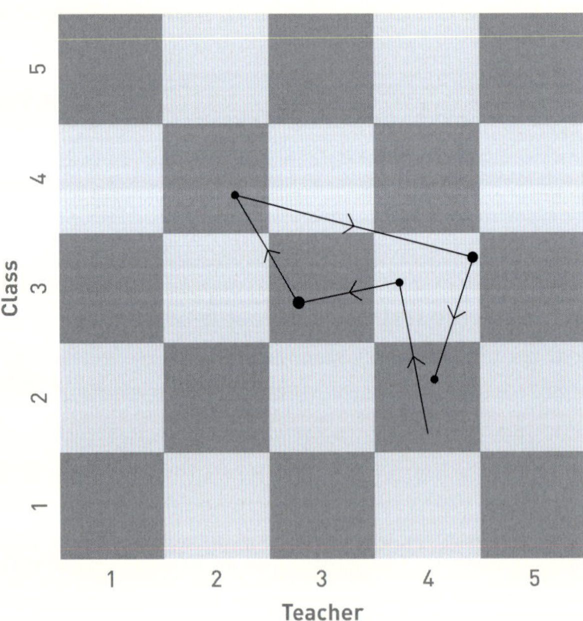

Figure 3.13 Emerging teacher–student interaction trajectories drawn by the space state grids method
Source: Mainhard et al. (2012).

Below are some useful resources on applying the SSG method when studying teacher–student relationships in a classroom setting.

- Mainhard, M. 'T'. et al. (2012). Mapping control and affiliation in teacher–student interaction with state space grids. *Teaching and Teacher Education*, 28(7), 1027–37.
- Queen's University, Department of Psychology: 'State space grids'.

Please read them carefully. You might like to apply the method when analysing classroom interactions if you have to observe them. When drawing the trajectories on the SSG, please ask these questions.

Questions

1. Are observed interactions *stable* (i.e. moving around only a small number of grids, staying in a specific state for a long time) or *variable* (i.e. moving around a large number of grids)?
2. What is happening when the interaction moves from one stable state to another state? Is the shift largely initiated by the teacher or is it students who drive the changes of interaction?
3. Is there any potentially vicious circle within the trajectory?

THE ACTIVE LEARNER

Working memory, metacognition and brain research

LEARNING OUTCOMES

By the end of this chapter, you will:

- Understand what is meant by working memory and executive functioning, and know that working memory and executive functioning enable metacognition
- Understand the role of fluid analogising in creativity and imagination, and know the four pillars of active learning
- Appreciate the importance of metacognition in active learning and know the three main phases involved
- Have an introductory understanding of the electroencephalogram (EEG) and the neural processes involved in metacognition, as well as the neuroscientific processes of error-related negativity (ERN) and feedback-related negativity (FRN), as they relate to active learning

Active learning, working memory and executive functioning

In this chapter, we build on and extend what you have already learnt in the previous three chapters about the science of learning and development, focusing on *working memory* and *metacognition* as important aspects of *active learning* in educational contexts. You will also meet Stanislas Dehaene's (2020) claim that there are four pillars of learning: (1) 'attention'; (2) 'active engagement'; (3) 'error feedback'; and (4) 'consolidation' – all of which encompass Hebbian learning (Chapter 2) and the dynamics of self-organisation (Chapter 3). Mention of error feedback, as one of the four pillars, means we will need to come to grips with some basic brain research methodology that uses the electroencephalogram (EEG), and that will introduce you to the neuroscientific terms of *error-related negativity* (ERN) and *feedback-related negativity* (FRN), both related to metacognition. We begin by revisiting Donald Hebb's account of learning via synaptic plasticity, which you studied in Chapter 2, and its implications for the active learner in school.

Hebbian synaptic plasticity and active learners in school

A most important message that teachers should bring to their practice in school is that education 'really does involve changing children's and teacher's brains' (Geake, 2009, p. 56). In Chapter 2 it was said that *teachers are physically changing the brains of their students* – teachers are quite literally 'brain changers'. But notice that it is not just children's brains that are being changed at school as a result of Hebbian synaptic plasticity, it is *also the brains of their teachers* as they learn and develop in their teaching role (Figure 4.1). There's an old adage or

Figure 4.1 The brains of teachers are physically changed through Hebbian plasticity as they perform their teaching role in school

motto in education that 'the best way to learn something is to teach it'. Change is occurring on a regular basis in teachers' brains, not least because teaching involves a lot of *repetition* and, as you saw in Chapter 2, the core of the Hebbian account of learning is cells exiting neighbouring cells 'repeatedly and persistently' (Hebb, 1949, p. 62). Teaching involves a lot of repetition and persistence – it's what helps to develop professional experience in teaching.

However, it is important to note two points: (1) not all learning is Hebbian learning; and (2) not all Hebbian learning occurs at school. With regard to the first of these two points, as just noted, Hebbian synaptic plasticity depends on the repetition of events, but learning can also occur as a result of one-off events. Common to both kinds of learning is the role played by *value* and *emotional salience*, which you encountered in Chapter 3, in learning about Edelman's account of synaptic strengthening and synaptic pruning, and how memory also involves a selective process driven by value and emotional salience. Generally, we remember best those things that have value and are emotionally salient.

There are experiences (point 1) that have high emotional salience, but do not need Hebbian repetition to be learnt and retained; these include experience of great joy but also of dread. Probably the most striking examples are those experience that result in post-traumatic stress disorder (PTSD), as a result of the fear or gross unpleasantness they engender. These include the kinds of experiences that arise in the context of war, but such experiences may also occur to all of us in situations that are frightening, such as almost being hit by a car when crossing the road. Thankfully, most of us don't need to have that kind of experience repeated to learn to take utmost care when crossing the road. Also, thankfully, these kinds of one-off events are rare at school, and most school learning requires Hebbian repetition. However (point 2), not all Hebbian learning occurs in the context of education, early childhood settings, primary school, high school or university; it is occurring in all aspects of our lives, as a result of *repeated regularities* in our everyday experiences and as part of our ongoing development.

One example of this can be seen at the very start of life's trajectory, in the way babies babble (Figure 4.2). As Geake (2009, p. 51) notes, 'Through fluctuations in their larynxes, young babies can babble with every different sound that a human is able to make. In doing this, babies are not, as sometimes claimed, universal linguists: their laryngeal outputs are random.' Needless to say, this ability to babble is genetically endowed. However, during the first half-year of life, the babbling swiftly exhibits 'use it or lose it' synaptic strengthening and pruning, so that 'by as early as 6 to 8 months of age, their repertoire of babbling sounds has become constrained to the phenomes of the parents' mother tongue(s) and accent(s)' (Geake, 2009, p. 51). The other side of that coin is that the

Figure 4.2 The chuckling, babbling infant can make every different sound that a human is able to make

lost non-mother tongue phonemes are mainly lost forever, which is why in trying to learn a second language in adulthood, it almost always has a 'wrong' accent.

The example of babies babbling points to another important issue related to Hebbian learning and the active learner at school. For some aspects of learning, there are critical *sensitive periods*, or *windows of opportunity*, that are lost if not activated when available at a given age. Thus it is widely recognised that it *becomes harder* to learn a foreign language as we get older. Could it be, therefore, that second-language learning in schools should start in early childhood education, which of course would require properly trained second-language teachers working in early childhood settings? At that point, the issue becomes political and one of funding; educationally, though, wouldn't it make sense from a science of learning perspective? The same issue might also apply to music education.

REFLECTION

To what extent do you think curriculum decisions about when children start to learn subjects should be guided by brain and developmental science rather than policy, funding and political concerns? What might you have learnt better if you had started earlier?

Another issue regarding the active learner at school, explained by Hebbian learning, is why school learning is so hard for most, if not all, students and why 'wrong' learning is so hard to erase. In a large, in-service teacher development training session, with primary and high school teachers, participants were asked to indicate with a show of hands what level of success in teaching a topic to a class of students they would accept as satisfactory. In teaching a lesson, is it satisfactory if a post-lesson test shows that 40 per cent of the students had learnt it? Not many hands went up. What about 50 per cent? A few more. How about 60 per cent? More hands went up. The majority of teachers opted for 80 per cent. The presenter asked, 'So what are you going to say to the parents of the 20 per cent who don't learn – why should it not be 100 per cent? The response was as spontaneous as it was clearly audible – that's unrealistic!

The point is that they were probably correct, but why? Now, it is important to stress that this issue is not to be confused with a different issue, that of 'enjoyment'. It is perfectly possible for students to thoroughly enjoy a lesson, but still not learn it clearly or retain it in the long term. It is not enough for teachers to entertain the students, although as we will see shortly, gaining their attention is paramount. The best way to know whether students have understood and grasped the lesson is to test it, not to assume that because they enjoyed it and paid attention, they have therefore learnt it. Teachers also need to encourage students to employ metacognition skills, which we will come to in the next section of this chapter.

From a Hebbian perspective, as you saw in Chapter 2, if learning is to occur there has to be a physical change in the brain: memory has to be consolidated in order to be retained. Students can experience the emotional joy of learning without that physical change occurring, and without it be being consolidated in memory. All sorts of causes may be impacting the failure to learn, but as teachers know all too well, and as constructivists have emphasised, one of those factors is *prior learning* – particularly 'erroneous' learning that occurred in the early years of pre-school and kindergarten and more broadly in the home and in social and cultural settings. The 'Hebbian model of neural plasticity not only can account for inefficiencies in learning, but it can also explain why erroneous learning is so hard to eliminate or counteract' (Geake, 2009, p. 55). The core problem is that what gets learnt tends to stay learnt because it has been physically instantiated in the brain at the synapses. Unlearning and re-learning requires a change at the synapses.

Working memory and executive functioning

Do the words 'learning' and 'memorising' mean the same thing, or is learning more than memory? If it is more, then what does this entail? Learning and memory are certainly closely related within educational contexts – which is why, in assessing students' learning, we check what they remember what they have learnt. To learn something at school, one has to remember it, and if an assessment test shows you have remembered it then you are said to have learnt it. But how well a student understands what they have learnt and whether they can apply it imaginatively, creatively and wisely goes beyond simply memorising.

Memory is necessary for learning but not sufficient, and that should be fully recognised in the assessment of what is learnt at school. In designing tests, pay attention to what the tests is actually testing, is it simply long-term memory (which of course is important) or do the tests also require the application of something deeper and more profound (Figure 4.3)? This is especially important as students become older. You may recall from Chapter 2 that Donald Hebb strongly advocated that university students should be *evaluated* on their ability to think creatively and come up with interesting and probing questions, rather than an ability to recite established knowledge or conventional wisdom.

Figure 4.3 Teachers need to closely examine what the tests given at school are actually testing, and whether it is what they claim to be testing

Thinking about your learning at school and university, to what extent has your learning required you to simply remember and recite established knowledge, and how much has it required you to be creative and come up with interesting and probing questions?

When learning about memory in Chapter 2, the focus was on long-term memory. This is what we normally mean by memory: the ability to recall what we have learnt and experienced over an extended time. That was the kind of memory lost by HM (studied by Brenda Milner, Hebb's doctoral student and later his colleague). Teachers want their students to remember what they have learnt in the long term, if only to do well in exams; however, arguably that is not the most important kind of thinking and learning that goes on in schools on a day-to-day basis. This learning is much more to do with students paying attention, with discernment (including comparing and contrasting), thinking imaginatively and creatively, and problem-solving. In fact, the ability to perform these skills is what many scholars mean when they use the term 'working memory', which is a key component of the 'executive functioning' of the brain, using a metaphor taken from the world of business management.

However, it is important to note that within cognitive psychology, working memory is often taken to be synonymous with *short-term memory*, which typically 'allows us to keep a phone number in mind: during the time it takes us to type it into our smartphone, certain neurons support one another and thus keep the information in an active state' (Dehaene, 2020, p. 90). Other scholars argue, however, that these two terms should not be viewed synonymously, especially within educational contexts. John Geake, an experienced Australian schoolteacher who became a leading educational neuroscientist before his untimely death in September 2011, said that 'temporary storage of information is an epiphenomenon, an outcome of what working memory is really all about' (Geake, 2009, p. 69). In saying it is an epiphenomenon, he means it is a secondary phenomenon, going so far as to say that 'the tag "memory" could even be misleading – the emphasis is very much on "working" … working memory is a central feature of executive functioning in the brain' (2009, p. 69).

The key point is that there is memory that is temporary, but working memory is more than that, and there is memory that is retained over a much longer time, as long-term memory. Whether something is retained in the long term has a lot to do with whether it has *emotional salience and value*. A single telephone number might be retained over a long period of time if it carries personal value and is thus emotionally meaningful. Moreover, as emphasised in Chapter 1, memory is highly distributed across the brain and is not fixed; it is always a reconstruction or reconstitution, bringing together different aspects of a single memory in a 'trick of timing' (Damasio, 1994, p. 95).

Geake actually aligns his claim that temporary storage is only an epiphenomenon of working memory with the notion of a dynamic **global neural workspace**, developed by Stanislas Dehaene and Jean-Pierre Changuex, which is conceived as 'a temporary conscious memory within which we can maintain, for a short period, practically any piece of information that seems relevant to us and relay it to any other module' (Dehaene, 2020, p. 160). This model,

global neural workspace: 'a temporary conscious memory within which we can maintain, for a short period, practically any piece of information that seems relevant to us and relay it to any other module' (Dehaene, 2020, p. 160)

which goes back to 1987, is grounded in the complexity or dynamic systems of nonlinearity and self-organisation. Geake says that in 'this model, working memory is essentially the workspace … Working memory is not just a temporary store, but is rather, in a sense, what the brain essentially does' (Geake, 2009, p. 70).

If you are finding this a bit confusing, it is clear that you are paying attention, which is very necessary for learning. What you are witnessing is 'science in action'. A very active search for understanding the functioning of the brain, as a whole, is taking place within brain science, and many of the emerging ideas are finding a place within a very new account of brain functioning, known as predicative processing, which you will encounter in Chapters 10 and 11. When it comes to brain science, we are living in exciting times. However, for our purposes in the context of education, what we basically need to appreciate is that there is a constant and highly dynamic flow of information coursing through the brain, which is constantly self-organising and involves both temporary and long-term memory.

Combining these ideas, Geake provides a very helpful definition of working memory in saying that it is a 'cognitive construct', a model 'of the neural process of combining information from the perceptual here and now with information from long-term memory, under attentional selection of what is relevant to the task in hand' (Geake, 2009, p. 72).

Creativity, fluid analogising and the four pillars of learning

Surely one of the greatest joys of teaching is to see our students develop in their creativity and imagination, in addition to them learning whatever it was they were meant to learn. But what is involved in being creative and imaginative? What is consciously happening when you are being creative and imaginative, and what is happening in your brain?

REFLECTION

In terms of your own conscious experience, what is involved when you are being creative, and is imagination an essential part of that? Select three different ways in which you are creative and then consider what are you actually doing in your thinking that enables you to be creative and imaginative.

Part of the answer to the second question about the brain is contained in the definition of working memory given above: a 'neural process of combining information from the perceptual here and now with information from long-term memory' (Geake, 2009, p. 72). There is a constant dynamic 'chatter' in your brain, across multiple brain regions and neural systems, that is monitoring your thinking, feelings and actions, comparing them with all that you are perceiving right now and all the experiences you have stored in long-term memory.

A key point is that brain function does *not* comprise *multiple intelligences*, as proposed by Howard Gardner in 1983. Neuroimaging research has consistently shown that brain function is both localised and highly distributed, so that the same neural systems 'are involved in many different cognitive abilities – spatial, verbal, language, logic, mathematics, memory. This commonality of brain functions provides a neural explanation for general intelligence over a multiple intelligence' (Geake, 2009, p. 94). There is a constant flow of information and also constant feedback between systems, which is being assessed for its relative importance by working memory, as a function of the *global neural workspace*. In short, creativity and imagination arise from the massive interconnectivity between the different neural systems in the brain, in response to immediate perception and mediated by the interface between working and long-term memory. There is no special part of the brain dedicated to creativity or imagination – that is what healthy human brains are doing all the time. And what healthy brains are doing in the global neural workspace is *analogising*.

The concept of fluid analogising

A long time ago, American pioneering philosopher and psychologist William James (1842–1910) wrote, 'A native talent for analogising is … the leading fact in genius of every order" (James, 1890, p. 530). In other words, the ability to use metaphor and analogy is the key to genius in every area of human endeavour. In 1964, in a book called *The Act of Creation*, Arthur Koestler (1905–1983) coined the term 'bisociation' to name what he considered to be at the core of creativity: the ability to bring together and combine different ideas that seem to be unrelated. Educational neuroscientist Usha Goswami (1992) has provided research evidence that young children's conceptual development is characteristically analogical, and it starts as early as three years of age. Steven Mithen (1996, p. 202) argues in *The Prehistory of the Mind* that what he called the achievement of *cognitive fluidity* 'transformed the modern human mind and behaviour', allowing modern humans to advance way beyond their earlier human ancestors.

Geake (2009, p. 96) argues that 'the *modus operandi* of creativity is analogising'. However, following AI researchers Melanie Mitchell and Douglas Hofstadter (1995), he adds that the kind of analogising required in creative thought is 'fluid analogising, where there is not one correct but several possible' (Geake, 2009, p. 95) options in framing the analogy. It is 'the judicious (albeit often instinctive) employment of fluid rather than exact analogies that enables effective categorisation and thus the assimilation of new knowledge' (Geake, 2009, pp. 96–7). As Hofstadter (1995) notes, 'categories are quintessentially fluid entities; they adapt to a set of incoming stimuli and try to align themselves with it. The process of inexact matching between prior categories and new things being perceived … is analogy-making par excellence' (Hofstadter, 2001, p. 499).

Geake makes the point that the notion of fluid analogising is especially appropriate in educational contexts because 'pedagogy entails fluid analogizing to deal with the imprecise categorical relationships of the real world' (Geake, 2009, p. 96). In other words, good teachers are those who are able to think of good analogies when explaining and clarifying tricky ideas. The same may be said of successful active learners: they are particularly good at creating and imagining good analogies in the process of understanding tricky ideas. These analogies do not need to be verbal; they may be largely or entirely pictorial – as in the use of diagrams and concept maps, for example.

The four pillars of learning

Stanislas Dehaene, a leading brain scientist with a particular interest in education, has identified what he calls 'four pillars of learning' (Figure 4.4). Because each of them plays an essential role in the stability of our mental constructions: 'if even one of these pillars is missing or weak, the whole structure quakes and quivers' (Dehaene, 2020, p. 145). The four pillars are attention; active engagement; error feedback; and consolidation. Dehaene (2020, p. 146) says, 'Teachers who manage to mobilise all four functions in their students will undoubtedly maximise the speed and efficiency with which their classes can learn.' The key point is that, from the perspective of brain science, the ability to think using creative and imaginative fluid analogies and attending to the four pillars of learning are the keys to learning success – both your own and that of your students. Let's look at the four pillars.

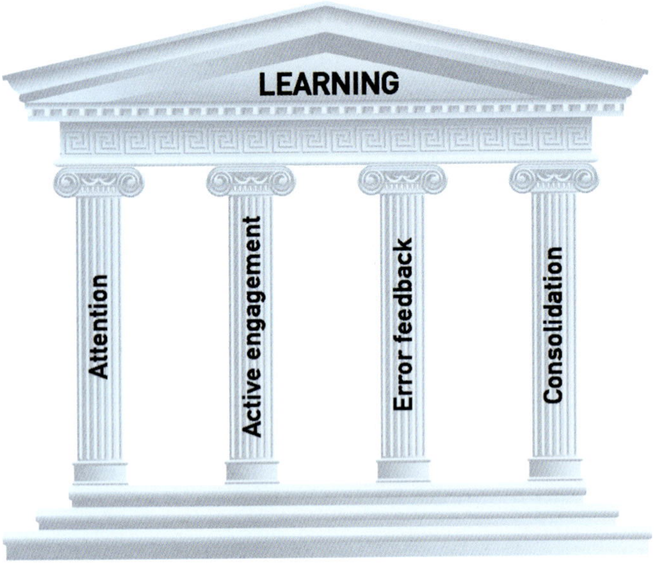

Figure 4.4 Stanislas Dehaene's four pillars of learning

Attention

The ability to attend to what is *relevant* is widespread throughout the animal kingdom; it is essential for survival. It also helps to solves another very important problem: the ability to deal with the saturation of information that constantly bombards our senses. The brain is alerting the organism *when* to attend and *what* to attend to. In the wild, if a predator approaches, an array of subcortical nuclei are strongly aroused, releasing neuromodulators such as serotonin, acetylcholine and dopamine and 'sending signals throughout the whole cortex' (Dehaene, 2020, p. 150). This has a very important message for teachers and parents: *pay close attention to each child's attention*. Students must pay attention if they are to learn, so a 'teacher's greatest talent consists of constantly channelling and capturing children's attention' (2020, p. 150). The basic message for teachers is not to try to speak to a child or a class unless everyone is paying attention, otherwise you are just wasting your breath.

Paying attention to what is relevant means *suppressing* what the brain is not perceiving as relevant, salient or meaningful. In paying attention, your students' brains are being highly selective, acting as a filter and suppressing what they consider to be unwanted information. Students' brains that have become disenchanted with school or with your classes have filtered out much of what you teach as unwanted information. Beyond the problem of student disaffection, there is a more general implication of filtering: the brain can make us 'blind to what it chooses not to see' (Dehaene, 2020, p. 155). Dehaene illustrates this with reference to the 'gorilla experiment', which he believes should be known by all parents and teachers because it provides a strong warning that the 'intrinsic limits of our awareness leads us to overestimate what we and others can perceive' (2020, p. 156). If you know the gorilla experiment, you will know what he is referring to. If not, please check it out on YouTube, using the search word 'gorilla experiment' – watch the video before reading the explanation.

Dehaene also points out that teaching involves *attending to someone else's knowledge* (2020, p. 173) and he speculates that no other species can teach like we do because, he believes, we are the only species with the ability to imagine what others are thinking and what *they* think others are thinking. This is called having a **theory of mind**, and it plays a very important role in school learning. Teachers constantly try to monitor what students don't know and students are aware that teachers pretty well know what they don't know. 'Once children adopt this pedagogical stance, they interpret each act of the teacher as an attempt to convey knowledge to them. And the loop goes on forever: adults know that children know that adults know that they do not know … which allows adults to choose their examples knowing that children will try to generalize them' (2020, p. 172).

theory of mind: 'the ability to attribute mental states to others, such as knowledge, intentions and beliefs' (de Waal & Preston, 2017); the mental process of monitoring and regulating one's cognitive activities

Active engagement

Active learning requires active engagement. Human brains basically contain two modes of learning. One mode, which is ubiquitous in education, is learning from what others (caregivers, teachers, lectures) are telling us or transmitting to us. One sees evidence of that in other species, particularly those that are self-aware, such as the great apes, dolphins and elephants, but it is modern humans who have exploited this ability over many hundreds of thousands of years. The other mode is active, involving creativity, imagination and fluid analogising as discussed above. It is the critical and analytical mode that tests out ideas and rejects them if they are found wanting. Both modes are necessary: the passive acceptance of what we are told has to be balanced by critical analysis if we are to be properly educated. Herein lies the problem of indoctrination, bigotry and fake news, which are founded on the first mode and seem immune to the second. As teachers, we need to provide a mix of the two. A passive learner will learn very little: 'Efficient learning means refusing passivity, engaging, exploring, and actively generating hypotheses and testing them on the outside world' (Dehaene, 2020, p. 178). Active learning also means being active in the learning process, having 'hands on' experience, 'learning through doing', which we come back to when considering the notion of embodied cognition in Chapter 9.

However, Dehaene is at pains to stress that this 'must not be confused with classical constructivism or discovery learning methods – which are seductive ideas whose ineffectiveness has, unfortunately, been repeatedly demonstrated' (Dehaene, 2020, pp. 181–2). He then adds that research has consistently shown that the 'most efficient teaching strategies are those that induce students to be actively engaged while providing them with thoughtful pedagogical progression that is closely channelled by the teacher' (2020, p. 184). The key point for teachers is that active learning has to be guided learning. Moreover, 'every teacher should keep in mind that all children learn using the same basic machinery – one that prefers focussed attention to dual tasking, active engagement to passive learning, detailed error correction to phony praise, and explicit teaching over constructivism or discovery learning' (2020, p. 186).

Error feedback

Active learning does not just require paying close attention to what is being learnt and active engagement with it; the brain is also involved in dealing with error feedback. Details of how it accomplishes this will come later in this chapter, when you learn the neuroscience notions of 'error elated negativity' (ERN) and 'feedback-related negativity' (FRN). You will return to these important concepts in Chapter 10 when studying assessment and learning, as well as when learning about the predictive brain in Chapters 10 and 11.

The key point is well known: *we learn by our mistakes*. However, we are now appreciating as never before that the brain is constantly predicting the inputs it will receive from the senses (sight, sound, touch, smell and taste) and adjusting 'these predictions according to the degree of surprise, improbability or error' (Dehaene, 2020, p. 203) that is actually given by the senses. In short, learning from our mistakes as a result of error feedback, which we experience on a regular basis, is a *conscious* manifestation of what the brain is doing *subconsciously* when it is dealing with error feedback.

In the context of education, when providing rapid and precise feedback on errors, teachers can considerably enrich the information available to their students to correct themselves. Indeed, John Hattie (2008) at the University of Melbourne, Australia, has convincingly argued that the *quality* of feedback students receive has a big influence on their academic achievement. Error feedback is also a key component of metacognition. This means that error should never be used as a means of labelling students as failures. Dehaene is very concerned that, 'All too often, schools use grades as punishments. We cannot ignore the tremendous negative effects that bad grades have on the emotional systems of the brain: discouragement, stigmatisation, feelings of helplessness' (Dehaene, 2020, p. 212).

Consolidation

When the British neuroscientist Graham Collingridge (mentioned in Chapter 2 in relation to the discovery of LTP) was asked by science education doctoral students in Korea what they should do to improve their learning, he said, 'Get more sleep' – surely not the answer these science students were expecting. Perhaps this is a message for all students and maybe their teachers, although it applies particularly in Asian contexts. The point is that sleep plays a key role in consolidating learning. As Dehaene (2020, p. 224) points out, 'While we sleep our brain remains active, it runs a specific algorithm that replays the important events it recorded during the previous day and gradually transfers them into a more efficient compartment of our memory.' And that does not just apply to humans – it is also found in rats.

Evidence for the role of sleep in rat learning comes from research conducted by brain researchers Matthew Wilson and Bruce McNaughton (1994), who discovered that what are called place cells in the hippocampus of sleeping rats spontaneously activate. Place cells are 'neurons that fire when an animal is or (believes itself to be) at a certain point in space' (Wilson & McNaughton, 1994, p. 226). Wilson and McNaughton's experiment found that when rats move through a corridor, their place cells fire in particular sequence. When the rat sleeps and the place cells are spontaneously reactivated, the firing sequence fire in the same order. The only difference is that during sleep the rat goes through the sequence much more quickly.

Although human beings are not rats, this experiment does seem to confirm what other research has found: that sleep plays a very important role in consolidating the learning process. For teachers, the key message is that students need to get sufficient sleep if they are to become successful, active learners. This seems to be especially applicable for adolescent learners, who are well known to experience a shift in their sleeping pattern, staying up later and also getting up later.

Supporting students in becoming metacognitive, active learners

The 'more you test yourself, the better you remember what you have to learn' (Dehaene, 2020, p. 215). The importance of students becoming *active learners* and *learning how to learn* has been known for a long time. In 1976, John Flavell coined the word 'metacognition' to capture some of the key processes involved. Subsequent research has taken this work forward, so that now, as some previous learning theories such as constructivism and discovery-based learning seem to be losing their appeal (Hattie, 2008), many in education view metacognition as one of the most important and productive learning strategies that teachers can use (Darling-Hammond et al., 2020). Given the importance attached to metacognition, it is clearly imperative that school teachers understand the processes involved, but perhaps it is even more important that they support their students as active learners and help them to understand how to employ metacognitive strategies.

What is meant by metacognition?

The Greek word *meta* when used as a prefix means 'going beyond'. In Chapter 3, it was said that DST may be viewed as a *meta*theory, in that it encompasses but also *goes beyond* many previous theories. In the case of **metacognition**, the prefix contains a self-referential connotation, in the sense that an individual learner is able to *go beyond* what they have learnt and reflect on the learning process itself. In short, humans know when they do not know and can do something about it. Metacognition is therefore often described as *cognition about cognition*, or *thinking about thinking*.

metacognition: the mental process of monitoring and regulating one's cognitive activities

It is tempting to assume that this must require advanced levels of thinking. That assumption would seem to be supported by Piaget's stage of formal operations, which he claimed is reached in adolescence and constitutes the end-point or final culmination of his stage theory. However, it is not only human adolescents that are able to engage in metacognition; it is also found in infants and even some non-verbal animals such as dolphins (Smith et al., 2012).

REFLECTION

Thinking back to your own education in school, can you remember being encouraged by your teachers to engage in metacognition as an active learner? If so, was it introduced systematically or was it just mentioned? Can you give examples?

Metacognitive strategies are those that learners use to *monitor* or *purposefully regulate* their thinking – for example, when a learner checks whether an arithmetic technique she used was accurate or when a learner seeks a more appropriate strategy when the pre-chosen method was found to be wrong. Robin Fogarty (1994) suggests that metacognition is a process that comprises *three distinct phases*:

1. Develop a *plan* before engaging with a learning task.
2. *Monitor* their progress while undertaking the learning task.
3. *Evaluate* the success/effectiveness of the strategies.

You can see these three phases in Figure 4.5. The questions in the boxes provide examples of what might be asked by a learner in each phase. Needless to say, teachers who teach the metacognitive learning process can incorporate these questions into lesson plans to facilitate their students' metacognitive thinking during the lesson.

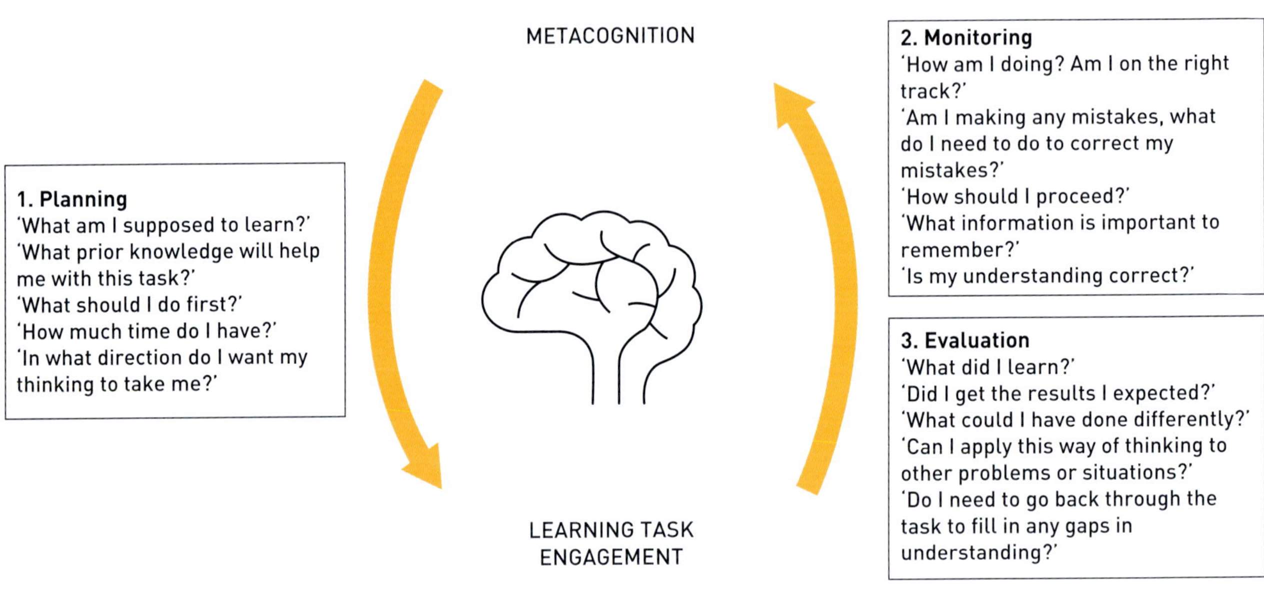

Figure 4.5 Three recursive phases of metacognition
Source: Adapted from US Department of Education Office of Vocational and Adult Education (2020).

You may notice there are a number of ways in which the above description of metacognitive phases has some similarity to ideas introduced in Chapters 2 and 3. For example, it is possible to see parallels between the three phases of metacognition and Esther Thelen's account of how a baby learns to grasp a rattle (Chapter 3). The baby has a 'plan' to grasp the rattle, the baby 'monitors' the success of plan, some attempts work, and others do not. The baby 'evaluates' the success and, as a result, gets better at grasping the rattle. Of course, the baby is not engaging with metacognition, but perhaps, at the neural level, metacognition is employing neuronal group selection, guided by value and salience that accounted for the baby's learning. Now that begins to sound like the basis of a good research question.

Also, in Chapter 3, you encountered the notion of recursive *feedback loops*. The way Fogarty described the three phases of metacognition is very linear and therefore misses the dynamics of feedback that is involved, though the diagram attempts to incorporate that using the two arrows. Feedback seems to be playing an important role in metacognition, although that needs a lot more research. From a dynamic systems perspective, the inter-relationship between planning, **monitoring** and evaluating doesn't go in the straight line, from beginning to end. Rather, all three are in play throughout the process, with each feeding off the others. Ideally, the plan is not static and it will be shaped and reshaped in response to the monitoring and evaluation. The evaluation isn't the end of the process: it will feed back into the fluidity of planning and monitoring. From the perspective of the science of learning and development, we need that fluid, ever-evolving dynamic, not only at the observable, behavioural level but also at the neural level.

Mention of these two different descriptive levels raises a question, if not a major problem. A key assumption of metacognition is that, as conscious human 'selves', we are able to reflect on our own thinking processes and generate reliable and accurate knowledge about our own learning. But during this century, as discussed in Chapter 1, we have become increasingly aware that much of our thinking and learning is occurring below the level of conscious awareness, and therefore is not available to conscious introspection. This is often referred to as the cognitive unconscious, although the term used in this book is 'subconscious' (the word 'unconscious' seems to suggest what happens if a boxer gets knocked out in a boxing match, or a state of coma, whereas 'subconscious' more directly connotes that much thought, learning and action, is occurring *below* conscious awareness).

As neuroscientist Chris Frith (2012, p. 2214) states, studies have confirmed that 'we have little or no direct conscious access to higher order cognitive processes. We may have access to the outcomes of these processes, but, through introspection, we get very little idea as to how these outcomes are achieved. In some circumstances, we have rather limited access even to the outcomes of decision-making processes. So, a word of caution is warranted about claims regarding our ability to self-report on our own thinking and learning processes. However, that does not mean researchers are not able to access what is happening subconsciously at a neural level, and we will consider this in the following section. First, however, a few words about metacognition and memory.

> **monitoring:** any activity aimed at evaluating or regulating one's own cognitions (Flavell, 1979), which includes checking errors, self-testing, assessing one's progress, etc.

Metacognition, context and memory

As you saw in Chapter 3, one of the important messages from a dynamic systems approach to learning and development is that *context matters*. This is especially relevant in relation to metacognition. A common misconception is that metacognition is a learning skill that can be separated from the context of subject knowledge. To the contrary, the extent to which we are able to consciously monitor, regulate and self-report on our own learning *is highly subject-knowledge specific*. A strong grounding in subject knowledge and the improvement of metacognitive skills must go hand in hand, with each feeding off the other through feedback.

Another related misconception is that you can teach metacognition in specially de-signed 'thinking skills' or 'learning to learn' lessons. Again, this implies that it is possible to separate metacognition from subject learning, or what phenomenologists have called the *object* of knowledge – the subject knowledge to be learnt. Phenomenologists have long insisted that it is *impossible to think without thinking about something*, and it is impossible to learn without learning about something. This important observation goes back to the nineteenth-century German philosopher Franz Brentano (1838–1917) and his notion of the *intentional object* (Brentano, 1874). Notice that the phenomenologist's notion of the *object* of learning (what the learning is *about*) is not to be confused with the behaviourist notion of behavioural learning *objectives* (the notion of setting observable and measurable outputs in learner performance).

REFLECTION

Do you think the phenomenologists are right that it is impossible to think without thinking about something, and impossible to learn without learning something? Give it a try: see whether you can just think without thinking about something.

An important implication of bringing metacognition and subject knowledge back together is that a student could be metacognitively strong in one subject area and not another. This means that in each subject area, teachers should not only teach the subject but also explicitly teach students how best to plan, monitor and evaluate their own learning as fully as possible.

In the process of enhancing their metacognitive skills in the context of subject knowledge, students are inevitably drawing on memory. As we have seen in Chapter 2, learning occurs at the synapses, where – hopefully – what is learnt is laid down and retained in long-term memory. You will recall that two biologically distinct forms of memory were discovered by Brenda Milner: declarative memory and non-declarative memory. These constitute two different forms of knowledge, given that what we have memorised is what we know.

Declarative *knowledge* is explicit knowledge of facts and ideas, including facts about oneself as a learner – one's own capabilities and preferences as a learner. What we are able to know about our declarative knowledge because we are consciously aware of it, can be *declared*, spoken about and discussed. Non-declarative *knowledge* is implicit, subconscious knowledge related to performance – what we are able to do automatically, without having to declare it. We know about it by being able to do it. Metacognition draws heavily on declarative knowledge, as can be seen in the questions in the planning, monitoring and evaluating boxes in Figure 4.5 above. However, as students acquire the skills of metacognition, through practice and repetition, much will be retained as non-declarative knowledge and performed automatically.

One final point about skills acquisition that is particularly appropriate for teachers: one of the best ways to help your students become metacognitive thinkers and learners is to do it yourself, and then to *explicitly model* it for your students. Tell them how you are approaching a subject-knowledge task and then, most importantly, let the students try it themselves. Don't make learning too difficult and don't make it too easy: students need a level of difficulty to develop, balanced by good modelling. As teachers, try to experiment with modelling metacognition with class groups and also individual students. Do your own research as a means to enhance your own metacognitive teaching skills.

Researching active learning and metacognition using EEG

Whether or not schoolteachers engage with research, as some do, all teachers should know how to read education research and to do so with a critical eye. In other words, when reading the claims made by researchers, teachers need to *comprehend* what is being said and *evaluate* the research findings (and ideally metacognitively monitor and evaluate their own learning processes when doing so). A lot of neuroscientific research that relates to learning and development uses the electroencephalogram (EEG) and much of it also employs event-related potentials (ERPs). So, let's start with the EEG and its potential for use in education research.

Introduction to the electroencephalogram (EEG) and its potential for use in education research

In Figure 4.6, a child is wearing an EEG cap. In this case, the cap is fitted with 64 electrodes that connect to the EEG, which amplifies the brain signals received by the electrodes one million times. Some research uses 32 electrodes, much more uses 64 (Figure 4.7) and sometimes 128 are used. Basically, the EEG is recording the neural activity (neural field created by the firing of neurons in the cerebral cortex).

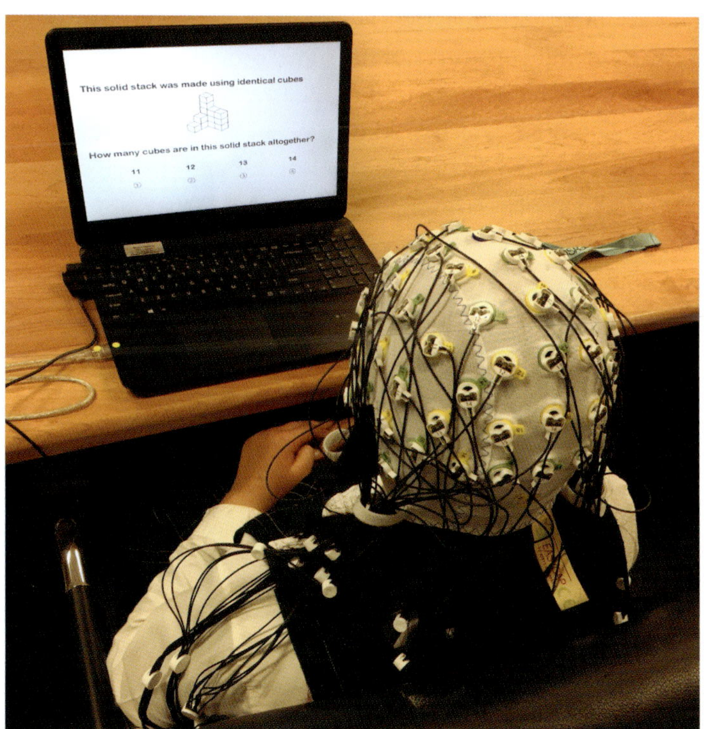

Figure 4.6 The EEG is recording the child's brain activity as she does a maths test

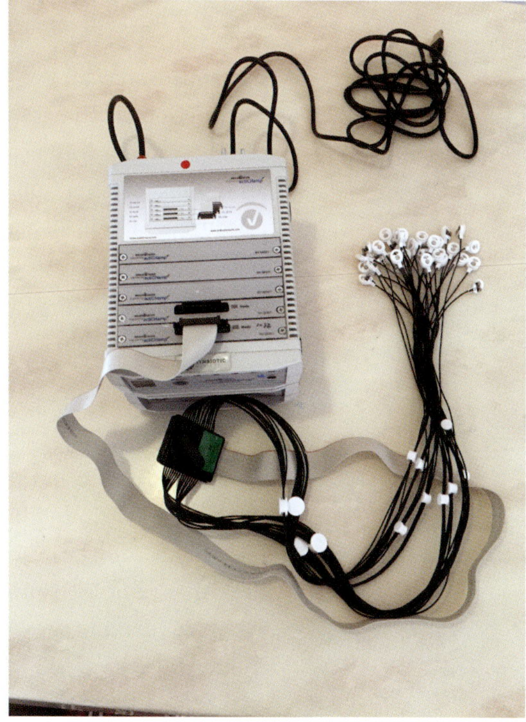

Figure 4.7 EEG showing electrodes

Nothing is being done to the brain: the electrodes are simply listening to the neural activity. Because EEG is a 'passive technology' (not in any way invasive) it is very suitable for use with young children and even babies, so very useful for education research. It is also portable, so ideal for research conducted in schools rather than in a laboratory. School-based use is still rare, but it is expected to increase significantly in the years ahead.

Having been amplified by the EEG, the signals from the brain go to a computer with a software program that turns the signals into a brain wave pattern, with each wave form representing one electrode and each signal representing activity over time (in milliseconds, abbreviation ms). Referring back to Chapter 2 and Hebb's postulate, what the EEG electrodes are actually thought to be capturing is a summation of the **post-synaptic potential** from a large cluster of neurons that are firing synchronously. That is because the post-synaptic activity occurs over a longer period of time (50–200 ms) than the **pre-synaptic action potential**, which is much quicker (10 ms or less) and less distributed, so not so easy to pick up by electrodes on the scalp and record.

Post-synaptic potentials come in two different kinds. As you learnt in Chapter 2, in a normally functioning brain, an action potential travels down the axon of the neuron to the synaptic terminal, where it releases a chemical neurotransmitter. The neurotransmitter then acts on the (post-synaptic) receptors of the receiving neuron. We can now add that this results in a change in what is called the membrane potential or transmembrane potential.

Basically, when a neuron is not firing and is at rest, it has what is called a **resting membrane potential**, which means that there is a difference in electrical potential between the interior and exterior of the neuron, measured in millivolts (mV). If the signal produced by the neurotransmitter has an excitatory effect on the receiving neuron (called *depolarisation*), it causes a reduction in the transmembrane potential, which is known as an **excitatory post-synaptic potential (EPSP)**; this usually occurs in the dendrites. If, on the other hand, the signal has an inhibitory effect on the neuron (hyperpolarisation), it is called an **inhibitory post-synaptic potential (IPSP);** this usually occurs on the cell body (Rowan & Tolunsky, 2003; Luck, 2005).

The main point to hang onto is that it is the combination of the ESPSs and IPSPs that induces currents of sufficient strength to be picked up by the electrode on the scalp.

Brain dynamics and the interplay of multiple spatial and time scales

When measuring brain activity related to metacognition using EEG, we are not working at the level of individual neurons, but rather at the level of large-scale neural networks. Figure 4.8 shows the relationship between three different special levels of operation and the corresponding timescales. At the microscopic level of individual neurons, the electrical synaptic currents are operating at a timescale of around 1 millisecond. Individual neurons cluster together for local networks at what is called the mesoscopic scale. Local networks are formed from thousands of neurons and their electrical activity (local field potentials) is operating at around 10 milliseconds. These then dynamically scale up to involve millions of neurons operating in entire brain regions as large-scale networks. It is these large-scale networks, operating at hundreds of milliseconds, that provide the brain-wave patterns that are recoded by EEG.

post-synaptic potential: change in the membrane potential of the post-synaptic terminal of a synapse

pre-synaptic action potential: depolarisation that occurs when the inside of the nerve cell fibre becomes positively charged

resting membrane potential: the difference in electrical potential between the interior and exterior of the neuron

excitatory post-synaptic potential (EPSP): a shift in membrane potential when the change of the receiving cell is *more* likely to fire its own action potential after receiving the neurotransmitter

inhibitory post-synaptic potential (IPSP): a shift in membrane potential when the change of the receiving cell is *less* likely to fire its own action potential after receiving the neurotransmitter

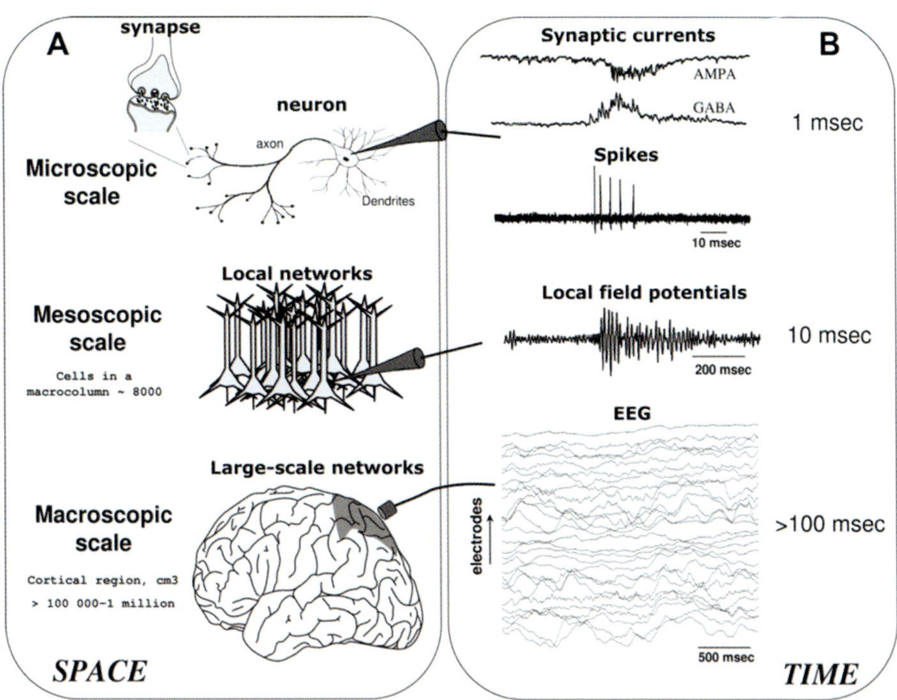

Figure 4.8 The brain viewed at different levels of organisation and corresponding timescales
Source: Le Van Quyen (2011).

It would be reassuring to tell you that we now have a clear and complete understanding of the multi-scale dynamics of the brain and nervous system, but this is far from reality. The brain is highly complex, it is the most complex system known to human beings. It has, however, become increasingly clear this century that if we are going to understand the multi-scale dynamics of the brain and nervous system, and how these different levels of operation self-organise and produce emergent patterns of behaviour, we need to apply the core concepts and mathematics of dynamic systems theory. Figure 4.8 is taken from one such study, by Michael Le Van Quyen (2011). For our purposes, there is no need to go into all the details. The main point is to be aware that research using EEG (when investigating the dynamics of active learning and metacognition) is operating at the macroscopic scale, large-scale networks) which is why it is providing data at a timescale of hundreds of milliseconds, and not the timescale of individual neurons.

Setting up the EEG and reading the waveforms

EEG waveforms are presented on a computer screen (or other monitoring device). In Figure 4.9, each waveform line displayed on the screen represents the recording from one electrode. So if the researcher is using 32 electrodes (see Figure 4.10) there will be 32 lines; if they are using 64 electrodes, there will be 64 lines. The screen identifies which waveform is coming from which electrode.

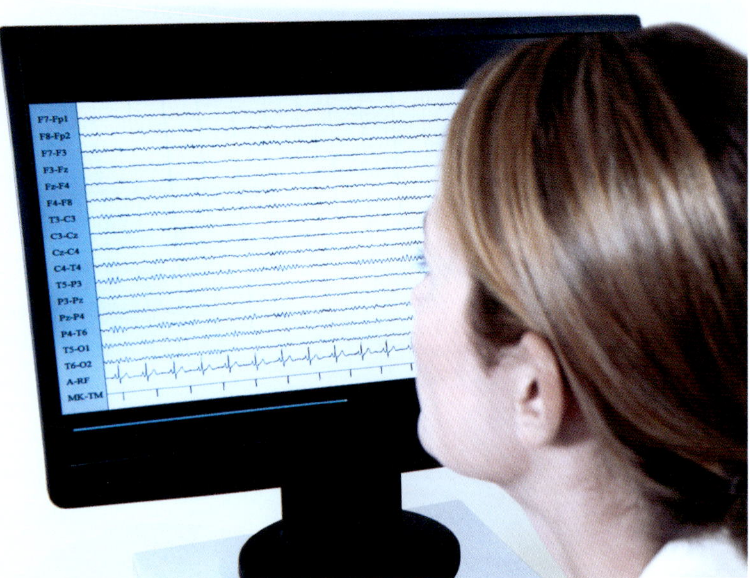

Figure 4.9 A 32-channel waveform, with each line representing the recording of the brain activity picked up by just one electrode (identified on left screen)

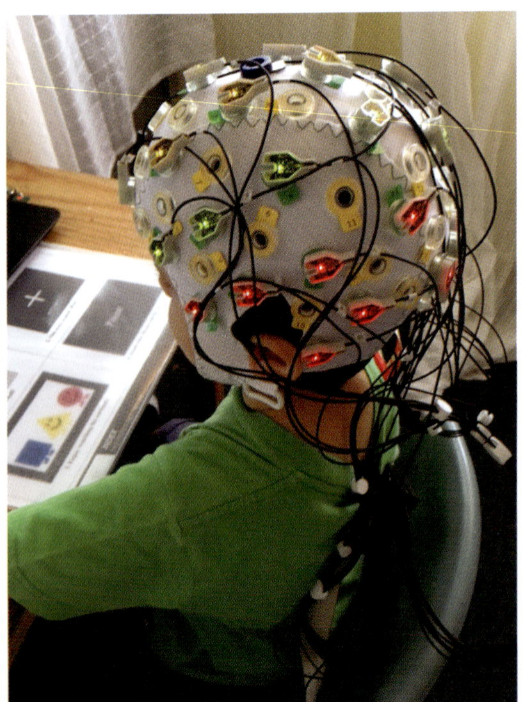

Figure 4.10 A cap showing some electrodes have good contact and some poor contact with the scalp

Each electrode is placed in a very specific area of the brain, which is determined by their position on the cap. The researcher therefore has to place the cap very accurately on the head of the research participant and each electrode has to make good contact with the scalp so that there is no 'impedance', as it is called. To ensure a good contact, a gel is used, which is inserted into the cap placement site with a syringe. When the EEG is turned on, after getting the cap carefully positioned, green or red lights are displayed at the terminals of each electrode (see Figure 4.10). Green is good – it means there is good electrical contact with the scalp. Red is bad (impedance) and has to be fixed by adding more gel. A must-not-forget part of the researcher's toolkit is shampoo and towels, so each participant can have a thorough hair wash after taking part.

Event related potentials

The use of EEG goes back to 1924, when a German psychiatrist Hans Berger recorded the electrical activity of the brain by placing an electrode on the scalp, amplified the signal and plotted how voltage changed over time. From those simple beginnings, EEG developed into the sophisticated technology used today. A major advance occurred in 1935 when it was realised that by using a simple averaging technique, it was possible to extract specific neural responses, associated with sensory, motor and cognitive events, from the wave patterns produced by EEG. However, it was not until the 1960s that this technology entered what might be called the modern era, when these specific electrical responses became known as **event-related potential (ERP)** because they are electrical potentials (changes in voltage) related to specific events.

Figure 4.11 provides the basic idea. What you see is the representation of an averaged waveform, which consists of positive-going and negative-going voltage *peaks* or *components*. These are referred to as P100, N100, P200, N200 and P300. The letters P and N are used to indicate positive and negative peaks, while the number indicates the timing of the peak within the waveform. Thus, P300 occurs 300 milliseconds after the onset of the waveform. You might notice that, in this diagram, positive is below the baseline, which is how it is often presented, though some researchers reverse this so positive is above the baseline.

Nowadays, EPRs play an important role in research on human thinking and emotions, and ERP studies are published regularly in child/human development journals. A small number of journals publish ERP research that is related specifically to learning and development in education, such as *Mind, Brain and Education*, and this is expected to increase as undergraduate and postgraduate studies in education become more focused on the science of learning and development. Already, some of this material is being made available in schools. Sometimes research results are discussed in school staff-development sessions and sometimes they are summarised in articles written for school publications that are sent to parents.

event-related potentials (ERP): a time-locked electrophysiological response that arises as a direct response to a specific event

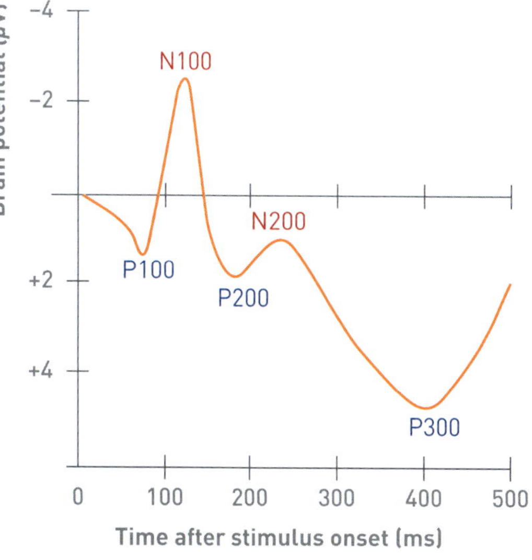

Figure 4.11 An averaged ERP waveform

An introduction to the Brodmann areas of the brain

We have already seen that the cortex of the brain can be divided into four lobes (Chapter 2), but if we need to be more specific in identifying areas of the brain – for example, where metacognition is occurring at a neural level – we need a better, more detailed map. Over the years, a number of mapping systems have been designed, often based on studies of patients with brain damage that has been linked to certain areas of the brain. In the early twentieth century, a German neuroanatomist, Dr Korbinian Brodmann (1868–1918) devised a map that divided the cortex into 52 areas, which are now known as **Brodmann areas**. Brodmann's map is not necessarily better than other such maps, but it is the one that is most commonly used.

Two copies are provided in Figure 4.12, one showing the outer cortex (left) and the other the inner cortex (right). Take a look at the left-hand diagram and see whether you can identify Brodmann area 44 (somewhere near the middle). That is also known as **Broca's area**, after Paul Broca (1824–80) and is well known for being implicated in the disability called expressive aphasia, the inability to *produce* speech, even though the patient understands language. We return to this in the following chapter. Now look at both images and see whether you can find area 10 (shown at the front of the brain in both diagrams) and area 32, located in the inner cortex (right-hand diagram) as these are implicated in aspects of metacognition (Burgess & Wu, 2013; Holroyd et al., 2004; Rouault et al., 2018).

Brodmann areas: functional areas of the cortex assigned by a German anatomist Korbinian Brodmann. Until recently, this was the most commonly used system of brain reference to facilitate communication among scientists regarding the regions implicated in different functions.

Broca's area: a region in the frontal lobe of the dominant hemisphere, usually the left for right-handed people. It is conventionally linked to speech production.

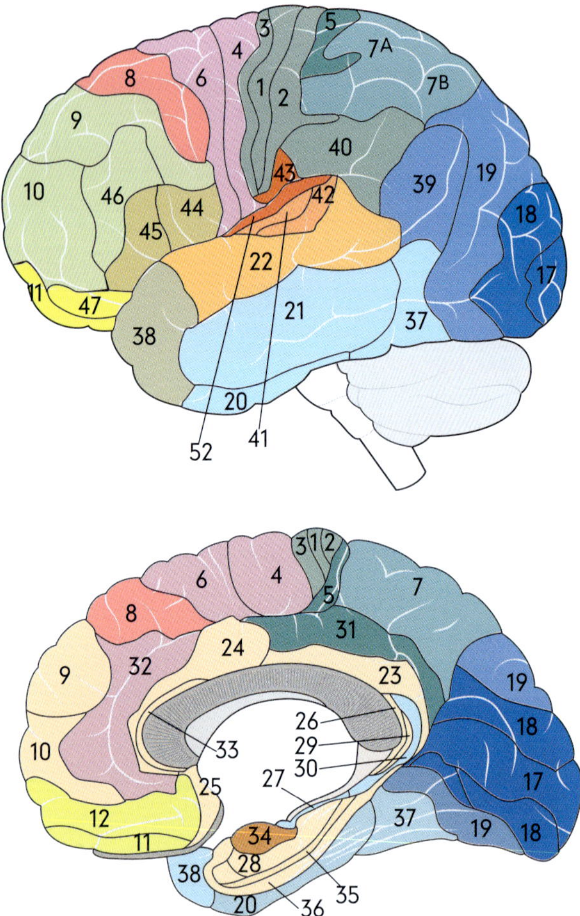

Figure 4.12 Brodmann areas
Source: Gage & Baars (2018).

Metacognition and monitoring

One of the three main phases of metacognition is *monitoring*. This is when the learner is reflecting on their progress and performance in response to their learning plan or specific goal, or, in the process of assessment, when answering a test question. One of the questions in the 'monitoring' box in Figure 4.5, asks: 'Am I making any mistakes, what do I need to do to correct my mistakes?' When researching the monitoring process in metacognition, researchers often focus on participants' response to error.

This century, the use of ERP technology is shining new light on what is happening in the brain when a learner is reflecting on and/or responding to error. This involves two processes: **error-related negativity (ERN)**, where no external (e.g. teacher) feedback is involved (Gehring et al., 1993); and **feedback-related negativity (FRN)**, where external feedback is given. Before being given the feedback (correct/incorrect), participants are often asked how confident they are in the answer they have given. Let's look at ERN and FRN in a little more detail.

Error-related negativity

Error-related negativity (ERN) occurs very quickly in the brain, within some 50–150 ms after the learner responds to a specific task (e.g. when a primary school student chooses 'k' not 'c' in response to '_ake with cake image').

error-related negativity (ERN): a negative deflection in an EEG waveform that is observed between 50 and 150 ms after perceiving that an error is made. It reflects the brain's activity as a monitoring system.

feedback-related negativity (FRN): an electrical brain signal measured with an electroencephalogram when a person engages with performance feedback (e.g. right vs wrong). It reflects the brain's activity as a monitoring system.

The ERN shows the brain spontaneously monitoring whether there are any errors in the decision; even when there is no explicit external feedback (e.g. teacher saying 'not correct'), the brain is doing it automatically.

Converging evidence has revealed Brodmann area 32, also known as the anterior cingulate cortex (ACC), as the source of the ERN (Holroyd et al., 2004). The ERN shows up on the waveform displayed on the computer in the region of N100, but it is designated as ERN because of its location, as indicated by the specific electrodes the signals are coming from. Moreover, it has been found that the relative magnitude of ERN (shown by the height of the waveform peak) actually predicts the degree to which learners learn from their own errors (Holroyd et al., 2004). In other words, this suggests that there are individual differences in our metacognitive abilities.

We know the location in the brain where ERN is generated, but what is going on? At a chemical level, it appears to be related to the neurotransmitter dopamine, which is known to play a major role in reward-related cognition, feelings of wanting and decision-making. Beyond that, there is still much to be discovered about how the brain is monitoring errors.

Feedback-related negativity

Feedback-related negativity (FRN) occurs when explicit feedback is provided to the learner – for example, when a teacher tells a student that the answer they have just given is not correct. This elicits a negative ERP component, which occurs within 200–300 ms from the moment when a learner received the feedback, so relatively slower than ERN. Nevertheless, ERN and FRN share much the same localisation in the brain (Brodmann 32), which means that the general error monitoring system in metacognition operates both when learners self-monitor without feedback and when external feedback is given.

REFLECTION

Can you think of different types of feedback you received in your classes? Do you think the ways to provide feedback lead to different approaches to learning taken by students? Also, would it be possible that two different students respond differently when they are given the same feedback from teachers? Why do you think such differences occur?

Armed with what we have just learnt about ERN and FRN, we can now briefly look at two very informative brain studies of metacognition.

ERN study with pre-verbal infants

RESEARCH LINK 4.1

Goupil, L. & Kouider, S. (2016). Behavioral and neural indices of metacognitive sensitivity in preverbal infants. *Current Biology*, 26, 3038–45.

The first, studying ERN, was conducted with very young *pre-verbal infants*. That is significant because the majority of studies of children's metacognition have relied on verbal self-reporting, but that is clearly not possible with pre-verbal infants. The only way researchers can get to their metacognitive processes – if indeed infants employ them – is by designing studies that record brain activity. They cannot rely on self-reporting.

But there is another related point to make here: perhaps research investigating older verbal children shouldn't rely on self-reporting either. For example, how can a researcher be sure that when children display poor metacognitive skills when self-reporting, it is not the result of poor metacognition, but rather poor self-reporting skills? In other words, observations from such studies might reflect children's limited capacity for explicit self-reporting, rather than limitations in metacognition (Goupil & Kouider, 2016).

There is also another major concern, mentioned in the first section of this chapter. Many neuroscientists emphasise that we actually 'have little or no direct conscious access to higher order cognitive processes' (Frith, 2012, p. 2214). In that case, the very thing that metacognitive studies are interested in (higher order cognitive processes) is out of reach because it is occurring below the level of conscious awareness. These concerns about self-reporting lead many researchers to use EEG and conduct ERP studies.

In 2016, two French researchers, Louise Goupil and Sid Kouider, found that infants as young as 12 months of age can 'estimate decision confidence', monitor their own errors and 'use these metacognitive evaluations' to guide subsequent behaviour. This means that, at a neural level, the 'core neural mechanisms of error monitoring' (ERN) observed in verbal children and adults is 'functional at 12 months of age' (Goupil & Kouider, 2016, p. 3043). In a second study, babies as young as one year of age demonstrate 'uncertainty monitoring' skills and are able, non-verbally, to ask for help in order to avoid making errors (Goupil et al., 2016).

RESEARCH LINK 4.2   ## FRN and error corrections in academic learning

Mangels, J. A., Butterfield, B., Lamb, J., Good, C. & Dweck, C. S. (2006). Why do beliefs about intelligence influence learning success? A social cognitive neuroscience model. *Social Cognitive and Affective Neuroscience*, 1(2), 75–86.

Turning now to FRN, a study by Jennifer Mangels and colleagues (2006) in the United States focused on how the assumptions that students bring to their learning influence the way they respond to negative feedback in the metacognitive process. The specific assumptions they investigated were related to intelligence and ability.

Two groups of students participated in the study. Half believed in the notion of innate ability and innate intelligence – the idea that we are born with these and they are therefore fixed for each individual. The other half believed that what we call ability and intelligence are not fixed, but result from learning experience and can change. It was found that these two groups responded very differently to explicit feedback after answering a range of general knowledge questions (history, literature, religion, geography, physical sciences, etc.), and indicating their confidence in the accuracy of their response.

When told they had got the wrong answer, those who believed in innate intelligence and innate ability displayed FRN only if the 'error' feedback was unexpected. In other words, they did not expect to be told (on the computer screen) that they were WRONG. For example, in response to the question 'What is the capital city of Australia?' the participant said 'Sydney'. They received error-feedback: WRONG. If they were expecting to be correct, this came as a surprise and on the EEG it produced an FRN component (negative spike on the screen) within about 200–300 ms.

Moreover, when being given the right answer – 'The capital city of Australia is Canberra' – this group of students demonstrated less sustained memory-related activity (i.e. smaller-sized left temporal lobe negativity) when attending to the correct answer. This implies a lack of attention: they were not engaging with effortful metacognitive learning when they were offered an *opportunity to correct* their misunderstandings. Perhaps not surprisingly, when called to do a second test (containing the questions they had got wrong) after some eight minutes, after taking off the EEG cap they made exactly the same error – still saying 'Sydney' instead of 'Canberra'.

By contrast, students who do not believe ability and intelligence are innate and therefore fixed for each individual, but instead see them as resulting from learning experience and able to change, registered FRN components whether they were expecting to be wrong or not and at a consistently large amplitude.

What this research reveals is that metacognition is dependent on certain *assumptions* that individual learners bring to their learning, and that this recognition needs to be incorporated in theoretical accounts of metacognition and, more importantly for education, into metacognitive practices and learning strategies in classrooms. Second, even at an individual level, researchers and teachers cannot assume that any given student is employing their metacognitive skill consistently; their assumptions about intelligence and ability may be impacting their attention to error.

APST and ACECQA curriculum specifications   STANDARDS

APST Standard 1.2 Understand how students learn
Demonstrate knowledge and understanding of research into how students learn and the implications for teaching

APST Standard 3.3 Use teaching strategies
Include a range of teaching strategies

ACECQA curriculum specifications
1.1 Learning, development and care
3.4 Teaching methods and strategies

Metacognition is widely recognised, both internationally and nationally, as one of the most important and productive learning strategies and teaching methods that practising schoolteachers can use. However, metacognition is not only a very important learning and teaching strategy; it is also a major research focus in education research. This chapter has provided an overview of how students and teachers can apply metacognition in the learning process, but it goes much deeper than that in describing how metacognition develops and how it is currently starting to be researched using brain imaging technology. Newly qualifying teachers were provided with a basic introduction to the use of encephalogram (EEG) research practices and what these are currently revealing about the nature of metacognition.

In addition, using examples of infant studies using EEG and what are called event-related potentials (ERPs), the chapter was able to address and rectify some common misconceptions about metacognition – for example, that it only develops in adolescents. In fact, EEG research shows that metacognitive abilities are present from infancy, so it is available for all teachers, including early childhood teachers, to provide appropriate teaching strategies that encourage young children's metacognitive learning abilities. Moreover, research findings regarding the brain's quick and subconscious processes underlying metacognition, and the relationship between subject knowledge and metacognition, can provide helpful pointers to the design of new teaching methods and strategies.

SUMMARY

- Not all learning depends on Hebbian repetition; learning can also occur as a result of one-off events, such as experiences that engender great joy or dread. Also, not all Hebbian learning occurs at school; it occurs in every aspect of our lives as a result of repeated regularities.
- Working memory is often taken to be synonymous with *short-term memory*, but temporary storage is only one aspect of working memory. Rather, it is best thought of as a central feature of executive functioning, including paying attention, thinking imaginatively and creatively, imagination and problem-solving.
- The ability to use metaphor and analogy is the key to genius in every area of human creative thinking, but the kind of analogising most required in creativity, especially in educational contexts, is fluid analogising, involving not just one correct but several possible and relevant comparisons.
- Stanislas Dehaene, distinguished French neuroscientist, argues that there are four pillars of successful learning: attention; active engagement; error feedback; and consolidation. When teachers employ all four pillars, they will maximise the speed and efficiency with which their classes can learn.
- While some previous learning theories such as constructivism and discovery-based learning seem to be losing their appeal, many in education view enhancing metacognition as one of the most important and productive learning strategies that teachers can use.
- Metacognitive strategies are those that learners use to monitor or purposefully regulate their thinking. These are found in infants and even some non-verbal animals, such as dolphins.
- Metacognition comprises planning, monitoring and evaluation of one's learning process. All three are in play throughout the process of metacognition, each interweaving with and feeding off the others.
- A common misconception is that metacognition is a learning skill that can be separated from the context of subject knowledge. Instead, a strong grounding in subject knowledge and the improvement of metacognitive skills must go hand in hand, each feeding off the other through feedback.
- According to Chris Frith (2012), there is little or no direct conscious access to higher order cognitive processes such as metacognition. We may have access to the outcomes of these processes but, through introspection (thinking about our thinking), we have very little idea about how these outcomes are achieved in the brain.
- EEG is a non-invasive recording technology of neural activity (the neural field created by the firing of neurons in the cerebral cortex). EEG electrodes are thought to be actually capturing a summation of the post-synaptic potentials from a large cluster of neurons that are firing synchronously.
- The event related potential (ERP) technique, using time-locked tasks and averaging wave patterns produced by EEG, extracts specific neural responses, associated with sensory, motor and cognitive events, from the wave patterns produced by EEG.

- The brain constantly monitors errors and external feedback. Error-related negativity (ERN) shows the brain spontaneously monitoring whether there are any errors in the decision, even when there is no explicit external feedback. Feedback-related negativity (FRN) occurs when explicit feedback is provided to the learner.

KEY POINTS FOR TEACHERS

- As a result of Hebbian learning, teachers physically change the brains of students, but they also physically change their own brains in their learning and metacognition when performing their teaching role, and in developing their professional expertise.
- Memory is necessary for learning, but research suggests it is not sufficient; the hallmark of sound learning is the ability to apply it creatively, imaginatively and wisely. This should be recognised in assessing learning at school. Teachers need to pay attention to what their test is actually testing.
- The best way to know whether students have understood and grasped the lesson is to test it, not to assume that because they clearly enjoyed the lesson and paid full attention, they have therefore learnt it. They may have significantly misunderstood it.
- All sorts of causes may be impacting the failure to learn. As teachers know all too well, and constructivists have emphasised, one of those factors is *prior learning* – particularly 'erroneous' learning that has occurred in lessons, and in the early years of pre-school and kindergarten, as well as more broadly in the home and in social and cultural settings.
- Contrary to *multiple intelligence* theory, as proposed by Howard Gardner, neuroimaging research has consistently shown that brain function is both localised, as required for multiple intelligence theory, and highly distributed. Teachers should be aware that the same neural systems are involved in many different cognitive abilities.
- Good teachers are those who are able to think of good analogies when explaining and clarifying tricky ideas. The same may be said of successful active learners: they are particularly good at creating and imagining good analogies in the process of understanding tricky ideas.
- Teachers who manage to ensure that their students attend to the four pillars of learning – attention, active engagement, error feedback and consolidation – will maximise the speed and efficiency with which they learn.
- Students must pay attention if they are to learn. Good teachers will constantly capture and channel their students' attention. However, all parents and teachers should be aware of the gorilla experiment, because it provides a strong warning that we are highly selective in what we perceive because the brain is filtering our perceptions.
- One of the best ways to help your students become metacognitive thinkers and learners is to do it yourself, and then *explicitly model* it for your students. Tell them how you are approaching a subject-knowledge task and then, most importantly, let the students try it themselves.
- Don't make students' learning too difficult and don't make it too easy – students need a level of difficulty to develop, balanced by good modelling.

- A common misconception is that metacognition is a learning skill that can be separated from the context of subject knowledge. On the contrary, the extent to which we are able to consciously monitor, regulate and self-report on our own learning is highly subject-knowledge specific.
- A key assumption of metacognition is that, as conscious human 'selves', we are able to reflect on our own thinking processes and generate reliable and accurate knowledge about our own learning. However, there are sound neuroscientific reasons to doubt that, not least because much of our thinking and learning is occurring below the level of conscious awareness. Researchers can gain some access to subconscious reactions by using brain imaging such as EEG.

Guided responses

REVIEW QUESTIONS

1. What key processes are involved in metacognition?
2. Is metacognition a form of high-level cognition that only emerges in adolescence?
3. Is metacognition a general learning skill that is unrelated to specific subject content?
4. How does the brain enable metacognitive monitoring?
5. What are the benefits of using EEG when studying students' metacognition?

FOOD FOR THOUGHT

1. The concept of metacognition in education goes back to the 1970, so it is not new. To what extent was your own learning in school influenced by metacognitive theory? Did you learn it as a learning strategy in your different subject disciplines? Did you learn it independent of subject knowledge, or did you not really come across it much at all? If you did encounter it, do you think it helped your learning? If you didn't encounter it, after reading this chapter do you think you missed something important?
2. Do you think that in studying as an adult learner, it is important to practise metacognitive skills or is it something you can bypass now? If you feel it remains important, based on what you have learnt in this chapter – particularly what you learnt about ERNs and FRNs – see whether you can improve on the three phases of metacognition proposed by Robin Fogarty (1994) and Figure 4.5.
3. In this chapter, it was pointed out that one of the important messages of a dynamic systems approach to learning and development is that context matters – it doesn't happen in a vacuum. Within Australia and the Australian education system, what contextual factors can you see operating and do these advance learning and development, or do they hold it back? Give reasons for your judgements.

RESEARCH ACTIVITY: USING METACOGNITIVE STRATEGIES TO SUPPORT STUDENT LEARNING

The importance of teaching metacognitive strategies has been recognised by a variety of stakeholders, both in Australia and in other parts of the world. Some of the key stakeholders have designed and disseminated resources to help practising teachers understand the importance of metacognition and gain strategies for their classroom teaching. Below are some examples of resources from well-recognised authorities. Please access these resources and think about the following two questions.

- Victorian Department of Education and Training (2021). *Professional Practice Note 14: Using metacognitive strategies to support student self-regulation and empowerment.*
- Education Endowment Foundation (n.d.). 'Metacognition and self-regulated learning: Seven recommendations for teaching self-regulated learning and metacognition'.
- Knaack, L. & Robertson, M. (n.d.). *Ten metacognitive teaching strategies: Helping students learn how to learn.* Nanaimo, Canada: Centre for Innovation and Excellence in Learning, Vancouver Island University.
- Yale Poorvu Center for Teaching and Learning, 'Encouraging metacognition in the classroom'.

Research activity links

Questions

1. Are the definitions of 'metacognition' introduced in each website consistent with the contemporary understandings gained from the science of learning?
2. Are the suggested metacognition strategies well aligned with the learning of subject knowledge?

THE DYNAMICS OF LITERACY AND NUMERACY LEARNING AND DEVELOPMENT

LEARNING OUTCOMES

By the end of this chapter, you will:

- Understand the emergence of language development as a human phenomenon, and in individuals
- Be able to identify the neural mechanisms that enable the emergence of literacy learning and development
- Be able to identify the neural mechanisms that enable the emergence of numeracy learning and development
- Appreciate how scientific research into the learning and development of literacy and numeracy is able to inform educational policy and classroom practice

Language development as a human phenomenon and in individuals

Have you ever stopped to think how remarkable it is that human beings have such an amazing ability to use verbal language, employing a massive array of different sounds and putting them together into words and groups of words that have meanings? It's not that other animals don't communicate through sound; they most certainly do. Not only mammals but birds signal to each other using sound all the time, and clearly these sounds have meanings – claiming territory and warning of danger, for example. Human beings are not alone in using sound to communicate; there are evolutionary antecedents for language, but the difference between all other creatures and human beings in the ability to use language is simply remarkable. So *when* did it begin in the story of human species development and *how* did it happen?

The origins of language in human social development

It is impossible to know for sure when language emerged, but archaeological evidence suggests that our species ancestors called 'archaic Homo sapiens and Neanderthals had the brain capacity, neural structure and vocal apparatus for an advanced form of **vocalisation**, which should be called language' (Mithen, 1996, p. 161), which would take the story of human language at least as far back as some 400,000 years. Notice the three factors identified here as essential for language: *brain capacity, neural structure* and **vocal apparatus**. The brain has to be massively endowed with neurons and neuronal connections (capacity), highly complex, dynamic and plastic so it has developed a rich and extensive array of neuronal pathways (neural structure) and the necessary physical vocal apparatus to make the sounds, including the pharynx, larynx, teeth, tongue and lips.

vocalisation: the act or process of producing sounds with the voice; also a sound thus produced

vocal apparatus: the organs, including the pharynx, larynx, teeth, tongue and lips, that produce sounds and speech

It is also postulated that language developed when humans began using *words to communicate about their social relationships* (an extension of grooming found in apes and monkeys) and their tool-making, plant gathering and hunting, which are also social events. On this basis, it seems all but certain that our own species (modern Homo sapiens), which emerged in Africa perhaps as recently as 100,000 years ago, already had highly developed forms of language.

But here's the problem: if language came quite late in the story of human species development (which goes back some two million years, to Homo habilis), this means that the neural mechanisms used in language were not 'built into' the brain from the start, as a result of evolutionary adaptation. They must have emerged later, by co-opting mechanisms that existed for other purposes. The word used for this is **exaptation**, first coined by Stephen Jay Gould (1991), who we will meet again in Chapter 7.

exaptation: a biological feature that has emerged later, by co-opting mechanisms that existed for other purposes, rather than being 'built into' the organism as a result of evolutionary adaptation

If we are going to understand *how* human language developed, an important starting point should be the recognition that *the brain recruited neural mechanisms not originally concerned with spoken language*, particularly those concerned with audio (sound) recognition and those that associate audial perception with meaning and potential action. The rabbit grazing in a field is highly alert to sound – is that particular sound a cause for taking flight, or can it safely be ignored? Gunnison's prairie dogs signal danger to kin by using different alarm calls *depending on the species of the predator* (Kiriazis & Slobodchikoff, 2006).

REFLECTION

Is it too much to claim that the ability of prairie dogs to signal the species of the predator means they have the rudiments of language? What do you think? If you think it is not language, what more is needed to make it language?

The human ability to use language has now evolved to the point where *language acquisition* has become as innate as the ability to perceive faces (Reid et al., 2017). We are born with it. As we will see later, although language acquisition is innate, literacy (reading, writing) is not. By about 28 weeks of pregnancy (third trimester), the foetus is already able to hear and is listening to the rhythm of language coming to them through the uterine wall; they even start to memorise it (Partanen et al., 2013). By birth, 'babies can tell the difference between most vowels and consonants in every language in the world' (Dehaene, 2020, p. 65). Of course, that doesn't mean that babies come into the world with language, only that they are born with the ability to acquire it.

Hebbian learning and language development in the early years

Having briefly considered the anthropology of language development, another starting point is to understand what is going on in the infant brain, as infants start to *grapple* with language. *Grappling* with language? That has a familiar ring to it. Do you remember from Chapter 3 the baby learning to *grasp* a rattle, grappling with it, trying to grasp it through repeated attempts? Could the development of words and the beginnings of language in an infant involve a very similar process? Or, to put it more formally, could the development of language in infancy result from classical Hebbian learning?

In Chapter 2, you met Donald Hebb and learnt about his core idea that learning occurs in the brain at the synapses and results from *repeated and coincident* firing of two synaptic connections. More specifically, something is learnt and memorised in the brain in response to a stimulus (seeing something, hearing something, etc.) as a result of electrical and chemical processes that signals across synapses, and that become more efficient as a result of repetition (repeated stimulation). That same Hebbian process (synaptic change through plasticity) occurs in the formation of neural pathways that 'learn', by becoming more efficient as a result of *repetition and feedback*.

As noted in the previous chapter, Hebbian learning through repeated stimulation is not the only kind of learning and memory; sometimes when the stimulation is very meaningful (either positively in moment of joy for example or negatively in trauma), no repetition is needed.

However, with regard to the kinds of learning that normally occur in school, Hebbian repetition plays the major role. Hebbian learning is also apparent in the way babies babble. As you learnt in the previous chapter, the vocal apparatus (fluctuations in their **larynx**) of babies allows them to randomly babble with every different sound a human can make. However, by the age of about six months, that repertoire of sounds has already been pruned

larynx: the 'voice box', located in the top of the neck

as a result of repeatedly listening to their caregivers' language ('mother tongue') and accents (Geake, 2009, p. 51), although these are often heard at a higher pitch than normal speech ('**motherese**'). Sounds that are heard are strengthened, while those not heard are weakened and lost, as the pathways in the brain for those sounds are weakened and pruned.

Hebbian *synaptic strengthening and pruning* are central to Edelman's notion of neuronal group selection that we met in Chapter 3, with regard to an infant learning to grasp a rattle. In the selective process, those synaptic firing patterns that are recognised as *valuable* become stronger and more frequent, while those that are not (those that have no value or salience in leading to successful grasping) are pruned. Strange as it might seem, it is precisely this *selective loss* through pruning that *opens the way to gain*, leading to ongoing language development and the intellectual development of the child.

Individual developmental pathways and early noun learning

Early language development 'consists mostly of learning object names' (Smith, 2013, p. 626) and, as just noted, the dynamic processes involved are *repetition and feedback*, and *synaptic strengthening and pruning*. The young child sees an object and is told the name. The process is repeated over and over: same object, same name. When the young child is able to grasp the object, it can be moved and rotated, such that it can be perceived in three dimensions. Hands and eyes work together, forming a perception–action loop that 'plays a role in real-time processes of stabilising visual attention on the object and supporting the binding of a heard name to the seen thing' (Smith, 2013, p. 623). In time, after numerous attempts, the child recognises the object and other similar objects, and is able to apply the correct name.

But does this mean each child is following the *same* developmental pathway to language development or is each child doing it their own way, but still ending up with much the same result? A dynamic systems account of language development emphasises that each individual child will have their *own pathway* to language competence. For example, in researching the relationship between infants' visual object recognition and the learning of the names of those objects, dynamic systems theorist Linda Smith (2013, p. 618) argues for a *pathways approach* to language development and defines a developmental pathway as 'the route, or chain of events, through which a new structure or function forms'.

Smith (2013, p. 626) concludes that children 'follow different developmental paths that depend on the specific tasks they discover and the intrinsic dynamics of their own systems, even if they end up, more or less, with the same set of human competencies'. For any given language (English, Chinese, Spanish, etc.), even though each child will follow their own developmental pathway, *all children learning that language nevertheless end up with the same names for objects and the same grammatical structure* – individually different pathways, but all leading to more or less the same set of vocabulary and language competence.

motherese: a term used in the study of child language acquisition to describe the way mothers often talk to their young children

REFLECTION

Think about the words infants first learn to say. How much do you think infants understand those first words? Talk it through with one or two classmates.

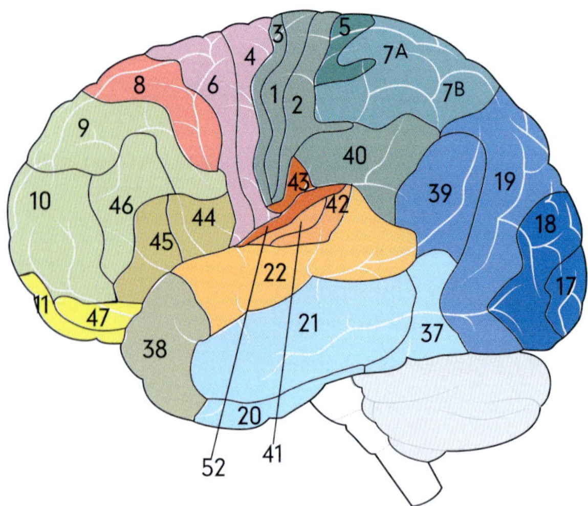

Figure 5.1 Brodmann areas. Note areas 44 & 45 (Broca's area) and 22 & 39 (Wernicke's area).
Source: Gage & Baars (2018).

lesion: damage to any part of brain

aphasia: 'an impairment of language, affecting the production or comprehension of speech and the ability to read or write' (National Aphasia Association, 2021)

expressive aphasia (or Broca's aphasia): an impairment in the production of language

Wernicke's aphasia: an impairment of language resulting in the loss of ability to understand spoken or written language

Wernicke's area: a region lying close to the auditory cortex in the temporal lobe of the dominant hemisphere. This area is traditionally considered to be a centre for the comprehension of language.

hemispherectomy: a neurosurgical procedure in which a cerebral hemisphere is removed, disconnected from the rest of the brain

The role of brain lesions in understanding language

Many of the early discoveries about the human brain have come from people who have suffered brain damage (**lesions**) and subsequently lost a particular ability. In regard to language, early discoveries were related to two different conditions that go by the general name of **aphasia** (or dysphasia). In the nineteenth century, the French neuroanatomist Paul Broca (1824–1880) studied two patients who had *lost the ability to speak words, though they could still understand them.* This is often referred to as **expressive aphasia (or Broca's aphasia)**. When they died it was discovered by autopsy that they had lesions in Brodmann areas 44 and 45, in the left hemisphere of the brain, which also became known as Broca's area (see also Chapter 4) – but see the note on brain specialisation, below. A different from of aphasia, often referred to as **Wernicke's aphasia**, was identified by the German neuroanatomist Carl Wernicke (1848–1905). This condition, in which a *patient loses the ability to understand spoken or written language*, results from lesions in both areas 22 and 39 in the left hemisphere, also known as **Wernicke's area** (Figure 5.1).

The study of brain lesions has assisted our understanding of language, but there are some important caveats that need to be noted. Most naturally occurring lesions, such as those resulting from a stroke, are not neatly restricted to small areas of the brain devoted to very specific tasks, such as recognising the orientation of lines in reading, but tend to be more widespread. Relating a particular lesion to a specific skill may therefore not be very precise and subsequent research is required to identify the actual site of impairment. Moreover, it cannot simply be assumed that if a lesion impacts an area that it is directly controlling that skill, the impact could merely be a side-effect of the lesion and the real control is elsewhere. It is often the case that damage to an area does not completely eliminate a skill, but rather substantially lowers performance.

The key message to bear in mind is that the acquisition of language 'requires a delicate balance between neural specialization and plasticity' (Gurunandan et al., 2020. p. 9715). For most people, language is located in the left hemisphere of the brain, but does that mean that if the whole of the left hemisphere is removed (as in a drastic treatment for crippling epilepsy called **hemispherectomy**), the patient will lose all language? The answer is no, especially if the operation is performed on a young child. The right hemisphere is able to compensate – at least in part, if not completely.

The issue of brain specialisation has been a thorny problem throughout the history of brain science. One view has held that the *brain is highly localised*, containing many specialised parts such that language and other main functions that occur in highly specific regions. One of the main early protagonists of this view was Karl Lashley, the doctoral supervisor of Donald Hebb. The opposing view has been that the different functions of the brain are not specialised and localised but are *highly distributed* and result from the global operation of the whole brain. In the case of Broca's area, for example, although it is often presented in textbooks as a clear example of localised brain function, neuroscience research 'has shown that the region is neither necessary nor sufficient for language' (Barrett, 2017, p. 167). As with many such dichotomies, the truth regarding whether brain function is highly localised and specialise or highly distributed and global probably lies somewhere in the middle, though currently the favoured view seems to be **localisation and specialisation**.

localisation and specialisation: the idea that brain is highly localised, containing many specialised parts such that language and other main functions occur in highly specific regions

A case study of language recovery after hemispherectomy

RESEARCH LINK 5.1

Boatman, D., Freeman, J., Vining, E., Pulsifer, M., Miglioretti, D., Minahan, R., Carson, B., Brandt, J. & McKhann, G. (1999). Language recovery after left hemispherectomy in children with late-onset seizures. *Annals of Neurology*, 46(4), 579–86.

A well-known case of language recovery after hemispherectomy is Michael Rehbein, who was longitudinally studied by Dr Dana Boatman (Boatman et al., 1999).

From the age of five, Michael began to experience uncontrollable seizures, as many as 400 a day. Following the doctor's recommendation, he had the left side of his brain removed at age seven. By six months after surgery, Michael's speech-understanding abilities were back to what they were before surgery. At that point, he was producing one- and two-word utterances spontaneously. The experience of Michael and his family was highlighted in Episode 2 of a documentary titled *The Secret Life of the Brain*, produced by the PBS.

Dr Boatman tested children like Michael as early as the first week after they had surgery. Within four to 16 days after surgery, these children started to show improved phoneme discrimination compared with their performance just prior to the surgery. Her results indicated that the isolated right hemisphere can discriminate consonants and vowels as early as four days after left hemispherectomy. Comprehension of spoken words and phrases reappeared later, but recovered fully within one year.

However, these children's expressive functions (i.e. speaking) still lagged behind. One year after surgery, naming remained impaired, and 'patient speech was limited largely to production of single words' (Boatman et al., 1999, p. 579).

Nevertheless, the recovery of higher-level comprehension and, 'to a lesser extent, expressive language functions is attributed to the plasticity of the right hemisphere, which appears to persist' beyond the early childhood (Boatman et al., 1999, p. 579).

Research link 5.1

Literacy learning and development

We now move from language to **literacy** learning and development. If language is quite a recent development in human evolution, then reading and writing (literacy) are *very* recent (a few thousand years) and not something all recent human societies and cultures developed. Moreover, within many present-day cultures, there remain people who, often through lack of educational opportunity, remain illiterate. This should reinforce the very important

literacy: the ability to both comprehend texts through listening, reading and viewing and compose texts through speaking, writing and creating

difference, noted above, between language acquisition and literacy. Whereas language acquisition is innate – we come into the world with the ability to acquire language – this is not the same for reading and writing. We do *not acquire them naturally*; they have to be formally learnt. In short, when it comes to literacy, *schooling is essential* (Figure 5.2).

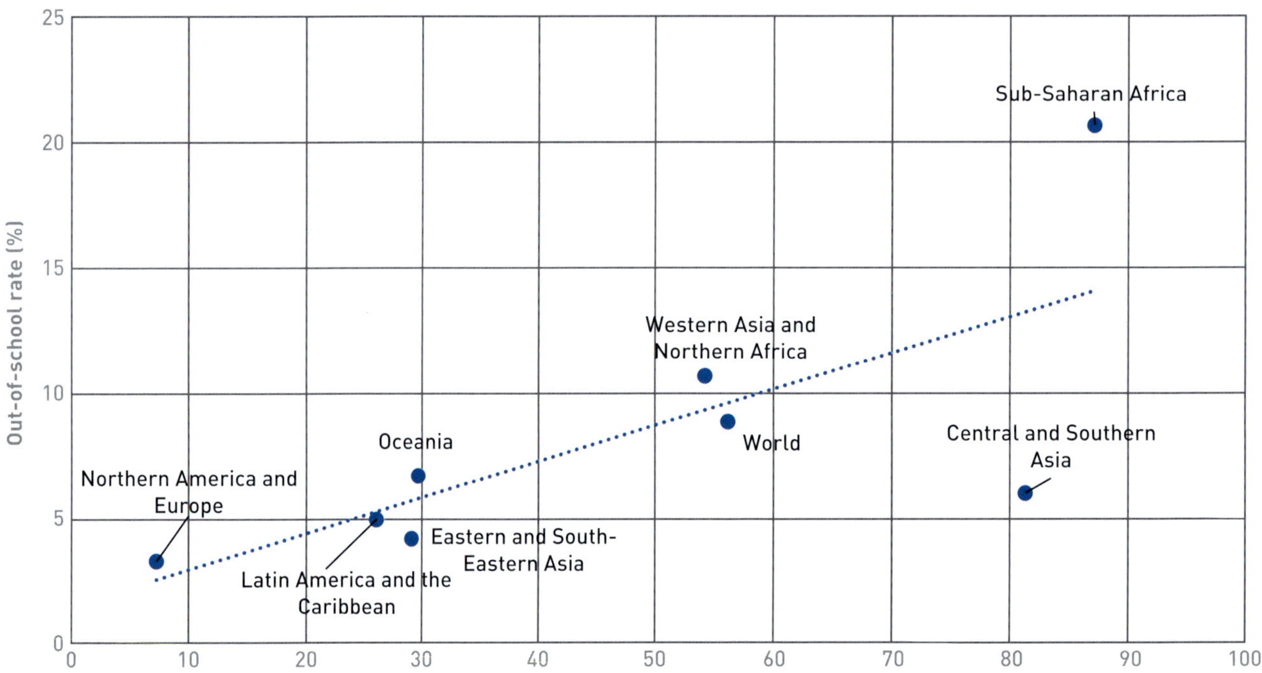

Figure 5.2 The relationship between out-of-school rates and the rates of children and adolescents not achieving minimum proficiency levels in reading
Source: UNESCO Institute for Statistics (2017).

We also noted earlier that the neural mechanisms used in language were not 'built into' the brain as a result of evolutionary adaptation; rather, language is an *exaptation*. A concept related to *exaptation* is **neuronal recycling**, used by Stanislas Dehaene to mean 'the partial or total invasion of a cortical territory initially devoted to a different function' (Dehaene, 2009, p. 147).

Literacy recycled a portion of the left hemisphere's cortical territory that originally evolved to *visually recognise and distinguish different shapes*. This cortical area is already present in the brains of monkeys and is able to recognise what Dehaene calls **proto-letters** (Dehaene, 2009, p. 137). Unbeknown to the many scribes who toiled to invent the different writing systems of each culture, they were actually discovering the preferred shapes of specialised neurons in the lower parts of the temporal lobes (Figure 5.3). Many of these preferred shapes 'closely resemble our letters, symbols, or elementary Chinese characters' (Dehaene, 2009, p. 137).

Basically, the reasons why your brain is literate, so you can read the black marks on this paper and make sense of them, are that you grew up in a culture that is literate and you have

neuronal recycling: 'the partial or total invasion of a cortical territory initially devoted to a different function' (Dehaene, 2009, p. 147)

proto-letters: shapes that resemble symbols, or elementary Chinese characters. Neurons on the inferior temporal lobe are particularly responsive to these shapes.

a brain that is *plastic*, with a *very efficient visual recognition system* that can recognise line junctions and other invariants in the shapes of written words or characters used by that culture. The black marks on this page are recognised by specialised neurons as symbols that you have learnt (remembered) when grappling with learning to read.

The story of Mr C and the brain's 'letterbox'

This also means that literacy can be lost. Imagine, as you are happily reading these words, that you suddenly find you can no longer read, that you cannot recognise the letters and words on the page. It seems highly unlikely, yet it is what happened to a man living in Paris in October 1887, who his doctor called Mr C. He could still see the letters and he knew they were letters, but he could not name them or turn them into words. He could still speak as normal and could write letters and words when the doctor gave him dictation, though after writing them he could not read what he had written.

Mr C was suffering from what his doctor called *pure verbal blindness* and what is now known as '**alexia** without agraphia' (literally meaning impaired reading without impaired writing), or **pure alexia**. When Mr C died four years later from a major stroke, an autopsy discovered that he had previously suffered a small stroke that had caused lesions in the region of Brodmann area 37 in the left hemisphere of the brain, while his right hemisphere remained intact. The case of Mr C provided the first evidence that the brain may contain a highly selective letter-reading area, so selective it seems that he could still recognise numbers. Very many cases of pure alexia have been reported since this first case, and we now have the imaging tools (especially MRI and fMRI) that are able to visualise a patient's lesion without the need to wait for the patient to die.

Using this technology, Laurent Cohen and Stanislas Dehaene identified what they believed to be a highly specialised region for letter recognition and word formation, which they named the **Visual Word Form Area (VWFA)**, located in the *left occipito-temporal* area (where the occipital and temporal lobes meet) within Brodmann area 37 (Dehaene-Lambertz et al. 2018). Dehaene (2009, p. 62) suggests that it is simpler to refer to this as the brain's 'letterbox' because, he says, it receives the incoming letters that we read before sending them to other parts of the brain concerned with pronunciation and meaning.

Altitalisch	Etrurisch	Altgriechisch	Phönizisch	Bedeutung
				a
				b
				g (k, c)
				d (t)
				e
				(Bau)
				z
				ē
				(h)
				(ch)
				th
				i
				k
				l
				m
				n
				x (s)
				o
				p
				(San)
				(Koppa)
				r
				s
				t
				v, u (y)
				f

Figure 5.3 Alphabet samples from old Italic, Etruscan, Ancient Greek and Phoenician

alexia: a partial or complete inability to read

pure alexia: impaired reading without impaired writing

Visual Word Form Area (VWFA): or brain's 'letterbox'; an area in the left occipito-temporal area (where the occipital and temporal lobes meet) within Brodmann area 37. It receives the incoming letters that we read before sending them to other parts of the brain concerned with pronunciation and meaning.

However, you will not be surprised to learn, given what was said above about the brain specialisation controversy that some scientists, including Cathy Price and Joseph Devlin (2003) in the United Kingdom, have argued for a more distributed model of letter and word recognition and that a precise VWFA or *letterbox* is a myth. Notice, however, that this is a very technical issue about the nature of brain complexity and how to interpret research data, especially lesion research data. A key question in the *localisation and specialisation* debate is 'What's the dynamic?' In other words, what's going on in that localised area of the brain (i.e. the letterbox area) and what's the balance between what's happening there and what's happening elsewhere, as part of the same overall process?

As always in science, it is important to appreciate that there are controversies – theories are often contested. However, that does not mean, as relativists insist, that all theories are of equal merit and there is no such thing as truth – that 'anything goes' and you can choose the theory that best suits your personal interest. That kind of relativism is totally opposed to science. Let's be clear: there is considerable agreement within science about how the brain learns to read. Disagreement allows science to progress – that's part of the excitement of science. Further research and analysis will eventually clarify the *degree* of specialisation in what is known as the VWFA, the brain's letterbox.

Some additional evidence was provided in a paper published in 2018. In a study of children who were just starting to acquire reading, it was found that the VWFA in the left hemisphere

> started to selectively respond to written words. In every child, it was then possible to go backward in time and ask what this region was doing prior to reading. We found that written words invaded a sector of visual cortex that was initially weakly specialized, slightly responsive to pictures of tools, and that lay next to a face-selective region. Reading acquisition did not displace those initial responses but blocked their development, such that face-selective responses became stronger in the right hemisphere. Those results provide direct evidence for how *education recycles the human brain* by repurposing some visual regions towards the shapes of letters. (Dehaene-Lambertz et al., 2018, pp. 1–2)

'Education recycles the human brain.' It is no exaggeration to say that teachers are literally *brain makers* and *brain recyclers*. That is a tremendous professional responsibility.

From the letterbox to sounds and meanings

Whether the process of letter recognition is highly specialised (a discrete letterbox) or somewhat more distributed, the key point is that brain research has consistently confirmed that 'in all cultures, the same area in the left occipito-temporal region is in charge of written word recognition' (Dehaene, 2009, p. 100). In every language, whether it is using an alphabet or individual characters as in Chinese, reading employs the same letterbox region of the brain (see the red area in Figure 5.4) and starts with a visual recognition process of previously learnt and invariant (not changing) letters, graphemes, morphemes and characters. This happens very rapidly and subconsciously, within about 200 milliseconds of first seeing the word or character. 'Beyond this point, the vision areas of the brain begin to work in tight conjunction with the spoken language networks' (Dehaene, 2009, p. 104), concerned with sounds and meanings. The overall process described by Dehaene (2009) is captured in Figure 5.4.

Basically, the brain area involved in converting letters into sounds involves the upper regions of the left temporal lobe and encompasses Broca's area, concerned with the articulation of words. However, as Figure 5.4 shows, the network involved in converting letters into sounds is very different from the network that analyses word meanings – a point that is very important for teachers, as it has a strong bearing on how to teach reading. Moreover, none of the semantic (meaning) areas is specific to reading – they all respond to spoken words and also images.

That said, it is one thing to know *where* the brain 'lights up' when engaged in converting written or spoken words into meanings, but quite another to understand *what* is actually happening. Although researchers have identified *areas* of the brain involved in meaning-making, the 'process that allows our neuronal networks to snap together and "make sense" remains utterly mysterious'

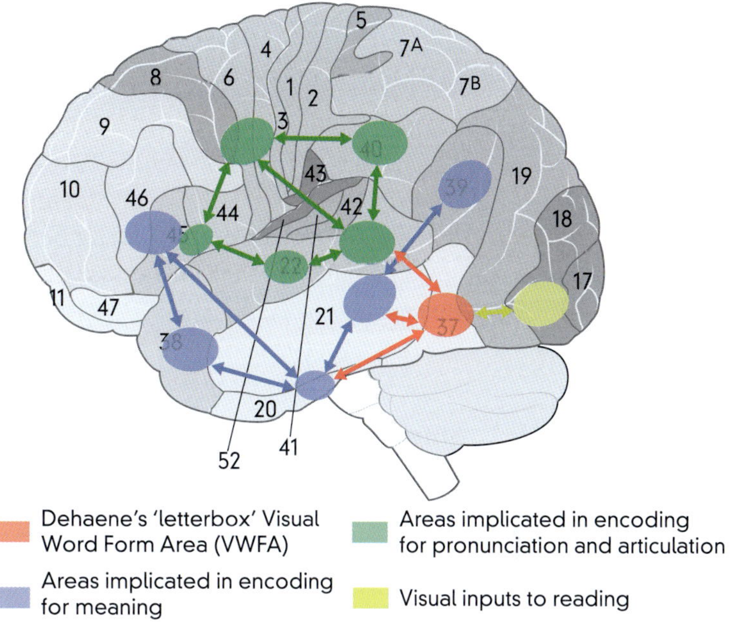

Dehaene's 'letterbox' Visual Word Form Area (VWFA)

Areas implicated in encoding for meaning

Areas implicated in encoding for pronunciation and articulation

Visual inputs to reading

Figure 5.4 Learning to read consists of creating an invariant visual representation of written words (red area) and connecting it to brain areas coding for speech sounds (green) and meaning (yellow). All connections are bidirectional.
Source: Brain image from Gage & Baars (2018). Concept adapted from Dehaene (2009, p. 63).

(Dehaene, 2009, p. 111). There is still a lot to discover about how human brains create meanings that are so essential to our everyday lives and to the human arts of storytelling, poetry and drama that enrich our lives and are introduced to children when taught in school. What we do know is that *multiple regions of the brain are involved in meaning-making* – it is highly distributed. The frontal and temporal lobes play a major role, but that may well be just the tip of a vast complex and dynamic network.

In Chapter 1, we mentioned that memory is highly distributed in the brain, that when we remember something the various parts of that *memory*, stored in different parts of the brain, are drawn together and reconstituted as a whole in what we have previously seen Damasio (1994, p. 95) call a 'trick of timing'. He makes a similar point with regard to *meanings in the brain*. Although the temporal and frontal regions are involved in meaning-making, it is likely that they are not where complete meanings are stored; rather, they act as 'convergence zones' (Damasio, 1989), which Dehaene says assemble 'dispersed fragments of meaning and bundles them together into articulated sets of neurons that constitute the genuine neuronal substrate of word meaning' (Dehaene, 2009, p. 112).

Learning to read: Three main stages

Learning to read is not easy. As adult expert readers, it is very easy to forget the grapplings and struggles involved. Even so, in just a few years of schooling, the child moves from seeing only a jumble of different marks on a page to recognising the marks as familiar spoken words that have meanings. How is that possible? Towards the end of the twentieth century, British researcher Uta Frith (1985, 1986) suggested there were three main *stages* in young children's learning to read.

In Chapter 1, we hoisted a warning flag about notions of developmental *stages* and *age phases*. Nevertheless, as we also noted, the notion of development stages can provide a rough approximation of children's developmental progress that can be helpful for teachers, so long as these warnings are heeded. Moreover, it is important to note that Frith's three stages of reading are directly related to the sequencing of events in the brains, and so are not like Piagetian constructivist stages. With these caveats in mind, her three main stages are:

logographic (or pictorial): children's tendency to recognise words as 'wholes' because they haven't as yet developed a set of alphabetic or orthographic spelling rules

Stage 1: The first stage in learning to read is **logographic or pictorial**. The child, at around age five or six, treats words as objects. The brain's visual system is picking up on visual clues, such as shapes of letters and recognising whole words as though they are pictures. Children at this age will typically recognise their own name and other words that they find visually stimulating. However, this is not *real* reading and is prone to error.

phonological: phonological awareness enables children to identify different units (e.g. syllables, rimes) in the words

Stage 2 The second stage is the **phonological** stage, when the child begins to break words that they see into their component letters and match them with spoken sounds, which can then be sounded out as a word. This transition to **phonemes** is *not* something that happens naturally or automatically – it has to be taught (Morais et al., 1979).

phonemes: discrete sound constituents of *spoken* words

Stage 3 The third stage is **orthographic**, which is when the child is reading fluently, no longer needing to systematically break words down into their component parts. Rather, the sequence of letters is recognised in one single act and the *meaning* of the word is grasped in parallel with recognising and pronouncing it.

orthographic: recognising a set of conventions for writing a language, including norms of spelling, word breaks, emphasis and punctuation

We will consider the educational implications of these stages in the final section of this chapter when we examine what has been called the *Great Literacy Debate* (Australian College of Educators, 2018) or more prosaically the *Reading Wars* (Castles et al., 2018; Dehaene, 2009), focusing in particular on the debate as it continues within Australia. First, though, we turn to the science of numeracy.

REFLECTION

How were you first introduced to reading? Can you recall the strategies that your teacher or parents used to teach you to read? Did they explicitly teach links between letters and sounds?

Numeracy development and learning

Perhaps we should begin by asking what is meant by numeracy? Is it, for example, simply the ability to recognise numbers and do basic number tasks, or does it encompass the ability to perform high-level mathematics? Is it simply concerned with performing mathematics skills or does being numerate include understanding its importance and role within social contexts? According to the Australian Curriculum, Assessment and Reporting Authority (ACARA, 2017):

> Numeracy encompasses the knowledge, skills, behaviours and dispositions that students need to use mathematics in a wide range of situations. It involves students recognising and understanding the role of mathematics in the world and having the dispositions and capacities to use mathematical knowledge and skills purposefully.

So, at least within the Australian educational context, **numeracy** is defined as the ability to perform and use mathematics with an appreciation of its social relevance and value.

numeracy: the ability to perform and use mathematics with an appreciation of its social relevance and value

Do babies possess a number sense?

This definition is helpful from a curriculum standpoint, but does numeracy also involve something more basic than this – something that even babies possess that enables these higher-level skills and abilities to be performed? To come straight to the point, are babies numerate, at least in some basic ways that provide a foundation for mathematical learning and development as they grow older? Do babies possess a **number sense** when they enter the world? If so, this would suggest that the origins of numeracy predate humanity, and can be found in other animals such as monkeys, for example. The idea that babies do possess a number sense was suggested in 1954 by Tobias Dantzig (1967). He believed that infants can 'recognize that something has changed in a small collection when, without his direct knowledge, an object has been removed or added to the collection' (Dehaene, 1997, p. 5).

number sense: a primitive ability innate to infants to appreciate quantities, numbers and their relationships

At this point, a clarification is needed between what is often referred to as **non-symbolic numeracy** and *symbolic* numeracy. Basically, non-symbolic numeracy is closely related to and arises from what we have called a number sense. It allows students to sense numbers and to estimate, compare and combine items such as dots on a page. Symbolic numbers are the digits and numerical symbols used in performing mathematics, starting with number symbols, 1, 2, 3 and so on, all of which are a human, social invention.

non-symbolic numeracy: the ability to discriminate the numerical magnitudes or perform numerical operations without involving numeric symbols

Over the past two decades, there has been significant growth in research investigating the numerical understandings of babies and infants. It is now generally agreed that non-human animals and preverbal children possess a primitive ability to appreciate quantities (Cantlon, 2012). Moreover, babies can recognise numerical changes in patterns of objects, whether these be visual such as collections of dots on a page or audial, as in repeatedly hearing sets of four identical syllables, with adjacent sets separated by pauses, for example 'tu, tu, tu, tu'. Pause, 'tu, tu, tu, tu' (Izard et al., 2009).

Indeed, these researchers have shown that new-borns (from 7–100 hours from birth) already recognise when sets of visual and audial objects share the same number (four dots, four tu's). They cannot discriminate the difference between sets of fours and fives, but they can if the gap is large, such as discriminating a set of four and a set of 12 objects (Izard et al., 2009). Discrimination among numbers becomes rapidly more precise during the first year. By six months of age, infants discriminate between sets with a 2:1 ratio (e.g. eight versus four dots or sounds).

From an educational standpoint, the idea that babies are born with a number sense flies in the face of what teachers were taught in the second half of the twentieth century, largely based on the constructivist theories of Jean Piaget (1965). Piaget believed that at birth the baby's mind is a numerical blank page. Over time, as a result of observing, internalising and rationally abstracting regularities found in the external world, numerical concepts are progressively constructed in young children's minds, sometime around the ages of four or five years. We will return to this important educational controversy about early numeracy development in the final section of this chapter, where we explain how Piaget got it wrong.

RESEARCH LINK 5.2

Infants' understanding of addition and subtraction

Wynn, K. (1992). Addition and subtraction by human infants. *Nature*, 358, 749–50.

In addition to recognising non-symbolic numerical magnitudes, infants can also perform simple arithmetic operations on small numbers of items. Karen Wynn (1992) studied whether five-month-old infants have a rudimentary grasp of addition and subtraction. Figure 5.5

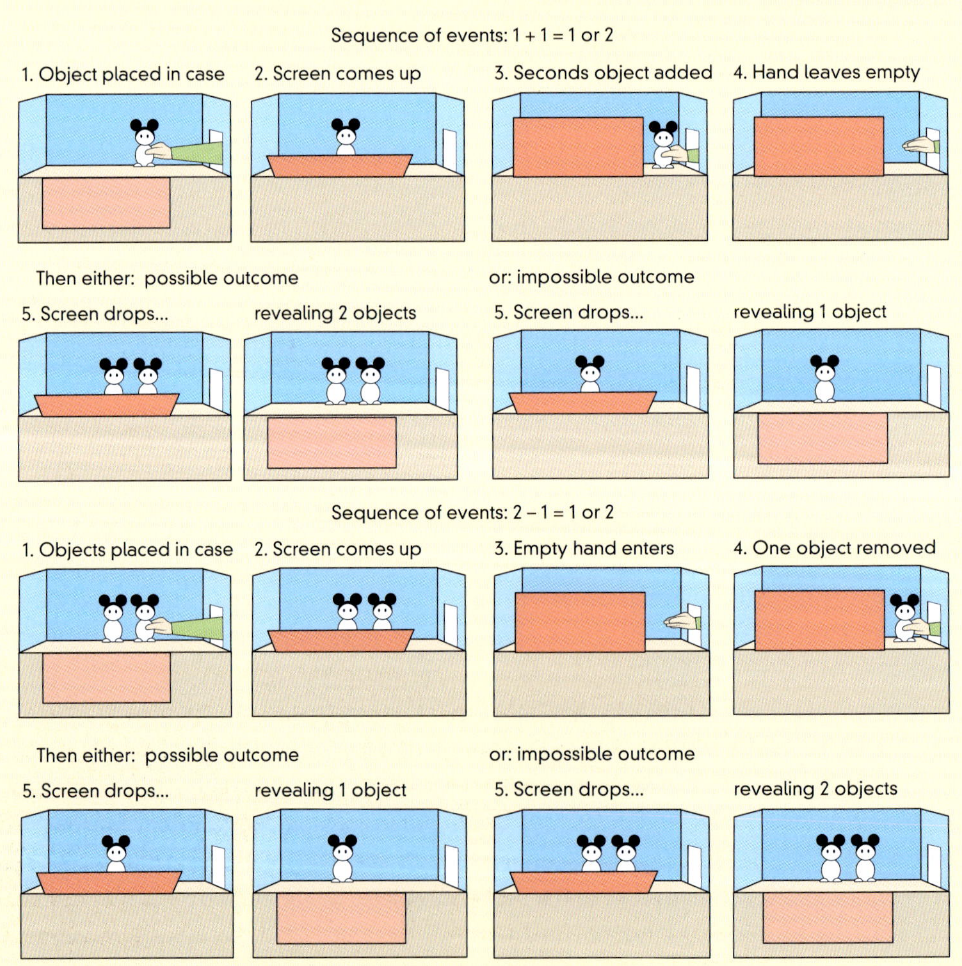

Figure 5.5 Infants' understanding of addition and subtraction
Source: Wynn (1992, p. 749).

illustrates the experiment performed by Wynn. In order to study infants' addition, five-month-old infants are shown a single puppet placed on a stage. Then a screen comes up and hides the puppet. The infants then observe a hand with another puppet that comes in and moves behind the screen, and the hand reappears empty after having been behind the screen. Then the screen drops, showing either the possible outcome of two puppets on the stage or the impossible outcome of one puppet on the stage.

Most of the infants looked longer at the impossible outcome (one puppet) than the possible outcome (two puppets), suggesting that they expected that 1 + 1 should equal 2 and they were surprised by seeing a single object. The results were consistent with subtraction (lower panel). Infants looked longer at the impossible outcome (2 − 1 = 2) than the possible outcome (2 − 1 = 1). Some may argue that this result does not imply infants can perform exact calculations. They may just sense that the numerosity of a set cannot remain the same when objects are added or removed. Therefore, they expect that 1 + 1 cannot be 1, but should be something else, without necessarily knowing the exact result for these operations.

In order to rule out this possibility, Karen Wynn showed the '1 + 1 addition' situation with outcomes of either two (possible) and three (impossible). The infants looked longer at the result of three items, showing that they were surprised when the addition of 1 + 1 resulted in 3. Therefore, five-month-old infants know 1 + 1 makes exactly 2, not 1 or 3. By nine months, infants can successfully add and subtract larger sets of objects, such as 5 + 5 = 10 and 10 − 5 = 5 (McCrink & Wynn, 2004).

REFLECTION

Why do you think Jean Piaget believed that at birth the baby's mind is a numerical blank page and children gain understandings of number later? Why do you think he was not able to see young children's number sense? What factors might have prevented his inability to see what subsequent researchers saw?

The social origins of numeracy

The social origins of symbolic numeracy are largely lost in the mists of time, although there are some archaeological artefacts that hold clues, such as the widespread use of *tally sticks* dating back to the Late Stone Age (or Upper Paleolithic) that lasted from about 50,000 to 12,000 years ago). Ancient tally sticks were usually long bones with notches carved on them, such as the Ishango bone dating back some 20,000 years, found in Africa. Split tally sticks were used as receipts in Medieval Europe. A length of wood was marked with a system of notches across its width and then cut in half. Matching the two halves provided proof of a transaction. Other clues about the origins of numeracy come from traditional peoples that do not have a written language and do not use modern tools and technology. Many such peoples use tally sticks, but on a day-to day basis often also use finger counting (as do many modern children and even adults) (Figure 5.6).

Figure 5.6 (Left) A baby is learning how to count using an abacus with decimal system.
(Right) A primary school child is relying on his finger counting while solving maths problems.

The decimal (base 10) system of counting we use today was inherited from the ancient Greeks. The ancient Babylonians, who were great astronomers, used a sexagesimal (base 60) system. Hence, 60 seconds in a minute, 60 minutes in an hour. The numerals we use today developed over time in India, which is also where modern rules for multiplication and division originated. Our numerals were further developed by Islamic scholars, who then transmitted them to the West. That is why our written numbers are referred to as the Hindu-Arabic numeral system.

Are literacy and numeracy related in the brain?

But how did numeracy and mathematics develop in the human brain? Clearly, as with language and literacy, numeracy is an *exaptation* and involves *neuronal recycling* – using neural mechanisms not originally concerned with numeracy and mathematics. So did numeracy piggyback on literacy, for example? That was the opinion of some scholars in the twentieth century, including the American philosopher and linguist Noam Chomsky, who famously believed that the origin of the mathematical capacity was 'an abstraction from linguistic operations' (Chomsky, 2006, p. 184).

Equally famously, Albert Einstein thoroughly disagreed. In *An Essay on the Psychology of Invention in the Mathematical Field*, Jacques Hadamard (1945, p. 142) cites Einstein as saying that words and languages 'as they are written or spoken, do not seem to play any role in my mechanism of thought. The psychical entities which seem to serve as elements in thought are certain signs and more or less clear images which can be "voluntarily" reproduced and combined.' Einstein is saying that, when doing mathematics, he is not mainly thinking in words but rather in images, which suggests that literacy and numeracy are using different networks in the brain. Moreover, as many teachers will testify, it is entirely possible for a student to be good at mathematics, yet struggle with reading and writing, and vice versa.

Marie Amalric and Stanislas Dehaene (2016) report studies they conducted, picking up on this controversy. Briefly, they used fMRI to scan the brains of 15 professional mathematicians and 15 academics who were non-mathematicians as they responded to mathematical and non-mathematical statements (Figure 5.7). Participants in the study 'had to decide whether the statements were true, false or meaningless' (Ansari, 2016, p. 4887). The mathematical statements were concerned with four main strands of mathematics: algebra, analysis, topology and geometry. The non-mathematical statements were on general knowledge of the natural world and history.

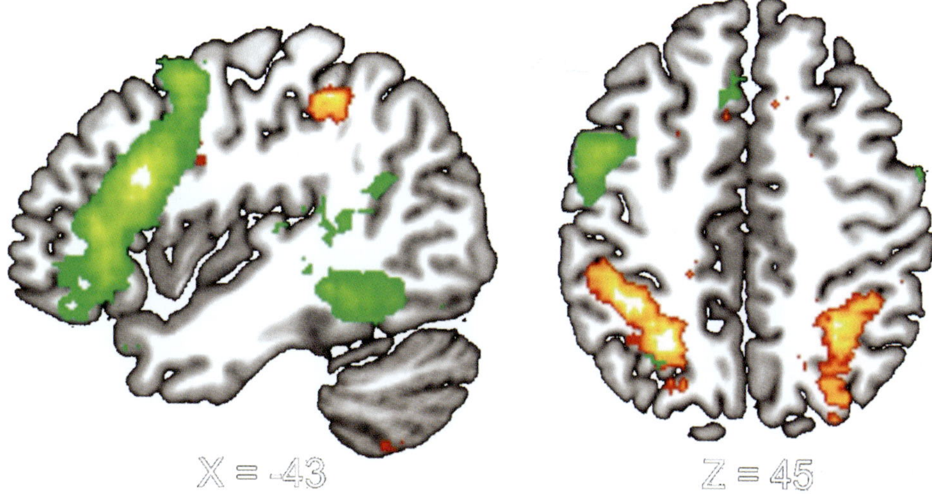

Figure 5.7 Brain areas associated with numerical processing are shown in red to yellow colours, and regions associated with language are displayed in green colours. Results support that neural circuits underlying numerical and complex mathematical processing are largely distinct from neural circuits engaged during language.
Source: Ansari (2016).

The study had two main aims. The first was to investigate whether high-level reasoning in mathematics is using and/or overlapping with the classical left-hemisphere regions of the brain associated with language processing. The results of the study provided clear evidence that 'high-level mathematical thinking makes minimal use of language areas and instead recruits circuits initially involved in space and number. This result may explain why knowledge of number and space, during early childhood, predicts mathematical achievement' (Amalric & Dehaene, 2016, p. 4909) later, as the child develops further. So numeracy and literacy are only marginally related in the brain *when the brain is engaged in mathematical thinking*.

Have you seen students who are good at mathematics but struggle with writing good, coherent essays? Why do you think students' performance is inconsistent across two different subjects? What advice should a teacher give to such students?

Are basic numeracy and advanced mathematics using the same cortical areas?

The second aim of the study was to investigate whether the professional mathematicians in the study are using the same areas of the brain when engaged with basic number recognition as they are when performing high-level mathematics. The second of these is particularly interesting, because there are many mathematicians who argue that 'number concepts are too simple to be representative of advanced mathematics' (Amalric & Dehaene, 2016, p. 4909). This research provided compelling evidence that 'advanced mathematics, basic mathematics, and even the mere viewing of numbers and formulas recruit similar and overlapping cortical sites' (Amalric & Dehaene, 2016, p. 4913). In other words, basic number sense and advanced mathematics are both using the same cortical areas of the brain, which are different from the areas used in literacy. The main areas involved are indicated in Figure 5.8.

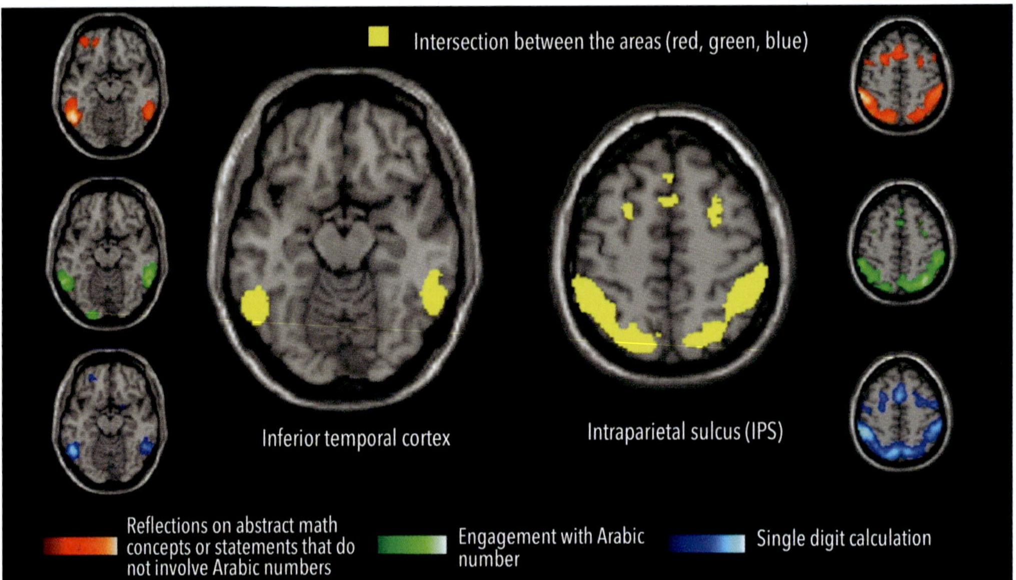

Figure 5.8 Overlap between areas involved in different levels of mathematical engagement
Note: The areas that are active when professional mathematicians reflect on highly abstract maths statements (in red) overlap with the areas we use when recognising numbers (in green), which we also use when running arithmetic calculations (in blue). The brain networks that initially respond to basic number sense and simple arithmetic are recycled for more abstract concepts.
Source: Adapted from Amalric (2017, p. 64).

Then what about mathematical genius? Are those who appear to have a very special talent for mathematics still using those same areas? And are they born with a special talent or does it perhaps come from constant rehearsal and repetition? Yes, they are using the same areas and much evidence is now pointing to the conclusion that, although there may be some genetic or other biological contributors to mathematical talent, they 'do not weigh much when compared to the power of learning' (Dehaene, 2009, p. 171). Those who are fond of mathematics devote a lot of time to it and also tend to befriend others who have a passion for mathematics. On the other hand, if a child develops anxiety about mathematics, they will try to avoid it and, without rehearsal and repetition, that child's learning of maths will be impoverished.

Brodmann area 39 and the story of Mr M

Imagine losing your number sense. Earlier in this chapter, we met Mr C, who lost the ability to read. Now meet Mr M, who lost his sense of numerical quantities. Although he can still read numbers, he can no longer understand the numbers he reads. His loss of number sense becomes very apparent when he is trying to do subtraction and he is doing what is called a bisection test, deciding which number comes between two others. When asked what is 3 minus 2, he says 2. When asked which number comes between 3 and 5, he says 3. Nevertheless, his rote memory for multiplication tables remains intact.

The lesion (brain damage) associated with Mr M's loss of number sense is in the lower parietal lobe of the brain, in Brodmann area 39, also called the 'angular gyrus'. Dehaene, who conducted the study of Mr M, notes that this cortical region is known to play 'a crucial role in the mental representation of numbers as quantities' (Dehaene, 1997, p. 189). It is also situated at the 'convergence of highly processed data steams stemming from vision, audition and touch – an ideal place for arithmetic because the number concept applies equally well to all sensory modalities' (Dehaene, 1997, p. 189). Although Dehaene admits that we still do not know much about what is happening in that area, and much more research is needed, he believes with many other scientists that this is where number sense is deposited in the brain. In short, it is mainly in the left hemisphere.

Applying the science of literacy and numeracy research to educational policy and practice

In this chapter, we have covered a lot of ground in coming to grips with the science of literacy and numeracy, but what does this have to do with educational practice – why should teachers care?

We end this chapter by considering three specific applications relevant to teachers, focusing on (1) what teachers should know about the dyslexia, (2) the central role of phonics in learning to read (addressing the so-called Reading Wars and, (3) why Piaget's stage-theory and constructivist account of numeracy in the early years, which has had a massive impact on early years educators, is now known to be wrong.

The problem of dyslexia

In Chapter 1, the problem of dyslexia was introduced to illustrate how, over the past 20 years, research is beginning to offer new scientific perspectives on this complex impairment. After reading the content of this chapter, we can now briefly answer three important questions: what brain areas are implicated in dyslexia? What might be the cause? And does this offer hope for a remedy?

grapheme: a letter or a group of letters representing a sound (phoneme) in a word

It is now generally accepted that while there are a number of symptoms, the core problem in dyslexia is **grapheme** to phoneme (letter-to-sound) conversion. Even more fundamentally, it appears not just to be a reading problem, rather it stems from a problem with phonemes that are the basic constituents of *spoken* words. It might seem a bit odd for a problem that manifests itself in reading to be due to a problem in speaking, but research has now established 'a solid link between early phonological abilities and the ease with which literacy will later be acquired' (Dehaene, 2009, p. 240). Having read this chapter, you will not be surprised to learn that a key area implicated in dyslexia is the letterbox area, or VWFA, in the left hemisphere. So the first key message for teachers is that dyslexia is a brain problem, not a problem with faulty social conditions or poor education, even if some children with dyslexia also suffer from those disadvantages.

What might have caused this brain problem? From a dynamic systems perspective, the idea of a single cause is highly problematic, as most developmental phenomena have multiple causes. One causal factor in dyslexia may relate to specific languages (for example, recognising Chinese characters requires much more rote memory compared with reading alphabetic languages, and reading English is considerably more difficult than reading Italian). However, across all languages and cultures, an important cause of dyslexia lies at the interface between vision and speech, located in the web of neural connections in the left temporal lobe.

Research suggests that genes play an important causal role, and that dyslexia runs in families, but another main culprit appears to be the process of cell migration, as the foetal brain is formed in the womb (introduced in Chapter 2). There is considerable evidence that the migration process in dyslectics went astray, as some neurons never reached their proper targets. A second key message for teachers is that dyslexia is not just a problem in learning how to read; causally, it is a problem of development – especially early brain development in the womb.

If this all sounds very bad, the good news is that there is considerable hope that comes by way of brain plasticity. As we saw in Chapter 1, within the brain there are millions of redundant pathways that can be used to compensate for pathways that have been damaged or not formed properly in development, if only we can find ways to use them. So the final key message in regard to dyslexia is that there is cause for cautious optimism, the brain has the capacity to remedy the dyslexia. Teachers should constantly bear in mind that each 'new learning episode modifies our gene expression patters and alters our neuronal circuits' (Dehaene, 2009, p. 257). In the act of teaching, that is what teachers are doing. This ability to reshape brains provides hope that, with new and improved intervention methods, we will be able to overcome dyslexia and other learning and developmental problems.

phonics: a language teaching method that explicitly demonstrates the relationship between phonemes and graphemes

whole-language approach: a literacy teaching approach in English, based on the belief that language should *not* be broken down into letters and combinations of letters and 'decoded'. Instead, learners should focus on whole meanings.

Is the systematic teaching of phonics important when learning to read?

The issue of how to teach reading – whether it should employ the systematic teaching of **phonics** or whether it should employ a *global* or **whole-language approach** – has been highly contentious, leading to what some have called the *Reading Wars*. In Australia in 2018, this hard-fought dichotomy produced what was dubbed the *Great Literacy Debate*, sponsored by the Australian College of Educators. On one side of the debate are those who advocate the importance of systematically learning phonics; on the other side are those who advocate the importance of meaning-making in reading, which they claim gets lost if children are drilled in phonics, which they sometimes disparagingly call 'barking at text'.

What you have read in this chapter on the science of reading would strongly support the importance of prioritising the learning of phonics, especially in the early stages of reading. We have stressed that the area of the brain involved in converting letters into sounds (the letterbox area) is very different from the area that processes meaning. We have noted that recognising letters in a word and turning them into speech sounds (converting graphemes to phonemes) radically recycles the young child's brain, and we have seen how this process forms the second *phonological* stage of learning to read, between stage 1, the *logographic or pictorial* stage, and stage 3, the *orthographic stage*, when the child starts to read fluently, no longer needing to systematically break words down into their component parts (Figure 5.9).

Figure 5.9 Kindergarten students are learning the alphabet and the sounds of letters using digital tablets

It is important to note that those who stress the important of the phonological stage 2, when young children learn to break whole words down into their constituent parts before reassembling them again into words, are not denying the importance of meaning-making or the eventual aim of word-fluency in stage 3. However, you can't simply skip from stage 1 to stage 3. Even though it is hard and disciplined work for young children, they have to go through stage 2. The pedagogical challenge for teachers is finding the best classroom strategies to assist children in transitioning through stage 2 to stage 3.

Piaget's error and early numeracy learning

As part of Piaget's (1965) comprehensive stage development theory, it was strongly affirmed that babies are born without any sense of number. Rather, he claimed, numerical concepts are *progressively constructed* in young children's minds, as a result of interaction with the world, around four or five years of age. This view was strongly supported by highly repeatable experimental evidence. The test he used to support this claim is referred to as **number conservation**. Before the age of around six years, young children fail the test.

number conservation: awareness that the number of objects remains the same, even if the objects are rearranged or hidden

A researcher shows a child two identical arrays of clay marbles (box 'a' in Figure 5.10) and asks the child whether the two arrays have 'the same' number of marbles.

Children at age four usually reply that the two arrays of four marbles are indeed 'the same'. The researcher then adds or subtracts a few marbles in one array and changes the way marbles are distributed, either closely located or widespread. The researcher then asks the same question: Do both arrays have the same number of marbles, or does one array have 'more'? If the arrays in box 'b' are presented, the same four-year-old typically answers there are now 'more' in the upper row. However, if the arrays in box 'c' are shown, the same child typically says there are more in the lower row. When the researcher repeats the same procedure with children of various ages, it is only by about age of six that a child correctly indicates that it is the array with the added material that has 'more'.

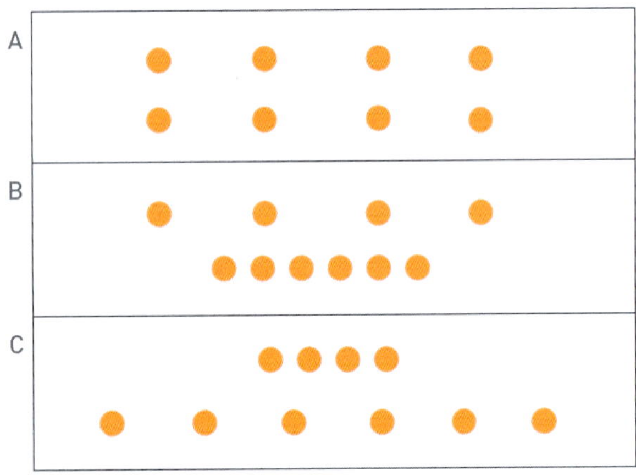

Figure 5.10 Number conservation experiment settings
Source: Adapted from Mehler & Bever (1967).

However, and significantly, it is now appreciated that although the experiment is highly repeatable (and was repeated in a number of variations by Piaget's supporters over may years, getting the same result each time), there are *problems with how it is conducted*, which start to invalidate the evidence it provides. For example, it is now noted that the test is highly reliant on language – conversation between experimenters and young children – so is it testing numeracy or literacy? Does the four-your-old child understand what they are being asked – the lines are longer and longer is more? In fact, when children are placed in contexts where their capabilities are probed with fewer words, as used in studies with animals, children's numerical understandings are observable much earlier in life than this experiment suggests.

For example, as far back as 1967, when the Piagetian paradigm was at its height, Jacques Mehler and Tom Bever showed that children's response to the conservation task dramatically changes if the task involves more attractive materials. Instead of using clay pellets, they used M&Ms – chocolate sweets. Also, when running the experiment with M&Ms, instead of asking the children to state which row had more in situation 'b' above, the experimenter asked them to 'take the row you want to eat, you can eat all the M&Ms in that row' (Mehler & Bever, 1967, p. 141). By simply changing the materials and instructions, they found that children aged three years and above were able to select the row with more M&Ms. Moreover, as noted above, Tobias Dantzig (1967) was already arguing, contrary to Piaget, that infants could recognise that something had changed in a small collection of objects when, without the infant's knowledge, an object was removed from or added to the collection.

REFLECTION

Why do you think children's performances differ between two experimental contexts? How do you think early childhood and primary teachers might use this insight that young children already possess an understanding of conservation?

There are two very important messages in this for prospective teachers. The first is that research over the past 20 years or so has confirmed that babies *do possess a limited and yet discernible number sense*. Piaget was wrong. Moreover, he was *known to be wrong* when his ideas were having such a strong influence on educational policy and practice, not only in Australia but also in the United Kingdom – for example, when in teacher education it was taught as fact and it underpinned a major UK government report, called the Plowden Report (1967), calling for major reforms to primary school education to come into line with Piaget's theories.

That raises a key issue for the study of education: how could the evidence produced by Piaget and his very many supporters have been so compelling, but the conclusion wrong? The idea that educational policy should be evidence-based is frequently repeated by politicians and policy-makers. However, in science, evidence is only as good as the quality of the research on which it is based. Teachers need to be constantly on guard when encountering evidence-based claims. Remember Piaget and number conservation – there was *plenty of evidence but in the end, it was wrong*, because the experiment was faulty. We will see the same problem arising regarding the reliability of evidence in Chapter 6 when we consider emotion and in Chapter 8, when considering moral development.

APST and ACECQA curriculum specifications   STANDARDS

APST Standard 1.2 Understand how students learn
Demonstrate knowledge and understanding of research into how students learn and the implications for teaching

APST Standard 1.5 Differentiate teaching to meet the specific learning needs of students across the full range of abilities
Demonstrate knowledge and understanding of strategies for differentiating teaching to meet the specific learning needs of students across the full range of abilities

APST Standard 2.5 Literacy and numeracy strategies
Know and understand literacy and numeracy teaching strategies and their application in teaching areas

▶ ▶   **ACECQA curriculum specifications**
1.1 Learning, development and care
1.2 Language development
1.7 Learners with special/additional needs
5.3 Numeracy, science and technology
5.4 Language and literacy

This chapter provides a frontline overview of recent research into the development and learning of language, literacy and numeracy that is highly relevant in advancing newly qualifying teachers' knowledge and understanding of these essential educational foundations. Given the focus on the science of literacy and numeracy, this chapter goes beyond what is generally available to newly qualifying teachers in their initial teacher education. In doing so, it adds considerably to their understanding of research into how students learn to read and the implications of this for the teaching of reading. It also provides them with an understanding, drawn from neuroscientific research, that literacy and numeracy in the brain are largely independent. Moreover, research suggests that advanced mathematics and basic numeracy are employing much the same areas and processes of the brain, so the learning of basic numeracy through to advanced mathematics can be viewed as a continuum.

The chapter provides sufficient grounding for newly qualifying teachers to address what has been called the 'great literacy debate' with informed perspectives about what is required in learning to read, drawing on the neurobiological evidence. This chapter also addresses the causes of dyslexia, a language impairment about which, arguably, all teachers should be sufficiently informed. The chapter builds on what readers have previously learnt about cell migration and brain plasticity to identify what is believed to be a major cause of dyslexia. It also encourages teachers to understand that there is hope for a remedy, using carefully considered teaching and learning strategies that can employ redundant pathways in the brain and thus bypass the causes of the impairment.

SUMMARY

- The history of human language dates at least as far back as 400,000 years. Biologically, three factors were essential for the evolvement of human language: brain capacity, neural structure and vocal apparatus.
- The neural mechanisms used in language were not 'built into' the brain from the start of human evolution. These neural mechanisms must therefore have emerged later, by coopting mechanisms that existed for other purposes (a process known as 'exaptation' or 'neuronal recycling').
- Early language development consists mostly of learning object names. The perception–action loop enables an infant to pay visual attention to an object and bind a heard name to the object seen by them.
- During early language development, each infant follows a distinctive developmental trajectory that is jointly shaped by the specific tasks in which they engage and the 'intrinsic dynamics' of their own neurobiological systems (Smith, 2013).

- Acquisition of language requires a delicate balance between neural specialisation and plasticity. For most people (who are right hand dominant), language is located in the left hemisphere of the brain.
- Although language develops as the young child hears their native language, literacy (the ability to read and write) has to be taught. Learning to read and write involves what is called the Visual Word Form Area (VWFA), or the letterbox area located in the left occipito-temporal area of the brain (Brodmann area 37). When children are taught to learn to read, the VWFA or letterbox area begins to respond selectively to letters, characters and written words by recycling cortical territory in the left hemisphere that originally evolved to visually recognise and distinguish different shapes. When learning to read, at the same time as invading this cortical territory the VWFA begins to form connections back and forth in feedback loops with areas specialised for speech sounds and meaning.
- Based on the work of Uta Frith, it is widely recognised that there are three main stages in learning to read an alphabet language. The first is a *logographic or pictorial stage* when the young child is recognising whole words as though they are pictures. The second stage is *phonological*, when the child begins to break words that they see into their component letters and match them with spoken sounds. The third stage is *orthographic*, when the child is reading fluently, no longer needing to systematically break words down into their component parts.
- Research has produced clear evidence that basic numeracy and 'high-level mathematical thinking makes only a minimal use of language areas of the brain and instead recruits circuits initially involved with space and number' (Amalric & Dehaene, 2016, p. 4909).
- It has also been shown that newborn infants already possess a basic number sense – a primitive ability to appreciate quantities. They can discriminate between whether sets of visual and audial objects share the same number or not. This is contrary to Piaget's constructivist belief that numerical concepts are progressively constructed in young children's minds as a result of interaction with the world, around the ages of four or five years.
- 'Advanced mathematics, basic arithmetic, and even the mere viewing of numbers and formulas recruit similar and overlapping cortical sites' (Amalric & Dehaene, 2016, p. 4913). Moreover, recent research has questioned whether there is such a thing as innate mathematical genius. It seems more likely that mathematical ability is closely related to an interplay between a passion for the subject and constant practice (Hebbian repetition).
- A knowledge of number and space acquired in the early childhood years of learning and development is a reliable predictor of mathematical achievement later in life.
- Scientific research into the learning and development of literacy and numeracy is able to inform educational policy debate and classroom practice. Research is also contributing to educators' understanding of impairments such as dyslexia.

KEY POINTS FOR TEACHERS

- A dynamic systems account of language development emphasises that each individual child will have their *own pathway* to language competence. Teachers should be aware that each child is doing it their own way, but still ending up with much the same result as other children.

- The key message for teachers to bear in mind is that the acquisition of language requires a delicate balance between neural specialisation and plasticity. For most people, language is mainly located in the left hemisphere of the brain, though it is also highly distributed using many parts of the brain.
- Research shows that, whereas language acquisition is innate, we come into the world with the ability to acquire language; this is not the same for reading and writing. We do not acquire reading and writing naturally – they have to be formally learnt. Schooling matters.
- The neuroscience of reading strongly supports the importance of prioritising the learning of phonics, especially in the early stages of reading. Young children need to break whole words down into their constituent parts before reassembling them again into words.
- Basically, the brain areas involved in converting letters into sounds involve the upper regions of the left temporal lobe and encompass Broca's area, concerned with the articulation of words. However, the network involved in converting letters into sounds is very different from the network that analyses word meanings – a point that is very important for teachers, as it has a strong bearing on how to teach reading.
- Teachers should be aware of the three stages of learning to read: the logographic or pictorial stage; the phonological stage; and the orthographic stage. Teachers can't simply skip from stage 1 to stage 3, as some educators have argued ('whole-language' or 'balanced approach'). Although it is hard and disciplined work, young children have to go through stage 2.
- Teachers should be aware that research over the past 20 years or so has confirmed that babies do possess a limited and yet discernible number sense. This flies in the face of what teachers were taught in the second half of the twentieth century, largely based on the constructivist theories of Jean Piaget (1965), that newborns possess no sense of number. Piaget was wrong.
- Some scholars in the twentieth century, including the American philosopher and linguist Noam Chomsky, believed that the origin of the mathematical capacity was an abstraction from linguistic operations. However, recent research suggests numeracy and literacy are only marginally related in the brain when the brain is engaged in mathematical thinking and not using words.
- Clues about the origins of numeracy come from traditional peoples that do not have a written language and do not use modern tools and technology. Many such peoples use tally sticks, but on a day-to-day basis often also use finger counting (as do many modern children and even adults). There is nothing wrong with finger counting.
- Research is contributing to educators' understanding of impairments such as dyslexia. In the case of dyslexia, it is not just a problem in learning how to read, it is primarily a brain impairment. Genes play an important causal role (dyslexia runs in families), but another main culprit appears to be the process of *cell migration*, as the foetal brain was formed in the womb.
- A key area implicated in dyslexia is the 'letterbox area', or VWFA, in the left hemisphere. In understanding that dyslexia is mainly a brain problem, teachers should also appreciate

that it is not a problem related to faulty social conditions or poor education, as sometimes claimed, even if some children with dyslexia also suffer from those disadvantages.
- The idea that educational policy should be evidence based is frequently repeated by politicians and policy-makers. However, in science, evidence is only as good as the quality of the research on which it is based. Teachers need to be constantly on guard when encountering 'evidence-based' claims.

REVIEW QUESTIONS

1. Are babies born with the ability to acquire language in the same way they can start to recognise faces from birth, or does language ability need to be taught?
2. Do all infants follow the same developmental trajectory when learning nouns, which forms a major part of language learning in the early years?
3. Which region of the brain is specialised for the recognition of letters and written words? What is happening in this region of the brain when a child starts to learn how to read?
4. What relationship is there between language and numeracy in the brain and is a high school student's ability to solve highly abstract mathematical problems very different from their basic number sense, learnt in early childhood?
5. How is the science of learning and development beginning to make an important contribution to understanding the language impairment known as dyslexia, which many teachers will encounter in their school?

Guided responses

FOOD FOR THOUGHT

1. Can you remember anything about your own process of learning to read words? Do you remember whether you learnt to read using phonics, breaking words down into their component parts and matching them with spoken sounds? See whether you can talk to (and perhaps record) adults or older siblings who knew you around the age five or six, when you started to read. What recollections do *they* have of you learning to read?
2. One of the messages you have been given in this textbook is that there is still a lot we don't know about brain and that within science there are disagreements about how to interpret data, for example. Do you think that weakens or strengthens your trust in science and what it has to offer education?
3. Does it surprise you to learn that babies are born with a basic sense of number and quantity, which is contrary to the view of Piaget that this has to wait until children are around the age of four or five years, when he claimed numerical concepts are progressively constructed in the child's mind as a result of interacting with the world? Do you think it is possible that both theories might be right – it is just a matter of different epistemologies (theories of knowledge) and teachers should be free to choose which they prefer? If so, how might these conflicting claims be reconciled? Or was Piaget simply wrong?

RESEARCH ACTIVITY: ENGAGING WITH THE GREAT LITERACY DEBATE IN AUSTRALIA

With what you have learnt in this chapter as background, including what has just been said, you can now dip into the Great Literacy Debate yourself.

It is suggested that you access the report the 2018 debate in the special edition of *Professional Educator* by the Australian College of Educators that contains the contributions of those who presented papers at a forum held in July 2018 (Australian College of Educators, 2018).

Pay particular attention to three papers: (1) Anne Castles affirming the importance of phonics (pp. 6–7); (2) Kathy Rushton, Robyn Ewing and Mark Diamond arguing for the priority of meaning-making, favouring whole language (pp. 14–17); and (3) Pamela Snow calling for a reading renaissance in education (pp. 37–40).

As an additional resource, check the websites below that aim to help teachers update their understandings on teaching phonics in their classroom.

Research activity links

- New South Wales Department of Education: 'Phonics'
- Victorian Department of Education and Training: 'Literacy teaching toolkit: Phonological awareness'
- UK government: *Learning to read through phonics*.

Question

In light of what you have learnt in this chapter about learning to read, compare and contrast what the authors of those papers are saying. What do you think are the key messages of this debate for teachers?

PART 2

THE EMOTIONAL, CULTURAL, MORAL AND EMPATHETIC BRAIN

In Chapters 1–5, a main focus was on the biologically embodied brain; in Chapters 6–8, the overriding framework is the *socially and culturally embedded* brain. A main connecting theme across these three chapters is the *neurobiology of emotion*. Concept map 2 shows the science and dynamics of emotion providing a platform for academic, moral and empathetic learning and development. In this model, there is no dichotomy between academic and social/cultural learning, as one often finds in education theory. From a neurobiological and dynamic systems perspective, *all learning and development is an emergent property of all the multitudinous processes occurring in the brain, working together in interweaving synchrony.*

The scientific study of emotion is recent in the West. From the onset of modern science in the seventeenth century, and throughout the Enlightenment and for most of the twentieth century, there was a strong prejudice against emotion. It was thought that to be rational, one must avoid emotion at all costs. It is now realised that human rationality is necessarily laced with emotion.

The story of that recent transformation is told in Chapter 6. It is not an absence of emotion in human rationality that is required, but rather a balance between rationality and emotion (emotion regulation). This has important implications for teachers' classroom

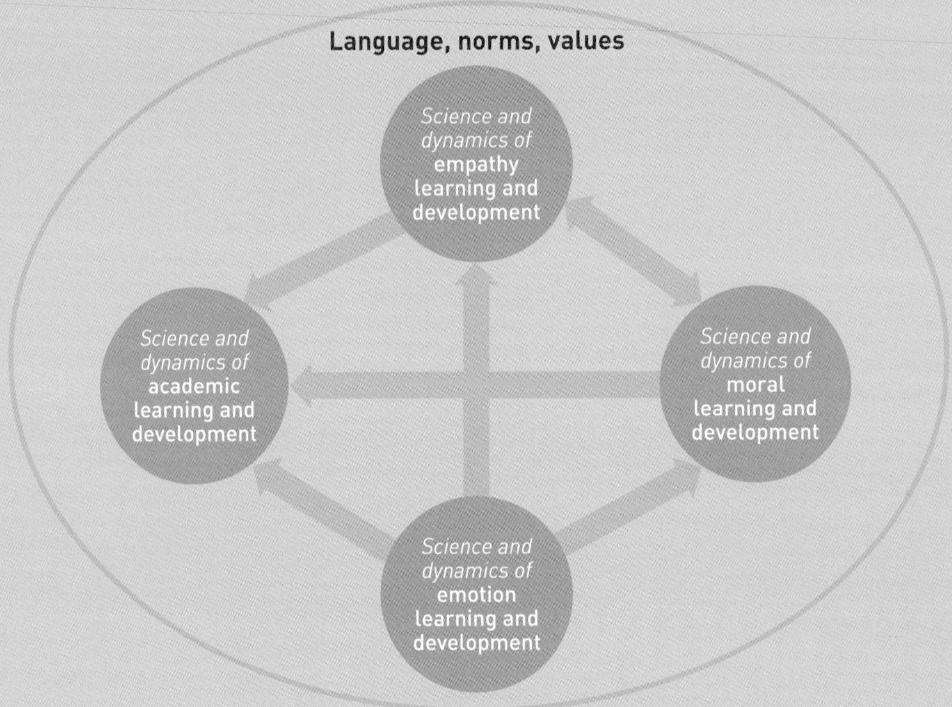

Concept map 2 Social and cultural context of learning and development

management. The chapter also examines the *classic view of emotion* and is critical of its claim that emotions are reliably and universally presented in *facial expressions*. This critique has implications for those teachers who believe they can reliably recognise what children are thinking and feeling by reading their facial expressions. It also provides an interesting case study of the contested nature of brain science, introduced in Chapter 5 – for example, with regard to emotion, whether it is modularised and *localised or* widely *distributed* in the brain.

Chapter 7 discusses how the dynamic interaction of culture, biology and context is instantiated in the brain. Building on the notion of the socially and culturally *embedded* brain, Chapter 7 introduces the idea that *culture is embrained* and the related idea that schools, classrooms and other learning environments are shaped by *neurodiversity.* In the process, it casts a critical eye on the standard concept in social science that learning results from socialisation. The 'classic' view of socialisation often incorporates the idea that children become *social and cultural creatures,* having somehow transcended their animal past in the process of being socialised. This chapter argues that this creates an unnecessary dichotomy between the biological and sociocultural. The chapter also introduces the notion that mirror neurons enable us to read of the intentions of others and examines whether mirror neurons are implicated in autism. The chapter closes with a detailed discussion of the social, cultural and political context of intelligence testing and its often unfortunate impact on education. It concludes that intelligence tests are not measuring genetically endowed intelligence, but rather providing a measure of social evolution and adaptation to modernity, known as the Flynn effect.

Chapter 8 focuses on the moral and empathetic brain, beginning with the claim that education is itself a moral endeavour, although it also notes that this insight is often missing in courses of teacher education. The chapter provides a detailed account of Laurance Kohlberg's paradigmatic formulation of moral development that dominated in education and in other disciplines for much of the second half of the twentieth century and into the twenty-first century. The chapter then introduces a dynamic systems account of moral development. In doing so, it considers the notion of a predilection to value within nature, which appears to be supported by the discovery that three-month-old babies are able to preference pro-social behaviour. The chapter also discusses the three gradations of empathy and their neurobiology. Finally, it discusses the neurobiology of moral thought and action before turning to the chemistry of care and trust, and the centrality of these in school communities.

THE EMOTIONAL BRAIN

Learning and development

LEARNING OUTCOMES

By the end of this chapter, you will:

- Appreciate the importance of emotion and emotion regulation in children and adolescents' learning and development, and understand that rationality and emotion, thinking and feeling are closely related in the brain
- Understand what is meant by the standard or 'classic' view of emotion and its claim that emotions are universal across all cultures
- Know that from a dynamic systems perspective, emotions are not things 'wired' into the brain at birth; they are processes that emerge through learning and development
- Know how this applies to emotion regulation, both generally and in educational settings
- Appreciate that the study of emotion is currently a very contested research field, including disagreements between those who favour the idea of dedicated circuits in the brain for emotion and affect and those who favour a distributed account

The emotional brain at school: Interweaving thinking and feeling

All teachers, from pre-school to high school, know that classrooms are filled with emotions. Young children starting pre-school or kindergarten may experience emotional trauma and cry when separating from parents; students in primary and secondary school may be anxious and freeze emotionally when required to give a presentation in front of classmates, while others may be so depressed or lethargic that they are not able to meet assignment deadlines.

REFLECTION

Do these kinds of examples ring true for you? Have you seen similar examples of emotions in classroom settings? What kinds of experiences at school had a strong emotional impact on you? Were they mainly social or were they related to learning? Did you feel you were able to manage your emotions at school?

The importance of emotion regulation in educational contexts

An important aspect of every child's learning and development at school involves the ability to respond to their experience of the world, whether physical or social, in a controlled and socially appropriate manner. Of course, in any given society or culture, at any given time, what is considered to be 'controlled' and 'appropriate' will be interpreted in terms of that society's conventional norms, which may be firmly or quite loosely defined. One can even see conventional differences across and within schools. Schools vary in the rules and values they prescribe and within any given school there may be considerable differences in how individual teachers interpret and apply those rules. When teachers start a new job at a new school, there is much to learn about the school and its students over the initial weeks and months.

One important aspect of workplace acculturation involves coming to grips with what the school expects in regard to conventions of acceptable behaviour – not only student behaviour but also teacher behaviour. This inevitably includes what is commonly called **emotion regulation**. In thinking about this, an important starting point is the recognition that, for all students and teachers', schools and classrooms are highly emotional settings. Moreover, emotional regulation is not just a matter of behaving appropriately in those emotionally challenging settings, it is also a matter of student and teacher wellbeing. As a teacher, in attending to your emotion regulation and helping students deal with their emotions, you are making a major contribution to the health of your brain and the brains of your students.

emotion regulation: the 'suite of executive functions directed at the modulation of emotional states and emotional impulses' (Woltering & Lewis, 2009, p. 160)

The importance of emotion regulation (Figure 6.1) is strongly emphasised in literature related to early years education – for example, the Australian Early Years Learning Framework (DEEWR, 2009). Emotion regulation is viewed as a significant developmental milestone in the early years, and there are constant warnings from experts that young children who fail to develop emotion regulation skills are likely to have difficulties forming meaningful relationships with others in later years (Eisenberg et al., 1993; Eisenberg et al., 2004). The teacher's role in the early years is therefore paramount (Denham et al., 2015), in assisting

Figure 6.1 A student's gestural expressions of his feelings

the development of emotion regulation of each child in their care. There are many suggestions on offer about how best to achieve this in early childhood settings, including the provision of what is called positive guidance – helping a pre-school child to name their own emotion by associating a bodily feeling with the related word, and talking about it with the child.

On the Western Australian Department of Education website, for example, a strong emphasis is placed on how the early years of emotional development contribute to the achievement of emotional maturity in later years. Indeed, 'Key skills underlying emotion are predictive of later success, wellbeing and mental health'. Moreover, 'self-regulation and resilience, the cornerstones of emotional maturity, are at the heart of children's learning ability' (Western Australian Department of Education, 2018).

The early years are clearly a very important time of emotional development, but so are the years of adolescence, following the onset of puberty. After what might appear to be the tranquil years of primary schooling, changes start to occur in the brain with the onset of puberty that can massively disrupt emotion regulation. The Victoria State Government of Education and Training advises all teachers to 'Promote mental health: social and emotional learning (SEL)', as this 'can help students learn the competencies and skills they need to build resilience and effectively manage their emotions, behaviour and relationships with others' (Victoria State Government Education & Training, 2020).

Mathematics anxiety in school

There are many different reasons why children and adolescents may experience negative emotional states at school, some of a social nature and others related to learning and performance. In particular, research indicates that a considerable number of high-school students experience **mathematics anxiety** – feelings of stress and worry when dealing with mathematics in school.

mathematics anxiety: feelings of tension, nervousness or worrisome thoughts about one's ability that impede engagement with numbers and mathematical problems

According to the Programme for International Student Assessment (PISA) study conducted by the OECD in 2012 and published in 2013, approximately 60 per cent of 15-year-old students from OECD member countries reported 'they often worry that it will be difficult for them in mathematics classes' and 'they worry about getting poor grades in mathematics' (Figure 6.2). Even worse, some of these students experience rather acute mathematics anxiety – for example, approximately 30 per cent reported that they feel 'very tense', 'very nervous' and 'helpless' when engaging with mathematics problems. Also, in all but nine participating countries, girls reported stronger feelings of mathematics anxiety than boys. Research link 6.1 provides you with some of the data on mathematics anxiety.

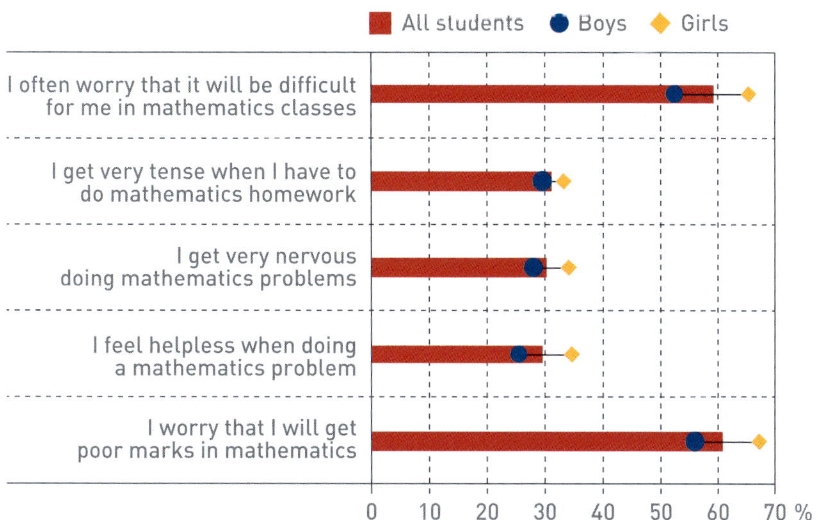

Figure 6.2 Percentage of students across OECD countries who reported that they 'agree' or 'strongly agree' with questions regarding mathematics anxiety
Source: OCED (2015).

Mathematics anxiety in schools

Organisation for Economic Co-operation and Development (OECD) (2013). *PISA 2012 results: Ready to learn. Students' engagement, drive and self-beliefs (Volume III)*. Paris: OECD.

According to cross-national comparison data, there is a significant association between 15-year-old students' average level of mathematics anxiety and their performance. Figure 6.3 'shows how countries where students report above average levels of mathematics anxiety are also countries where students tend to perform less well in mathematics' (OECD, 2013).

However, it should also be noted that in countries that show the highest mathematics performance internationally (China, Singapore, Hong Kong, Taiwan, Korea), students' mathematics anxiety is above the level of the OECD average. This means maths anxiety itself does not directly lead to poor maths performances. Other mediating factors may also exist. A competitive and high-stakes academic culture in Asian countries may create more anxiety about meeting the maths performance expectations, which are closely related to students' career pathways and jobs. Some research has suggested that it is not worrying and anxious feelings that impede students' maths performance, but rather the compromised 'attention' and 'working memory' resulting from such feelings (Maloney & Beilock, 2012). If correct, it suggests that those who are equipped with well-balanced emotional regulatory skills may be able to overcome detrimental impacts from their anxiety.

The PISA 2012 data do not include data from young students, but traditionally it has been believed that mathematics anxiety develops in junior high school or at the end of primary school, resulting from the increased difficulty and complexity of the mathematics curriculum

▶▶

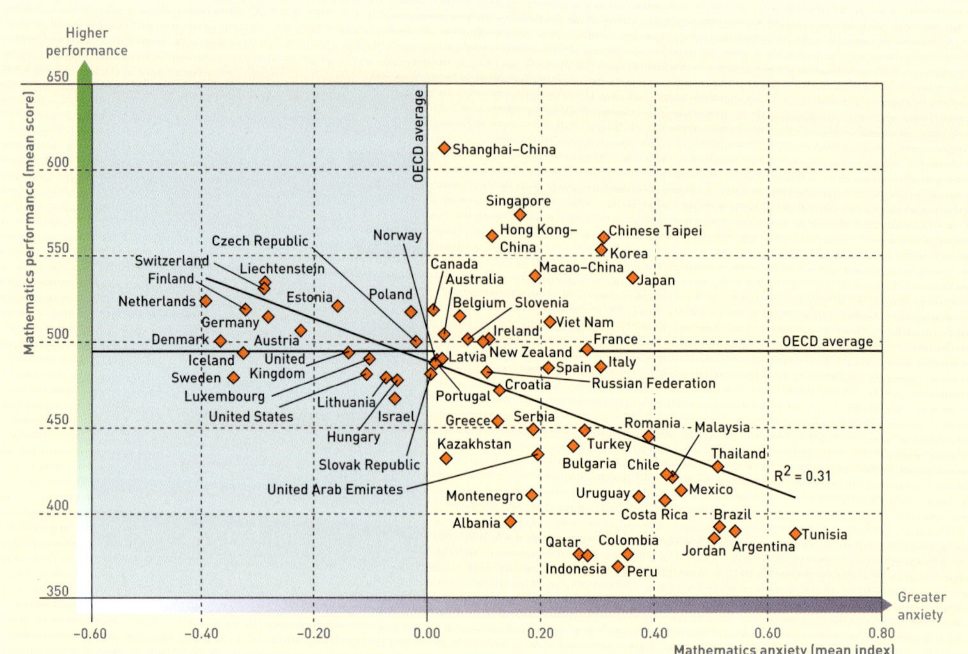

Figure 6.3 Association between mathematics performance and mathematics anxiety across countries
Source: OCED (2015).

(Maloney & Beilock, 2012). However, more recent studies show that maths anxiety emerges as early as the first years of primary school (Ramirez et al., 2013).

Some laboratory experiments show that the following strategies help students experiencing anxiety to maintain their learning and performance with maths tasks:

- Help the student interpret their physiological arousal in a positive manner – for example, by telling them that arousal may help them improve their performance (Jamieson et al., 2010).
- Give students opportunities to express or reappraise their feelings (e.g. writing) and encourage them to reframe the physiological reaction (Park et al., 2014).
- Provide resources that can help maths-anxious parents positively participate in math-related interactions with their children, without transmitting their own negative and fearful feelings (Berkowitz et al., 2015).

The Western myth of emotionally detached human rationality

Although, as just noted, there is ample recognition in policy documents that schools need to assist children and adolescents in their emotional development and emotion regulation, it is often not clear how they understand emotion, even though some attempt a preliminary definition. For example, we are told that, 'Emotional reactions are the brain's way of keeping us safe from danger and socially connected' and that they 'enable the body to return to a less aroused state' (Western Australian Department of Education, 2018). This is helpful in saying what emotions do, but *what exactly are they*? What is happening in the brain when we experience emotions, and are emotions always viewed in such a positive way as 'keeping us safe from danger and socially connected', as this document claims? Moreover, aren't emotions a problem – isn't being emotional the opposite of being rational, thinking emotionally and not objectively? Don't adults often tell youngsters to stop acting emotionally?

Over the past quarter of a century, brain science has consistently shown that 'human beings are fundamentally emotional creatures' (Immordino-Yang, 2016, p. 28), but that flies in the face of a pervasive myth that human beings are essentially rational, not emotional, creatures. To be emotional, we have been consistently told in the West, is to be irrational. Rational thinking and emotion are opposites, and emotion is so toxic it should be avoided at all costs. So where did this idea that human rationality and emotion are incompatible come from? How did this prejudice against emotion arise?

Like many things 'Western', its origins can be traced back to ancient Greece, but a more recent and major source was the onset of 'modern science' in the seventeenth century. Particularly influential was the rationalism of René Descartes (introduced in Chapter 1), the French philosopher, mathematician and scientist who had a profound influence on much European thinking until quite recently. From his perspective, there is no place for emotion in mathematics or natural philosophy, it simply muddies the human mind. Hence, for more than 300 years in the West, the notion of the impartial, detached logical, objective scientist, untainted by emotion, prevailed.

Lost in that image of the scientist was any acknowledgement that scientists are almost always passionate about their work and thoroughly, emotionally committed to the quest for truth. In short, science is a thoroughly human activity. In the mid-twentieth century, scientist and philosopher Michael Polanyi (1958) aimed to bridge the Enlightenment gap between science and humanity by drawing attention to the personal aspect of science and to what he called the *intellectual passions* of scientists. His key message was that science is not dehumanised; rather, it is thoroughly human and often very personal. By the end of the twentieth century, brain science was starting to reveal that it was not only the Enlightenment image of science that was wrong; human rationality had also been misunderstood.

Towards a more holistic view of rationality and emotion

In fairness, it wasn't just the Enlightenment that promulgated the idea that the human mind is 'a battlefield where cognition and emotion struggle for the control of behaviour' (Barrett, 2017, p. 81), nor was it just a Western view. Lisa Feldman Barrett argues that it is part of deeply held and widespread *classic view* that has 'been around for millennia in various forms' (2017, p. xi). It is premised on the belief that human 'rationality makes us special in the animal kingdom' (2017, p. 81) and, 'Without rationality, you are merely an emotional beast' (2017, p. xi).

Barrett (2017, p. xi) mentions Plato, Aristotle, the Buddha, Descartes, Darwin and Freud as past believers in what she calls the classic view. She names Paul Ekman (who you will meet later) and the Dalai Lama as current influential thinkers who 'offer up descriptions of emotions rooted in the classical view' (2017, p. xi). Moreover, she says that the idea of the detached rational person still persists in much psychology and in economics. Indeed, she says, the idea of the 'rational economic person' making cool-headed decisions in the marketplace, has been so influential that, 'Every economic crisis in the last thirty years has been related, at least in some part, to the rational economic person model' (2017, p. 80).

However, as mentioned in Chapter 1, in 1994 brain scientist Antonio Damasio published a book called *Descartes' Error,* which challenged the Enlightenment view that reason and emotion are incompatible and, moreover, the then prevailing scientific view that they are produced by two entirely different neural systems in the brain. As a result of studying many patients with brain lesions to the prefrontal lobe, he argued that 'feelings are a powerful influence on reason, that the brain systems required by the former are enmeshed in those needed by the latter, and that such specific systems are interwoven with those that regulate the body' (Damasio, 1994, p. 245). This was a very radical hypothesis, interweaving rationality and emotion, thinking and feeling. As you can imagine, it stirred a lot of emotions in very many people, who were opposed to what they saw as his attack on human rationality.

The dreadful story of Phineas Gage

In fact, the first scientific clue that there was something terribly wrong with the Enlightenment belief that rationality and emotion should be kept apart, and the scientific view that they are separated in the brain, came in the nineteenth century. It resulted from a terrible accident involving a very unfortunate worker that caused a major brain lesion, and it should have alerted science that there was a need to reassess the relationship between rationality and emotion; however, the clue was largely missed, and the most crucial evidence was lost.

Imagine this if you can. In 1848, workmen are building a railway track and, in order to break through rock that stands in the way, they are using explosives. Holes are drilled deep into the rock and explosives packed into the hole. Phineas Gage is leading a team of workers and is tamping down the explosives into the hole. His metal rod makes a spark when hitting the rock, causing an explosion, and the rod shoots out of the rock like a missile. It goes up through his 'left cheek, pierces the base of the skull, traverses the front of his brain and exits at high speed through the top of his head' and 'lands more than a hundred feet away, covered in blood and brains' (Damasio, 1994, p. 4) (Figure 6.4).

Enough to kill poor Phineas, you may think, but in fact he is still fully conscious and able to walk and talk. He is a popular leader and his stunned workmates carry him to a nearby road, sit him upright in an ox cart and take him to the local hotel. Phineas is able to get out of the cart with just a little assistance from his workmates. The owner of the hotel gives him a chair to sit on in the hotel entrance and calls for the local doctor. Dr John Harlow had been a schoolteacher before becoming a doctor and was still in the early years of his new profession.

Without the aid of antibiotics (they would not become available for another century), Harlow skilfully cleans the wound with disinfectants. Within a couple of months, Phineas

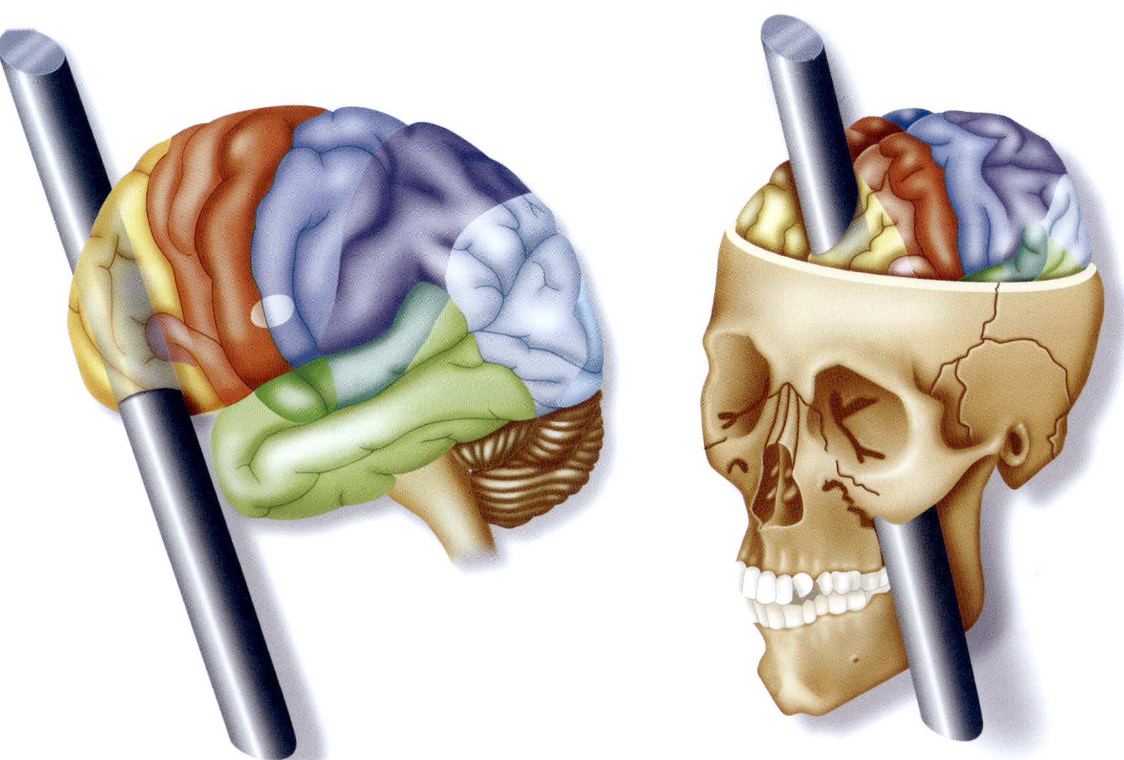

Figure 6.4 A recent digital image of where the metal rod went through the skull of Phineas Gage

recovers fully, or so it first appears. At least he is able to go about life normally, except there's a problem. From being the caring, thoughtful, temperate and orderly man that he was before the accident, his character has changed to be the direct opposite – uncaring, intemperate, disorderly and, as his doctor notes, 'fitful, irreverent, indulging at times in the grossest of profanity which was not previously his custom' (cited in Damasio, 1994, p. 8).

His friends would say 'Gage is no longer Gage': he simply wasn't the person he was before. He returned to work on his former job, but his change of character led to him being dismissed. He tried many other jobs, but without success: people avoided him. What had gone wrong? Well, he had a big hole in the head and part of his brain was missing, but he could still do most things he could do before the accident. Above all, he could still speak, despite the loss of the front part of the brain. All his language abilities remained largely intact, even if his use of language could be very abusive.

Phineas died in May 1861, some 13 years after the accident. Very strangely, the medical world paid almost no attention. He died in San Francisco, far from where the accident occurred. Although he had suffered a major brain lesion, there was no autopsy and his brain wasn't removed for scientific examination. He was buried and all the evidence was lost. Dr Harlow wasn't informed that Phineas had died and only found out five years later. He was shocked and he asked the family whether the skull could be exhumed for scientific examination (Figure 6.5). So why was this major medical event overlooked?

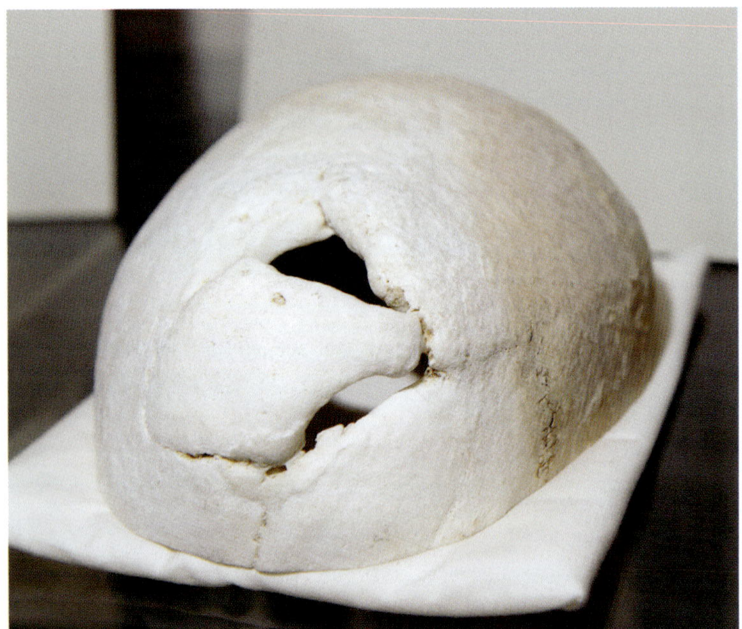

Figure 6.5 The skull of Phineas Gage, which Dr Harlow had exhumed from his grave, six years after he had died

In retelling this story at the start of his book *Descartes' Error*, Damasio (1994) recalls that it occurred at much the same time as the major focus was on understanding the brain and language, particularly the discovery mentioned in the previous chapter of Broca's area in 1861 (the same year that Phineas died) and Wernicke's area in 1874 some years later. However, Gage's injury was not relevant to that research focus; the only problem Phineas had with language after the accident was with *what* he said; his language abilities were intact. Another reason was that the divide between those who favour specialisation and localisation in the brain, versus those who favour the idea of distributed function and global action was just getting underway. It is still with us today, as we saw in the Chapter 5 and will see again in this chapter.

The work of Paul Broca and Carl Wernicke appeared to provide evidence for localisation and specialisation, and the same could have been said about Phineas' lesion, although in fact Damasio recalls that it was somewhat perversely used by the opposing camp to argue that if the brain is specialised, his lesion should have impacted his language and body movements (motor control). But maybe the real reason his injury was not fully investigated was that it was pointing to an uncomfortable truth: that there is a neurobiological link between the brain, who we are as persons, and how we act and interrelate in social contexts. Damage to certain areas of brain will very likely change your identity, how you behave and how you are perceived – in fact, who you are.

'The child who constantly causes behavioural problems in class is actually a brain that is functioning in a certain way that is not acceptable to his teachers and the school.' Is this a good definition of student misbehaviour? Should teachers adopt this when dealing with misbehaving students? What do you think? Talk it over with friends.

Damage to the prefrontal lobes: Emotion and rationality

Thankfully, although his damaged brain was lost to science, Dr Harlow at least ensured that Phineas's skull was preserved. It is now kept in a museum at Harvard University in the United States. In the 1970s, Hanna Damasio (Antonio's wife) launched an investigation into the actual trajectory the tamping iron took as it went through his skull and which areas of the brain would therefore have been damaged. Using modern imaging techniques, it was found that 'in spite of the amount of brain lost, the iron did not touch the brain regions necessary for motor function or language' (Damasio, 1994, p. 32). What it did remove was part of his prefrontal cortices – more precisely, what is called the **ventromedial prefrontal cortex (VMPFC)** region, which is the underbelly of the frontal lobe (ventral comes from the Latin *venter*, meaning belly). It was this damage that correlated with his drastic change of character, including his inability to abide by normal social rules and courtesies, and his failure, after the accident, to order his life productively.

However, you do not have to experience such a traumatic accident to have prefrontal lobe damage. As a neurosurgeon, Antonio Damasio began to encounter patients whose behaviour resembled that of Gage and who also had prefrontal lobe damage. One was a patient called Elliott, who had a very large brain tumour in the front of his brain that had to be removed by a surgical operation. The operation was a success, but it left Elliott with a devastating condition. Although in many ways he was much as he had been before the operation, something was missing: he had lost the ability to order his life rationally – his decision-making was totally erratic. Elliott, the thoughtful and reliable man was no longer Elliott. But what had gone wrong?

Damasio spent many sessions with Elliott struggling to work out what was missing. One day Elliott said something Damasio found astonishing. He said his feelings had changed since he had the operation. 'He could sense how topics that once had evoked a strong emotion no longer caused any reaction, positive or negative' (Damasio, 1994, p. 45). What he had lost was the ability to interweave emotion and rationality. He had become an Enlightenment thinking man – thinking without emotion – but that hadn't made him rational; instead, he had become totally irrational.

What is this thing called emotion? The standard or 'classic' view

If thinking and feeling, rationality and emotion, are intimately interwoven in healthy brains, what exactly are emotions and is there any way of knowing what emotions other people are feeling? Is there a way for teachers to sense what emotions their students are feeling? Can we read it in their faces or body language, for example? The belief that we can read emotions from facial expressions forms part of what Lisa Feldman Barrett calls the **classical view of emotion** (Barrett, 2017, p. x). In this section we will introduce the research claims regarding the recognition of emotion in faces. In subsequent sections of this chapter, we will consider alternatives to the standard or classical view and will also cast doubt on the veracity of facial emotion recognition.

Taking a critical approach is important. If much of what is currently believed about emotions is wrong, including that they can be reliably expressed by our faces, we will need to reconsider how emotion *should* be viewed in education. It also raises critical questions about the rapidly growing industry of computer-assisted *facial recognition of emotion*, particularly its use in education, because this is based on the standard view, as you can see in Research link 6.2.

ventromedial prefrontal cortex (VMPFC): a broad area in the lower (ventral) central (medial) region of the prefrontal cortex; a key region supporting the decision-making process

classical view of emotion: the contentious view that basic discrete emotions have unique, biological foundations that are universal because they are passed down from our evolutionary, biological past

RESEARCH LINK 6.2

Can we build computer systems that reliably read students' emotions from their facial expressions, or are there problems?

Computerised facial recognition (Figure 6.6) is big business, potentially bringing considerable benefits but also a host of concerns – including educational concerns. For example, you can see something of the research being conducted on recognising student emotions during online teaching by reading Wang et al.'s (2020) article 'Emotion Recognition of Students Based on Facial Expression'.

Figure 6.6 Facial emotion recognition system

This research is based on the belief that 'Facial recognition is one of the most powerful, natural, and universal signals for human beings to convey their emotional states' (Wang et al., 2020, p. 1). It is also based on the belief that teachers have a right to monitor the emotions of their students during online teaching. Both these claims are problematic, and many teachers will feel they are especially worrying. In this chapter, you will see *why* this research is problematic *methodologically* (Barrett et al., 2019), but in education the idea that teachers can use computer technology to monitor what are believed to be their students' emotional states clearly raises *major ethical concerns*, at least in Western educational contexts.

The above article provides a good example of current research being undertaken. It is suggested that you complete your reading of this chapter before checking two recent articles raising concerns about computers allegedly reading emotions. One, published in the journal *Nature* in 2020 is titled 'Why Faces Don't Always Tell the Truth About Feelings' (Heaven, 2020). The other, published in *The Guardian* (Schwartz, 2019), is titled 'Don't Look Now: Why You Should Be Worried About Machines Reading Your Emotions.'

Research link 6.2

Identifying emotions from facial expressions, the classic view

Let's start with some pictures of emotions. Take a look at the six pictures in Figure 6.7. Each picture is said to portray a facial expression of an emotion. By just looking at the pictures, without any clues, can you see what emotion each face is portraying? Okay, just one clue: the smiling face is showing the emotion we call happiness. What about the others? It would be good if you could pause in your reading to do this exercise, ideally with a couple of classmates. In that case, can you all agree about what emotion each face is portraying?

Figure 6.7 Some facial photographs representing basic emotions, similar to those used in basic emotions method studies
Source: Concept adapted from Barrett (2017, p. 5).

How did you go? These are actually pictures of actors posing those emotions, so not actual photographs of emotional faces. They are similar pictures to those used in a very well-known research study that dates back to the 1960s. The study was first designed by three researchers in the United States, Silvan Tomkins, Carroll Izard and Paul Ekman, though the research method is still actively used to this day. The actual pictures used can be found on page 5 of Barrett's (2017) book *How Emotions are Made: The Secret Life of the Brain*. The six emotions pictured are said to be anger, fear, disgust, surprise, sadness and happiness. Now can you see which picture goes with each emotion?

The research method that they called *emotion recognition* was based on the belief, which goes back to Charles Darwin (1965), that the expression of different emotions on the face is universal: 'All people, everywhere in the world, are said to exhibit and recognise facial expressions of emotion without any training whatsoever' (Barrett, 2017, p. 4). If you studied

Research link 6.2 above, you will have seen this claim repeated when the researchers said that 'Facial recognition is one of the most powerful, natural, and *universal* signals for human beings to convey their emotional states' (Wang et al., 2020, p. 1) (emphasis added).

universality of emotional facial recognition: the contentious view that people from around the world exhibit and recognise facial expressions of emotion without any training whatsoever

If this claim for the **universality of emotional facial recognition** is correct, you should have had no difficulty in matching the six emotions you were given with the six faces. However, experience suggests that it is not as easy as that. Even when given the six emotions, some students find it difficult to agree which is which. Perhaps the easiest one is happiness, which you were given as a clue. It seems well matched to the smiling face, but in certain social situations – for instance, when in a formal meeting with a very senior colleague –might we not *smile in anger* when responding to someone who has just insulted us?

And, if you think critically about it, this research is actually making an astonishing claim: that wherever in the world you are born or grow up, you should be able to recognise the same emotions expressed on the faces of American actors. Does that sound right?

REFLECTION

If you have the benefit of a multicultural class grouping, this would be an excellent opportunity to discuss this universalist claim. Is it true that all peoples, everywhere in the world, express emotions using the same facial expressions? Can all humans recognise and differentiate the same emotions in facial pictures? What do you think?

Pick the emotion that best matches the face.

Figure 6.8 A second example of emotion recognition test
Source: Concept adapted from Barrett (2017, p. 5).

If you think it doesn't sound right, you will be surprised to learn that this claim for universality is backed up by half a century of what is said to be firmly supporting research evidence. In education, we are frequently told about the importance of evidence but, as suggested in the previous chapter, evidence can be problematic and should always be viewed critically. So let's take a closer look at the methodology used by the experimenters. One approach used multiple-choice options (Figure 6.8). The participant in the research is asked to circle which of the emotions the face portrays. As you know, in this case the 'correct' answer is said to be happiness.

Another approach used by the researchers gives two images and a short storyline (Figure 6.9). The participant is then asked to pick the face that best matches the story below.

In recounting the history of this research, Barrett points out that it had a massive impact, revolutionising the scientific study of what was called *emotion recognition*: 'Using this method, scientists showed that people from around the world could consistently match the same emotion words (translated into the local language) to the posed faces' (Barrett, 2017, p. 7). Barrett notes that one of the original researchers, Paul Ekman, even went to a very remote tribe in Papua New Guinea and ran the experiment, in their language (Ekman, 1972, 1980). Sure enough, these traditional people could correctly match the pictures with the words or storylines. They also tested the experiment in Japan and Korea with equally convincing results. Surely this body of confirming evidence must support the universalist claims of the researchers. Barrett is highly critical of this research, despite the hundreds of studies that have been conducted that used this method, all pointing to the same conclusion: that emotion recognition is universal.

However, in thinking about this research, you may also have begun to sense that there may be something not at all right about it. If so, why is it still trusted? Well, one answer is that it fits neatly into the standard or classic view that emotions are universally produced because they come from our evolutionary, biological past and are hard-wired into us at birth. According to this account, all peoples on Earth are the inheritors of universal emotions and facial expressions that are said to be 'reliable, diagnostic fingerprints of emotion' (Barrett, 2017, p. 7). In the battle for survival in a hostile world, the classic view argues, emotions played a core role in giving our ancient ancestors in Africa an *evolutionary advantage*. As a result of natural selection, emotion became a fixed component of our biological inheritance, fixed into the brain as discrete emotion circuits that are activated (switched on) whenever we become emotionally aroused. But is this correct? We will return to this question later.

Pick the face that best matches the story below.

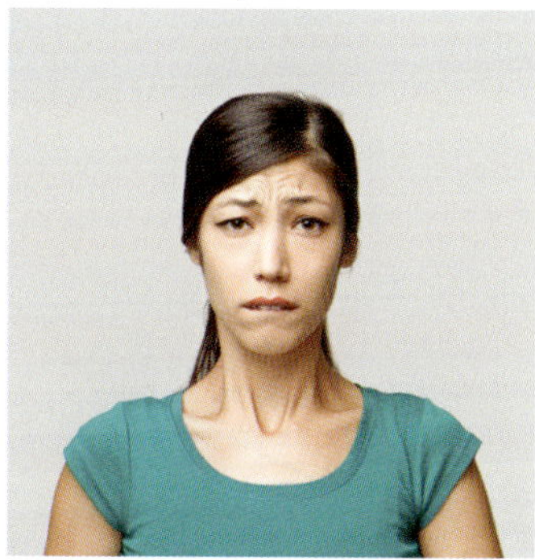

Her mother just died, and she feels very sad.

Figure 6.9 A third example of emotion recognition
Source: Concept adapted from Barrett (2017, p. 5).

The notion of the triune brain

Part of this account of how evolution has shaped the brain is the claim that the brain is comprised as three evolutionary distinct layers. The model goes back to neuroanatomist Paul MacLean, who first suggested his notion of the **triune brain** in the 1960s, although he later set it out in detail in his 1990 book, *The Triune Brain in Evolution*. According to MacLean, the most primitive layer of the model, which was allegedly inherited from reptiles, accounts for our most basic survival instincts. Following that, the first mammals developed what he called the limbic system, which he argues is the brain's basic emotion system. Finally, the cortex evolved with the higher mammals. With further evolutionary development in humans, the cortex is the seat of our human rationality – which, it is claimed, makes us special in the animal kingdom (Figure 6.10).

MacLean's identification of the limbic *structure*, which you met in Chapter 2, was arguably his greatest contribution to brain anatomy, although whether it comprises a discrete *system* concerned with emotion and emotion regulation is a very different issue (the identification of a distinct limbic *structure* in the brain doesn't imply that it is also a distinct *system*), even though MacLean claimed it was. According to the classic view, when we experience

triune brain: Paul MacLean's controversial idea that the human brain evolved in three distinct stages that are now represented in the current structure, comprising a primitive reptilian brain and then a limbic brain that appeared with the first mammals, both of which were wrapped around by a cortex that evolved with higher mammals

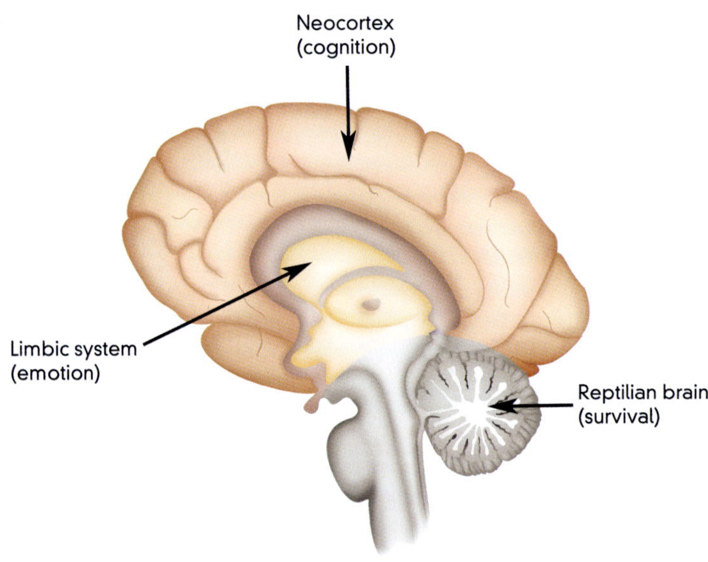

Figure 6.10 MacLean's notion of the 'triune brain'

an emotionally alarming event, ancient emotional circuits in the limbic system of the brain are triggered and switched on. This triggering is what causes your heart rate to increase, your sweat glands to activate, your breathing to speed up, your voice and body posture to change and your face to show the unmistakable tell-tale signs that you are experiencing emotion.

Certainly, to varying degrees, all these things may happen to us when experiencing emotion. But is this because different and discrete biologically ancient emotion circuits in your brain are kicking into action and firing in response to events – or could there be an alternative explanation for what is happening in the brain when emotions are experienced? And are emotions actually universally the same across all cultures, as researchers have claimed, based on their research findings? Perhaps it is time to start rethinking our understanding of emotion.

Teachers often say that they watch the faces of their students for clues to how each child is feeling and whether they are understanding what they are being taught. Do you think faces can reliably be read in that way, or might teachers misread what they see? Why might that be educationally problematic when dealing with students?

A dynamic systems approach to emotion and emotion regulation

Classrooms are filled with emotions and every teacher knows that helping children to regulate their emotions is a crucial part of their professional role. But what exactly are emotions if they are not evolutionarily fixed circuits in the brain, and what is involved in their regulation? What's the difference between a child who is able to regulate their emotions in school and one who seems unable to keep control? Can emotion regulation be taught in school, or is it something a child is born with – just part of a child's character or temperament?

Emotions as dynamically emergent and constructed

By the end of the twentieth century, a new perspective on emotion was being suggested within the framework provided by a dynamic systems account of learning and development.

In Chapter 3, you were introduced to the dynamic systems view of development as 'messy, exploratory, opportunistic and context sensitive', as well as the idea that development is an 'emergent, self-organising process with feedback' and the view of children as 'complex, emergent, self-organising beings'. Once we begin to apply these dynamic systems concepts to emotions, a very different account of what emotions are and how they might be regulated comes into view. For example, take a look at what dynamic systems theorist Alan Fogel and his colleagues wrote in 1992:

> We postulate that emotion is not felt experience alone, nor a pattern of neural firing, nor an action such as smiling. Emotion is the process that emerges from the dynamic interaction among these components as they occur in relation to changes in the social and physical context ... From a dynamic systems perspective, coherent emotions can be conceived of as relatively stable patterns that are continually constructed by a complex and dynamic process of interaction among the components. (Fogel et al., 1992, p. 129)

Emotions, they say, are *dynamically emergent*, and are *constructed* by a *complex and dynamic process of interaction* among *many components* in response to *changes in the social and physical context*. Two years after this quotation from Fogel and colleagues, Esther Thelen and Linda Smith (1994, p. 320) said, from a dynamic systems perspective, that emotions are 'emergent, self-organised processes' and, just like cognitions, they are 'fluid, context-sensitive, non-linear, and contingent'.

In a dynamic systems account, emotions are not pre-programmed *things* in the child's brain, with dedicated and fixed circuits that switch on in response to emotional triggers; rather, they are emergent self-organising *processes*. One could say that there is no such *thing* as emotions – they are not things at all, they are processes. As a result, they are *not fixed* in the brain, but are fluid, in the same way that all of our thinking and learning is *fluid*, which in turn means that emotions can be *changed and improved*. From a dynamic systems perspective, each child's brain creates their own emotions; emotions emerge in response to events, not simply as a here and now felt experience, but informed by their past experience – what Edelman calls 'the remembered present' (Edelman, 1989). The *dynamic processes* involved in producing emotion are *much the same in each child*. That much is universal, but the result is individual differences and variability in how those emotions are felt and expressed.

You will probably recognise that this is similar to the point about individual brains mentioned in Chapter 2: the neurobiological processes in the brain are the same in all human beings (and also very similar to those in many other animal species); however, that sameness of process produces individual difference. Thus, 'no two human brains are, have ever been or ever will be, identical' (Geake, 2009, p. 46) because of the multiple individual difference in experience. In other words, *individual difference is a product of similarity of process*. The same point applies to emotions. Emotions emerge in and are constructed by each individual brain employing the same dynamic processes, but that results in emotionally different individuals.

Let's briefly recall the six-month-old infant learning how to grasp a rattle in Chapter 3. What we observe is the infant making a large variety of random bodily movements, arms and legs moving in all directions that correspond with random neuronal firings of in the brain. When the infant touches the rattle, the particular firing pattern that gave rise to that contact is recognised by the brain as emotionally good, valuable and salient, causing a chemical reaction that reinforces that firing pattern. Through constant trial and occasional success, each selective success further strengthens that firing pattern. In this way, success is built on success so that, over time, the infant will learn to grab the rattle easily and put it straight into their mouth – another developmental learning skill mastered.

Similarly, within social contexts, as infants 'engage in social discourse, they produce actions, perceive their felt consequences, and view the reciprocal activities of the social partner, often in a game of mutual matching and turn-taking' (Thelen & Smith, 1994, p. 320). In the same way that relatively stable categories of perception and action emerge through the process of neuronal group selection (selecting and strengthening only those perceptions and actions that have value), so relatively stable categories of emotional feelings emerge in each individual child, associated with successful perception–action strategies.

The interweaving of thinking, feeling and body regulation

In stark contrast to the notion that rationality is required to control wild primitive emotions, as represented in the classic view of emotion and the notion of the triune brain, feeling can act as a powerful influence on reason. In other words, it is not only that reason can regulate emotion and feelings; the lesson of Phineas Gage and so many others with prefrontal lobe damage is it also works the other way around: *emotion can act as a regulator of reason*. The loss of emotion within human thinking and reasoning leads to an unregulated life.

Earlier in this chapter, we saw that Damasio (2018) provided a holistic view of rationality and emotion, thinking and feeling while also linking these to body regulation, technically called *homeostasis*. He said that 'feelings are a powerful influence on reason' and 'the brain systems required by the former are enmeshed in those needed by the latter'. Damasio also said that these 'systems are interwoven with those that regulate the body' (Damasio, 1994, p. 245).

Damasio is saying that the brain systems required for emotion and feeling are enmeshed in those that are involved in reasoning; moreover, they are interwoven with those that regulate the body. Elsewhere, he says that body signals of emotion are not simply reactive, occurring after an event; you can have a kind of bodily gut (visceral) feeling and other bodily sensations that something might be wrong in advance of an event. Damasio (1994, p. 173) calls these bodily feelings **somatic (body) markers** and suggests that they act as emotional warning signals that focus attention on the possible negative outcomes of an intended action. This, he found, was missing in his patients with prefrontal damage.

This points to an important message for teachers in managing emotions in classroom settings. We mentioned above how important it is that teachers help children and adolescents to regulate their emotions, to keep them appropriate and balanced (Figure 6.11). However, Damasio points out that it is a two-way process: emotions not only need regulation, they also act as regulators. The emotions of each and every child in a teacher's care are enmeshed in rationality and act as a control on irrationality. Moreover, they do more than that. As previously noted, one Australian education document says that, 'Emotional reactions are the brain's way of keeping us safe from danger and socially connected' (Western Australian

somatic (body) markers: bodily feelings that act as emotional warning signals focusing attention on the possible negative outcomes of an intended action

Department of Education, 2018). In other words, emotion reaction and regulation are very much related to each child's survival and wellbeing.

In the same year that Damasio produced his book, *Descartes' Error*, Thelen and Smith were also taking a holistic stance in arguing, as noted above, that emotions result from emergent, self-organising processes in the brain that are very similar to those that produce cognition and action. Emotions are not anomalies in the brain; rather, they are part of the way each individual's embodied brain functions, and how it develops. Also working within the framework of a dynamics approach to human development, Steven

Figure 6.11 An adult trying to help a boy's emotion regulation

Woltering and Mark D. Lewis refer to developmental pathways of emotion regulation, and make the important point that these 'trajectories can be seen as dynamically unfolding, self-organizing cascades, and skill acquisition may be different for each child and for any given child in diverse contexts' (Woltering & Lewis, 2009).

Woltering and Lewis (2009) draw on a distinction in the literature between two forms of emotion regulation: reactive and deliberate. We will use their actual words to make the distinction:

> **Reactive emotion regulation** is typically characterized as fast, stimulus driven, and containing implicit evaluations of objects or events that can be aversive or rewarding. The immediate feeling stopping you from taking the candy is also involuntary and strongly bodily in nature. It is in close alignment with Damasio's (1994) somatic marker hypothesis, which states that such visceral, bodily reactions are an integral part of a regulatory system that forms the basis of decision making and action. Our feelings, perceptions, appraisals, and actions can be regulated and influenced by relatively automatic processes over which we do not exert intentional or conscious control. (Woltering & Lewis, 2009, p. 161)

> **Deliberate emotion regulation** can be characterized as slower, reflective, and more sensitive to context and strategy (e.g. reappraising the situation, by thinking 'I have had B before. It is not too bad'). These regulatory processes minimize the unwanted state of fear or anger through planning and reappraisal, and they are more reflective and deliberate than those described for reactive control. Deliberate control is not necessarily conscious, but it is generally intentional and presumes executive control of attention. (Woltering & Lewis, 2009, p. 161)

Reactive regulation is not only found in infants and young children, and thereafter into later development; it is also shared across different species. Woltering and Lewis (2009) suggest its control is fast, involuntary and strongly bodily in nature – the result of visceral

reactive emotion regulation: relatively automatic processes of modulating an emotional state over which we do not exert intentional or conscious control

deliberate emotion regulation: slower, reflective (but not necessarily conscious) modulation of emotional states

(gut) feeling that things are not right. By contrast, deliberate regulation, which begins to emerge around the age of two or three years, is slower, more reflective and context and strategy sensitive, in response to feeling fearful, angry or perhaps over-anxious about taking a dreaded maths test!

When regulation is lacking, the results can be dramatic. Imagine a young child who is five years old. He comes from a high socio-economic status home. He is in a café, playing with a balloon, waving it all over the place. His mother becomes concerned that he could harm someone, especially his infant sibling if he puts the stick of the balloon into her eye. So she takes it away from him. The child flies into an immediate rage: the pupils of his eyes are wide open, he is crying and falls on the floor of the café lounge, kicking his legs, which brings his behaviour to the attention of everyone in the café. This child in this instance is clearly lacking reactive or deliberative regulation, although by this age he should have been capable of both. We can compare that with his friend when he was only three years old. Having fallen over when running on a path, he was very emotionally upset. However, after being picked up and given a hug, he deliberatively said, 'I think I need an ice-block, then I will be okay.'

Earlier, we said all teachers know that helping children to regulate their emotions is a crucial part of their professional role. Performing that role can be greatly aided by appreciating the difference between reactive and deliberate emotion regulation. In addition, Woltering and Lewis identify four recommendations, arising out this distinction that are particularly appropriate for teachers:

1. [It is important to] start teaching these regulatory skills early, as successful later development depends on prior skill acquisition.
2. The methodology of teaching should target not only just deliberate regulatory skills, but also incorporate those supporting reactive regulation.
3. [It is important to recognise that] in-the-moment emotion regulation requires the ability to coordinate reactive and deliberate control.
4. Recognizing individual differences in children's styles of emotion regulation highlights the need for an individualized approach to emotional issues. Thus, for example, for 'some children, the brain systems mediating coordination between reactive and deliberate styles … may not have matured or developed effectively. For these children, it may not be effective to rely on deliberate learning strategies such as reasoning, whereas other children may thrive on such an approach.' (Woltering & Lewis, 2009, p. 166)

REFLECTION

See whether you can imagine a child or adolescent who is not managing emotion regulation well in preschool or school. Given what Woltering and Lewis (2009) advise, what practical strategies could teachers adopt in order to encourage and nurture that child's (or adolescent's) emotion regulation?

Emotion regulation development in adolescence

Casey, B. J., Duhoux, S. & Cohen, M. M. (2010). Adolescence: What do transmission, transition, and translation have to do with it? *Neuron*, 67(9), 749–60.

With regard to emotion regulation, adolescence (approximately spanning from ages 10 to 19) is of critical importance for several reasons. It is well recognised that this period is not only associated with significant biological and physical changes but also growing academic pressures and expectations of social and financial independence. 'These challenges are often accompanied by increased emotional reactivity and stress' (Ahmed et al., 2015, p. 12). Many developmental neuroscientists suspect that ongoing brain development temporarily hinders adolescents from regulating their emotions effectively, 'putting them at greater risk for anxiety and stress' (Ahmed et al., 2015, p. 13).

Over the past few decades, neuroimaging studies have provided evidence of ongoing structural and functional brain development during adolescence. An important point is that structural development within a brain 'does not occur linearly over time within brain areas, with quadratic and cubic trajectories often evidence, nor does it occur uniformly across multiple brain regions' (Ahmed et al., 2015, p. 13). Instead, as we saw in Chapter 2 (see Figure 2.19), different brain regions that enable emotions, including the regulation of emotion, develop at different rates. Moreover, the connections between those regions are constantly changing. Therefore, it is suggested that a functional imbalance between different regions may result in less effective emotion regulation in adolescence.

One widely known study on adolescents' sensitivity to emotional cues and information was conducted by Hare et al. (2008), who studied 80 participants (aged between seven and 32 years) in order to see specific changes in the brain and in behaviour in adolescents relative to both adults and children. Further, they examined changes in activity over time, not just one-off transient event, with repeated exposure to threatful stimuli, then assessed whether these changes were related to what the participants reported about the anxiety levels experienced.

This research showed that adolescent participants would likely have an increased amygdala response to threat cues (fearful faces), compared with children and adults. (See Chapter 2, Figure 2.9, for the position of the amygdala in the brain.) We will shortly see that the role of the amygdala in emotion is contested, but the research by Hare et al. reported here found that the participating adolescents displayed increased amygdala activity compared with the maximums found in other age groups. See Figure 6.12(A), which shows increased activity up to age around 18 years; the maximums then reduce for older ages, making a triangular shape of distribution.

This non-linear (not in a straight line) pattern of distribution is very similar to findings from other neuro-developmental studies (Ernest et al., 2005; Galvan et al., 2006; Guyer et al., 2009). As shown by Figure 6.12(B), this 'heightened response in amygdala activity is age dependent and specific to adolescents relative to children and adults' (Casey et al., 2010, p. 757). Also, as shown in Figure 6.12(C), participants showed significant individual differences in the amygdala's habituation level, with some easily habituated to repeated exposure

to the threat stimuli (i.e. fearful face) while others remain sensitive. Most importantly, the extent of the amygdala's habituation correlated with participants' everyday anxiety collected via self-report ratings.

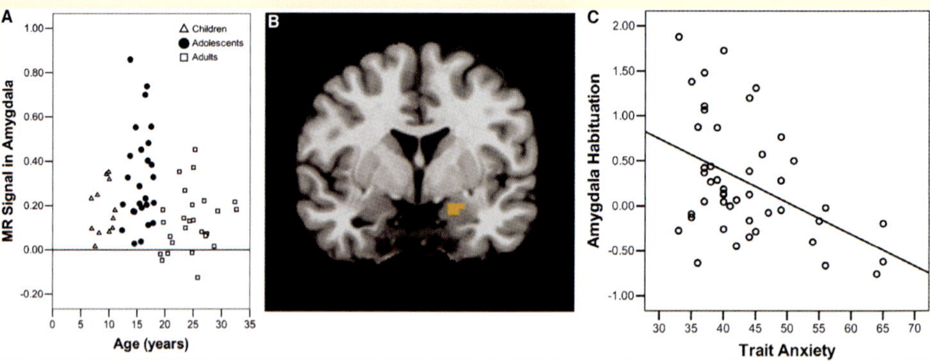

Figure 6.12 Developmental and individual differences in amygdala response to threat
Source: Casey et al. (2010), adapted from Figures 2 and 4 of Hare et al. (2008).

Casey et al. (2010) suggest that 'initial emotional reactivity to potential threat, as indexed by elevated amygdala activity, is *typical of*, or *normal* for adolescence', but not being able to reduce this heightened activity 'over time, with no impending threat, is *atypical* or *maladaptive* and may be indicative of risk for anxiety' (Casey et al., 2010, p. 757).

It should be noted that throughout this chapter our focus has been on emotions, although sometimes we have used the word 'feelings'. Within the literature on emotion and the brain, you need to be aware that you will often see a distinction drawn between emotions and what is called **affect** (where the emphasis in on the first letter 'a'), which refers to the feelings that are constantly occurring in your embodied brain, whether you are aware of them or not, characterised on a scale from pleasant to unpleasant or positive to negative.

The *extent* to which the feelings or emotions are experienced is called **valence**, which is the value associated with any given feeling. For example, a feeling of contentment will be characterised as a positive, pleasant valence. Another related term is **arousal**, which may be low or high. Feelings of sadness would be an example of unpleasant valance with low arousal. However, having made these distinctions, it should also be admitted that these terms are often used interchangeably in the literature. What matters for teachers is the recognition that not all feelings are emotions; some feelings are simply baseline, whereas emotions are complex states or mental constructs.

affect: a collective term referring to broad range of feelings that people experience. Emotion builds on affect.

valence: the extent to which feelings or emotions are experienced, from pleasant to unpleasant, and the value associated with any given feeling or emotion

arousal: a property of affect; a basic feeling that we experience, which can range from calm to agitated

The contested nature of emotion research

Clearly, a dynamic systems account of emotion and emotion regulation stands in stark contrast to the standard or classic view. Gone is the idea that emotions exist as distinct neural circuits or firing patterns in the brain, with each emotion having its own distinctive underlying pattern. Although we experience emotions as distinct states of being, that doesn't necessarily mean they are distinct processes in the embodied brain. Gone is the idea that emotions are built into the brain, waiting to be triggered whenever we experience the related emotion. On the contrary, the basic neurobiological processes that create emotions are closely related to body regulation. Emotions are emergent, self-organised processes that respond to perceived changes in physical, social and cultural contexts. Moreover, they are fluid, context-sensitive and non-linear. Gone is the notion that emotions are universal, the same for all peoples of all cultures. They are not universal; rather, they are highly dependent on culture and upbringing.

Revisiting the evidence for facial recognition of emotions

If that is the case, how is it that research into the facial recognition of emotions has been so successful in supporting the standard or classic view? Part of the reason is surely to do with how that research has been conducted. If you think back to how the pictures of six faces were first presented to you, only one clue was given to get you thinking on the right track. Other than that, you had to suggest what emotion the face was representing. Most people find that difficult. However, it became much easier when you were given the six emotions and all you had to do was match the emotion with the face, as a kind of multiple-choice exercise. Now why was that? Could it be that once you are given the emotions you have been *primed* to give the answer the researcher is looking for? And why would the researchers be looking for a particular answer? Could it also be that it *fits nicely into the theory of universal emotions* they are wishing to uphold?

In her book, *How Emotions are Made*, Lisa Feldman Barrett (2017) says that when research participants are not given a short-list of emotions to choose from to freely label the six pictures, 'the subject's success rate plummets'. She reports that, 'In one of the first free labelling studies ever conducted, subjects named the faces with the expected emotion word (or synonyms) only 58% of the time, and in subsequent studies the results were even worse' (Barrett, 2017, p. 45). Now that doesn't sound like a convincing success rate if we are all meant to be able to accurately read the emotions expressed on other people's faces, including the faces of the children we encounter in schools.

Another problem with this research is that each of the six emotions chosen for this research is highly variable. Emotions don't take a single physical form; they have multiple variations and they also show variation across cultures. If one studies an emotion such as fear, for example, there is no one kind of fear that can be represented in one kind of wide-eyed facial expression. Observe a young child when watching what they perceive to be something scary in a movie, the child may put their hands over their eyes, but if they see a snake on the path ahead, what kind of facial expression will they exhibit? And what about when they are about to fall off a chair? Perhaps you might like to try an experiment in a mirror: see how many different 'fearful' expressions you can create.

The emotion of fear and the amygdala in the brain

Fear may take many different forms and be expressed in many different facial expressions, but is it all related to one particular part of the brain, or is distributed across the brain? Barrett (2017, p. 17) notes that 'fear in the brain is an instructive example because for many years, scientists have considered it a textbook case of the localisation of emotion to a single brain area – namely the amygdala'. You will see the amygdala referred to in Research link 6.3, above. In fact, each brain has two amygdalae, one on each side of the brain branching into the temporal lobe as part of the limbic structure.

Research into the role of the amygdala has been the source of ongoing dispute in brain science with regard to emotion. It is believed to be the clinching argument in the claim that emotions are localised – a core belief of the standard or classic view. Although it is clear that the amygdala is certainly implicated in emotions such as fear and anxiety, as indicated in Research link 6.3, and other primary emotions (e.g. anger, joy surprise, sadness, disgust), this does not mean that emotion is localised to the amygdala, it is also distributed across the brain.

As is so often the case, brain lesions have been the source of what we know about the brain, including whether there is a correlation between fear and the amygdala. Many of these have been inflicted on animals in research laboratories. In the 1930s, experiments were performed on monkeys, removing their amygdalae by surgery and then observing their behaviour. It was found that these monkeys no longer avoided other animals such as the snakes that they would normally keep well away from. It was concluded that this was due to a lack of fear, caused by the removal of the amygdalae. But would this also apply to human beings and, if so, is it a total lack of fear or only partial? Evidence comes from people who have a very rare and debilitating disease that gradually erodes the amygdalae during childhood and adolescence, called Urbach-Wiethe disease. One is a patient called SM.

SM was tested on multiple occasions when she might be expected to experience fear, but with no success. Researchers even tried hard to teach her to fear, but also without success. It seemed as though there was nothing she feared, which was taken to be sufficient proof that fear is localised to the amygdala. And if that emotion is localised, surely that would equally apply to other emotions? But then it was found that she could recognise fear in certain body postures and in voices. Moreover, there was one thing she certainly feared – asking her to breathe in air that had been laced with additional carbon dioxide. So the evidence wasn't as clear cut as first supposed.

Much more telling counter-evidence came from the study of a pair of identical twins who suffered from the same disease as SM. One of the twins had exactly the same outcome as SM, but the other twin did not. They shared the same DNA, suffered the same brain damage and grew up together in closely similar environments, yet one twin had a generally normal response to fear. In her case, brain plasticity resulted in other brain circuits compensating for her loss of her amygdalae. In pointing to this evidence, Barrett says, 'These findings undermine the idea that the amygdala contains the circuits for fear. They point instead to the idea that the brain must have multiple ways of creating fear, and therefore the emotion category "Fear" cannot be necessarily localised to a specific region' (Barrett 2017, p. 18). She is not saying that the amygdala is unimportant in regard to emotion; on the contrary, brain regions including the amygdale are important; however, she claims that 'they are neither necessary nor sufficient for emotion' (2017, p. 19).

Agreement and disagreement in the brain-based study of emotion

Barrett's position on this issue and a number of others is situated at one end of a spectrum regarding brain specialisation, as mentioned in Chapter 5. One end holds that the brain is highly localised and specialised; the other end argues that brain function is highly distributed and results the global operation of the whole brain. That tends to be Barrett's position in regard to emotion within the brain. However, she is not denying that brain regions involved in specific functions exist, only that there is much more to the story than these particular regions. The brain, she believes, must be viewed as a *constructed whole*, with all areas working together in synchrony, and in harmony with the whole organism and its internal interoceptive bodily regulation.

However, she is certainly not alone in stressing that brain function and bodily regulation are interwoven – that is what is meant by the *embodied brain*. It is also what we have seen advocated by Damasio when claiming that feelings 'depend on a delicate multicomponent system that is indissociable from body regulation' (Damasio, 1994, p. 245), and is also stressed by dynamic systems researchers. For example, we have seen Fogel and colleagues refer to emotions as 'continually constructed by a complex and dynamic process of interaction among the components' (Fogel et al., 1992, p. 129).

However, Barrett (2017) tends to position herself as unique in many of the issues she discusses. Moreover, it is important to be aware that within the field of academic debate, when laying out her position she is not short of critics, such as Ralph Adolphs. Although agreeing with various aspects of Barrett's account, he is nevertheless concerned with other aspects, such as her 'concept of concepts' (Adolphs, 2017, p. 33). For those who enjoy the cut and thrust of academic debate, the argument between Barrett and Adolphs in the journal *Cognitive and Affective Neuroscience* (2017) will prove quite a treat.

That is not something with which we need to engage here, in the context of education. However, one key point for teachers that arises out of this debate on the neuroscience and dynamics of emotion is that science is often contested. We will see other examples of hotly debated issues as we consider the social and cultural brain in Chapter 7 and the moral and empathetic brain in Chapter 8. We will also see that these aspects of the embodied, embedded brain also involve emotion.

<table>
<tr><td>**APST and ACECQA curriculum specifications**</td><td>**STANDARDS**</td></tr>
</table>

APST Standard 1.1 Physical, social and intellectual development and characteristics of students
Demonstrate knowledge and understanding of physical, social and intellectual development and characteristics of students and how these may affect learning

APST Standard 4.3 Manage challenging behaviour
Demonstrate knowledge of practical approaches to manage challenging behaviour

APST Standard 4.4 Maintain student safety
Describe strategies that support students' wellbeing and safety working within school and/or system, curriculum and legislative requirements

ACECQA curriculum specifications
1.3 Social and emotional development
1.4 Child health, wellbeing and safety

► ► This chapter provides a frontline overview of recent research into the development of emotion that is highly relevant in advancing newly qualifying teachers' quality engagement with their students. Given the focus on the science of emotion, this chapter goes beyond what is generally available (e.g. classical view of emotions) to newly qualifying teachers in their initial teacher education. In doing so, it adds considerably to their understanding of research into how emotions are created by the brain. It also provides them with an understanding, drawn from neuroscientific research, that human rationality requires emotions. Moreover, research suggests that emotion regulation involves both reactive and deliberate processes, so the learning of regulation skills requires the balancing of both processes.

The chapter provides sufficient grounding for newly qualifying teachers to address some of the challenging behaviour of students, with informed perspectives about why such behaviour emerges, drawing on the neurobiological evidence. This chapter also addresses the causes of learning anxiety, about which all teachers should be sufficiently informed, in order to create engaging and safe classrooms, while supporting students' wellbeing.

SUMMARY

- Schools and classrooms are filled with emotions, and one important part of a teacher's role is to help children regulate their emotions. This is especially important in the early years, which influence the formation of meaningful relationships in the later years. Emotion regulation is closely associated with student wellbeing.
- For more than a quarter of a century, brain science has consistently shown that human thought and rationality are necessarily laced with emotion, but that contradicts a previously held belief that to be rational, one has to studiously avoid or systematically suppress emotions. In the West, this widely held belief, which is still heard today, goes back to the Enlightenment and ultimately to ancient Greece.
- Lesions to the prefrontal lobe are now known to severely compromise the ability to interweave emotion and rationality. Without emotion, human rationality ceases to be rational, with those who suffer such legions likely to be limited in real-life decision-making and social interaction.
- What has been called the 'classic' view of emotion posits that emotions are fixed into the brain as discrete emotion circuits that are triggered whenever we become emotionally aroused. As part of this view, much evidence has been assembled to support claims that emotions are reliably expressed by the human face and that these distinct emotional facial expressions can be recognised by all people everywhere, regardless of culture. However, there are strong reasons to think that the classic view is wrong, and this research is unreliable.
- A dynamic systems account of emotion and emotion regulation provides a stark alternative to the standard or classic view of emotion that it is 'wired in' by evolution. Rather, emotions are dynamically emergent and are constructed by a complex and dynamic

process of interaction among many components in each individual brain, in response to changes in the social and physical context.

- Emotions are created in the brain and, just like cognitions, they are fluid (not 'wired' in or waiting to be 'triggered') and context sensitive. From a dynamic systems perspective, teachers are encouraged to be aware of the highly dynamic and changing nature of emotion and emotional expression when helping children with their emotion regulation.
- There are two kinds of emotion regulation. *Reactive* emotion regulation is fast and strongly bodily in nature (gut feeling), and largely automatic and subconscious. Deliberate emotion regulation is slower and, as the name implies, reflective and deliberative. It is recommended that teachers start teaching these regulatory skills early, as later development appears to depend on prior skill acquisition.
- Research into emotion is ongoing and remains highly contested, especially between those who advocate the localisation of emotion in the brain and those who believe it is not localised but instead is distributed throughout the brain. Arguably, the details of these research contests in neuroscience are not important within education, although they do alert teachers to the contested nature of science. Scientific theories and research outcomes are not fixed in stone; they are always revisable and often contested.

KEY POINTS FOR TEACHERS

- The children teachers encounter in classrooms are both rational and emotional beings. All learning environments are filled with emotions – emotions are not abnormalities in classroom settings, but are to be expected. However, that doesn't mean that emotions should run wild. In maintaining a good learning environment, teachers should accept emotions, but also strive to keep them in productive balance.
- Working with children and adolescents in school inevitably involves working with thinking and feeling persons, which has implications for lesson planning and the way assessment tasks are designed – don't just focus on thinking; interweave learning and assessment with feeling.
- Within the research literature on emotion and the brain, a distinction is often drawn between emotions and what is called *affect*, which refers to the feelings that are constantly occurring in embodied brains. Teachers should be aware that not all feelings are emotions: some feelings are baseline, constantly occurring in the embodied brain, whereas emotions are complex states or mental constructs.
- Research shows that the early years are clearly a very important time for emotional development, but so are the years of adolescence, following the onset of puberty. After what might appear to be the tranquil years of primary schooling, physical changes start to occur in the brain with the onset of puberty that can massively disrupt emotion regulation.
- Lesions to the prefrontal lobes can significantly impact a child's behaviour. As teachers, we should always be aware that personality and behaviour are products of the brain, and children in our care may have experienced previous brain damage, either in the form of neglect or abuse, or physical damage.

- There are many different reasons why children and adolescents may experience negative emotional states at school, some social and others related to learning and performance. Research indicates that a considerable number of students experience 'mathematics anxiety', feelings of stress and worry when dealing with mathematics in school.
- Teachers help children and adolescents to regulate their emotions, to keep them appropriate and balanced. However, in doing so, they should also be aware that this is a 'two-way process': emotions don't just need regulation, they also act as regulators.
- In dealing with emotions in educational settings, teachers should avoid a 'one size fits all' approach; this needs an individualised approach. For example, there are differences between children in the rates and extent of maturity in brain systems mediating coordination between reactive and deliberate regulation, so strategies that work with some children may not work with others.
- Helping children to regulate their emotions is a crucial part of the teacher's professional role. Performing that role can be greatly aided by appreciating the difference between reactive and deliberate emotion regulation. Reactive regulation is typically fast and stimulus driven, while deliberative regulation is slower, reflective and more sensitive to context and strategy.
- Research strongly suggest that it is important to start teaching regulatory skills early, as successful later development depends on prior skill acquisition (Woltering & Lewis, 2009). Moreover, in assisting students in dealing with their emotions, teachers should target both reactive and deliberate regulatory skills.

REVIEW QUESTIONS

Guided responses

1. Which region of the brain was damaged when Phineas Gage had an injury? What social problems did Phineas have as a result of that lesion?
2. Can we identify people's emotions from their facial expressions, regardless of their cultural or ethnic backgrounds?
3. Why is it important to regulate emotions and how can teachers help children develop their emotion-regulation skills?
4. Does emotion regulation always occur as a result of a conscious, deliberative process?
5. Where in the brain is the amygdala located? What roles does the amygdala play when we face a perceived threat?

FOOD FOR THOUGHT

1. One core message that brain science consistently repeats is that humans are fundamentally emotional beings. Do you think the current education system, particularly what generally happens in schools and classrooms, sufficiently embraces this understanding?
2. A dynamic systems account of emotions stresses that they are not pre-programmed *things* in the child's brain, with dedicated and fixed circuits that switch on in response

to emotional triggers as often supposed. Rather, they are emergent processes and each child's brain creates their own emotions – emotions emerge in response to events, informed by their past emotional experiences. In what ways can schools and teachers help to enhance each child's emotional experience, so it more richly informs the child's future emotional experiences and also aids their present emotional regulation?

3. Do you think it might be problematic that technology is currently being designed on the basis of the theory of universal emotional facial recognition, which would allow lecturers to monitor students' facial expression when conducting online classes? Do you see any problems with the underlying theory, and would you be happy to be taught in such an online class where the lecturer is able to make judgements about your putative emotional state?

RESEARCH ACTIVITY: CATERING FOR STUDENTS' EMOTIONS IN CLASSROOM LEARNING

Mary Helen Immordino-Yang, a former schoolteacher, is now a well-known affective neuroscientist working in education. In a paper jointly written with Antonio Damasio, she calls emotions 'the rudder that steers thinking' (Immordino-Yang & Damasio, 2016, p. 28). She provides a number of guiding strategies to help teachers accommodate and support the development of emotional learning in classroom contexts. The main recommendations are:

1. *Foster emotional connection to the material.* This means designing 'educational experiences that encourage relevant emotional connection to the material being learned' (Immordino-Yang, 2016, p. 101) by involving students in deciding which specific topics to present and making room for students to relate the material to the life of themselves. Teachers need to create classrooms where 'rich emotionality is played out, that valuable emotional memories are accumulated, and that a powerful and versatile emotional rudder is developed' (2016, p. 102).

2. *Encourage students to develop smart academic intuitions.* 'From a neuroscientific perspective, intuition can be understood as the incorporation of the nonconscious emotional signal into the knowledge being acquired' (2016, p. 102). Students must be 'offered adequate changes for the development and feeling of experience-based intuitions about how and when to use the academic material' (2016, p. 103). Without the development of sound intuitions, 'it is likely that the students will not remember the material in the long-term, and that even if they remember it in an abstract sense, they will have difficulty applying it to novel situations' (2016, p. 103).

3. *Actively manage the social and emotional climate of the classroom.* Students need 'an atmosphere of trust and respect. It is here that the classroom climate and the social relationships between the teacher and students have a crucial contribution to make' (2016, p. 103). A carefully timed dose of humour or incentive can help students 'to feel safe in expressing and learning from their mistakes, and in building social cohesion among the students and between the students and the teacher' (2016, p. 103).

As the research activity, read the following three publications, then choose *two* of her three recommendations and, with reference to the ideas gained from the readings below, apply those two recommendations in *designing a 30- to 40-minute lesson* in your preferred subject area (e.g. maths, science, history) and targeting a chosen student age (K–12).

Research activity links

- Interview with Professor Mary Helen Immordino-Yang: Varlas, L. (2018). Emotions are the rudder that steers things. *ACSD*, 1 June.
- Immordino-Yang, M. H. & Faeth, M. (2010). The role of emotion and skilled intuition in learning. In D. A. Sousa (ed.), *Mind brain & education: Neuroscience implications for the classroom*. Bloomington, IN: Solution Tree Press.
- Immordino-Yang, M. H. & Knecht, D. R. (2020). Building meaning builds teens' brains. *ASCD*, 1 May.

THE SOCIAL AND CULTURAL BRAIN

Learning and development

LEARNING OUTCOMES

By the end of this chapter, you will:

- Appreciate the dynamic interaction of culture, biology and context in the learning process
- Understand that culture is embrained and that culturally diverse settings, including schools, classrooms and other learning environments, are shaped by neurodiversity
- Know what is meant by mirror neurons, how they were discovered and how they may provide important insights into how we relate to others and read their intentions in social and cultural contexts
- Understand what is meant by the reification of intelligence and the Flynn effect – the idea that IQ scores are a measure of adaptation to modernity and social development

The biologically, socially and culturally embedded brain

From the moment each child is born, and even when still in the womb, their brain is *relationally embedded* (needing relationships for survival and wellbeing) in dynamic and constantly shifting physical, social and cultural contexts.

In the seventeenth century, at the start of the scientific revolution, the English poet John Donne (1572–1631) (Figure 7.1) captured this idea when he wrote, 'No man is an island, entire of itself'. The words of the poem are given below, using modern English, because his actual words were written in the English of his time and place, and expressed within his cultural context. Donne is saying that any person's death diminishes him, because he is immersed in what it means to be human.

Figure 7.1 John Donne (1572–1631)

No person is an island entire of itself,
Every person is a piece of the continent,
A part of the main.

If a clod is washed away by the sea,
Europe is the less,
As well as if a promontory were,
As well as any manor of your friend's,
Or of your own were.

Any person's death diminishes me,
Because I am involved in humanity.
Never send to know for whom the bell tolls;
It tolls for you.

'No man is an island, entire of itself'

– John Donne

This is very much in line with the seemingly contradictory idea that *individuality is inherently relational*. Much more recently, dynamic systems theorist Alan Fogel (2008) has pointed out that, when we say human beings are fundamentally relational, we are also admitting that, as individuals, we are 'inherently incomplete'. People must, he says, 'find themselves in the other, become who they are through the other' (2008, p. 59). Which means that, at every moment on our trajectory through life, we discover our 'self' in relation to others.

However, Fogel also notes that in much conventional thinking, relationships are conceived as *linkages between distinct individuals*, each with their own independent identity. In schools, children are often viewed as separate individual selves, who are nevertheless *drawn* into relationships with others. These includes their parents, their extended family, family friends, peers and teachers at school, and social acquittances beyond school. Through these multiple social encounters, we are often told, children develop and *learn as a result of socialisation*. But is the idea of linkages between separate individuals the best way to think of human relationships, and is the notion of learning as socialisation entirely appropriate in educational contexts?

The 'classic' notion of learning as socialisation

From a dynamic systems perspective, individual children are always and inevitably nested in relationships (unless intentionally or accidently isolated from others) and the claim that each child's brain is socially and culturally embedded recognises that the relationship is always *dynamically interactional*: the child, society and culture are always intimately relational, each impacting the other, each creative of the other, at every moment. However, from this perspective the idea that relationships are *links* between individuals looks very *mechanical*, like the links of a chain, or between the cogwheels in a clock. By contrast, the notion of the embedded brain is *biological* (not mechanical) and *ecologically relational*.

In Chapter 6, you met what Lisa Feldman Barrett identifies as the *classical view of emotion*, the belief that human 'rationality makes us special in the animal kingdom' (Barrett, 2017, p. 81), which she says is deeply embedded in Western thinking. Anthropologist Tim Ingold (2008) notes how human beings have *classically* been viewed as *social and cultural creatures that transcend their animal past*, in the process of *being socialised*. For many sociologists and anthropologist, he says, learning and development arise out of processes of socialisation, which begin in the home, where babies and infants learn to recognise and interact with members of their immediate family. According to this **classic view of socialisation**:

> The new-born child … comes into the world as an entirely *asocial* being – equipped, to be sure, with certain innate response mechanisms, but without any of the information that enables adults to function as persons in the social world. Socialization … is the process whereby this information is taken on board. Among other things, the child acquires rules for categorizing and positioning other people in the social environment, and guidelines for appropriate action towards them … . Furnished with the rudiments of the kinship system, the child can then begin to participate in social life. The original asocial infant has become a social being, a *person*, equipped to play his or her part vis-à-vis other persons on the stage of society. (Ingold, 2008, p. 112)

Ingold identifies this theory of socialisation as deeply Western, by which he means the enduring belief that 'there is *more* to humans than their biology – that although each of us may start out, at birth, as a biological organism, wholly ignorant of society and culture, we nevertheless end up as persons with specific social identities and cultural competencies … *humans are supposed to grow out of biology and into culture*' (Ingold, 2008, pp. 112–13, emphasis added).

Biological versus social – a false dichotomy

From a science of learning and development perspective, the idea that we somehow grow out of our biology and into culture, out of the physical, natural world into the world of society, simply establishes a false and totally unnecessary dichotomy. Why can't we be both? And why can't we be viewed as *both biological and sociocultural* from the moment we are born – and even before, as we start to learn and develop in the womb?

socialisation (classic view of): the process whereby a social child acquires rules for categorising and positioning other people in the social environment and guidelines for appropriate action within the social world (Fogel, 2008)

social constructivist (or constructionist) theories: sociological theories of knowledge according to which knowledge is constructed through interaction with others. Social constructivists tend to ignore contributions from biology.

The same separation of biology from the social and cultural is often found in **social constructivist (or constructionist) theories**, which inform educational debate. As noted in the previous chapter, Barrett calls her biological account of emotion a *constructionist theory*. She identifies similarities with social constructionist theory, except that she notes, 'Social construction tends to ignore biology, however, as irrelevant to emotion' (Barrett, 2017, p. 33). Social constructionist theories, she says, are concerned primarily with social circumstances in the world outside you, without acknowledging that social and cultural experience is inscribed in the connections between neurons and across neuronal networks.

Human brains (and those of many other animals) are inherently social, just as they are also inherently biological. The social and the biological belong together, there is no need to set up a dichotomy between them, as one finds in much social and cultural theorising. Until recently, for example, much cultural psychology paid little or no regard to the biology of the brain in its theorising of culture and cultural development. However, writing from the perspective of what is now called *cultural neuroscience*, leading cultural psychology scholar Shinobu Kitayama (Kitayama & Park, 2010, p. 125) concludes, 'No longer is it possible to demarcate the domain of culture as separate from biology and ignore the latter in the analysis of the former ... the brain is the quintessential biological organ'.

social determinism: a contentious view that social research should systematically exclude biological explanations from its account of human behaviour, and only employ social factors such as social interactions and constructs

Viewing the biological and social as inherently interwoven is in stark contrast to the core claim of **social determinism**, which you will sometimes encounter in educational theory and research. For example, there is some evidence that Vygotsky's ideas were indebted to social determinism, even though he was not at all opposed to brain science. One quite recent discussion of Vygotsky states that, 'despite its insights and sophistication ... Vygotsky's psychological theory remains marked by a mechanistic social determinism stemming from a dualism of "natural" versus "cultural"' (Jones, 2016, p. 7). Social determinism does not just drive a wedge between the biological and social; many of its advocates also insist that cultural research should *systematically exclude* biological explanations from its account of human behaviour, and only employ social factors such as social interactions and constructs. But aren't highly social animals such as chimpanzees also thoroughly biological – and doesn't that also apply to human beings?

RESEARCH LINK 7.1

Can animals be taught – do they possess a zone of proximal development?

Vygotsky, L. (1978a). Interaction between learning and development. In M. Gauvain & M. Cole (eds), *Readings on the development of children*. New York: Scientific American Books, pp. 34–40.

When Vygotsky's ideas started to become available outside of the Soviet Union, one of the key notions that was particularly attractive to teachers was the idea of a *zone of proximal development (ZPD)* – the idea that in development and learning there is always a gap between what a child can do and what they on the verge of being able to do. Many educators saw the teacher's role as bridging that gap. Jerome Brunner (1915–2016) used the analogy of scaffolding (seen on building sites), providing a temporary structure so that the child can cross the ZPD and move on in their learning and development.

However, it is not generally realised that Vygotsky used the notion of the ZPD to *demarcate human beings from all other animals*. He was particularly critical of the work of Gestalt psychologists such as Wolfgang Kohler (1887–1967), who were finding very convincing experimental evidence that chimpanzees are very clever in solving problems. Vygotsky's attitude to other animals was very similar to that of Descartes, who viewed them as machines without a mind. One wonders what Vygotsky would have made of the discovery by Jane Goodall in the 1960s that chimpanzees use tools and *they are taught* to use tools by observing and imitating their parents and older siblings?

Writing in the context of his own time and place, Vygotsky claimed that primates could not be taught, and their intellect could not be developed, *because* they had no zone of proximal development. The full quotation from Vygotsky is as follows:

> However, Kohler failed to take account of an important fact, namely, that primates cannot be taught (in the human sense of the word) through imitation, nor can their intellect be developed, because they have no zone of proximal development. A primate can learn a great deal through training by using its mechanical and mental skills, but it cannot be made more intelligent, that is, it cannot be taught to solve a variety of more advanced problems independently. For this reason, animals are incapable of learning in the human sense of the term; *human learning presupposes a specific social nature and a process by which children grow into the intellectual life of those around them.* (Vygotsky, 1978b, p. 39)

For a more recent understanding of animal learning, you may like to check what we now know about crows and their problem-solving abilities (Jelbert et al., 2018).

Research link 7.1

From learning as socialisation to the notion of socialised learning

Arguing from a dynamic systems approach to learning, Ingold says that 'we need to replace the theory of learning as socialisation with a socialised theory of learning' (Ingold, 2008, p. 114). This alternative emphasises the dynamic interplay between the learner and the multi-causal and multi-directional social and cultural settings in which learning occurs. Whether these are traditional social and cultural settings or formal classroom learning environments, learning is always more than the 'internalisation' of a pre-existing and fixed body of knowledge that is assumed to exist independently of the learner.

He identifies three immediate concerns with the notion of learning as socialisation. First, the idea of the child moulded through socialisation invokes a very 'passive' view of childhood, in which the rules and conventions of their society are passed down to them from above, by adults. It is therefore also 'decidedly adult-centred' (Ingold, 2008, p. 113): adults have a job to do, which is to socialise the child into the ways of thinking of their own particular culture or society. On the contrary, 'children are agents with purposes and perspectives of their own' (2008, p. 113) and every child is immersed in the society and culture they are born into, from the very beginning. Every encounter physically impacts their brain, forming synaptic connections and synaptic pathways through Hebbian plasticity.

We should note that a strength of both Piaget and Vygotsky is that they viewed the child as an active participant in their own development. Nevertheless, Vygotsky also believed that the 'true direction of the development of thinking is not from the individual to the social, but from the social to individual' (Vygotsky, 1986, p. 36). From a dynamic systems perspective, this appears very linear and one-way, and it also sets up yet another false dichotomy. Surely, on the contrary, social relationships and social interactions are always multi-directional, and even very young 'children are agents in the socialisation of their parents as much as parents are agents in the socialisation of their children' (Ingold, 2008, p. 113). In similar ways, *children socialise their teachers*: 'Whether we are talking about an interaction at home between mother and infant, or in a western schoolroom between teacher and pupils, these contexts are *negotiated*' (2008, p. 113).

Second, the process of socialisation is *not a prelude* to the child's entry to into the world of society and culture, as the classic view supposes – a transition from the margins of society into full-blown, socialised membership. Rather, 'every infant embarks on life from the centre of the social world and begins at once to interact with other people in his or her surroundings' (Ingold, 2008, p. 113). Third, *learning is an ongoing, lifelong* process. There is no point at which it could be said to end – such as when children can be said to have become socialised, as the classic view supposes.

REFLECTION

Can you think of specific examples to support Ingold's claim that children are agents in the socialisation of their parents as much as parents are agents in the socialisation of their children, and that children socialise their teachers?

Culture and neurodiversity in educational settings

Teachers are often told that they should strive for 'culturally inclusive teaching and learning' (NSW Department of Education, 2020), but what does that mean and, moreover, what might be involved neurobiologically?

What is meant by culture?

We first need to recognise that the term 'culture' is defined in many different and sometimes misleading ways. For example, currently in everyday conversation, the term 'culture' is often used to create 'essentialist categorizations of entire groups with labels based on racial and ethnic membership' (Heath, 1997, p. 113). Essentialism assumes there is some basic *essence* that defines and demarcates one groups from another, perhaps based on geographical location (e.g. East Asian, Western), religious affiliation or ethnicity. For example, in education-related documents we often see terms such as 'Asian learners' (e.g. Biggs, 1994; King & Bernardo, 2016) and 'Muslim students' (e.g. Niyozov & Pluim, 2009).

The problem with using labels such as 'East Asian' and 'Muslim' is that it ignores a vast array of national, social and cultural differences within these putative essentialist groupings, not to mention the considerable differences between individual people within these groups. Among East Asians, for example, there are different nations (e.g. Korean, Chinese, Japanese), each with its own distinctive and defining language, history, customs and traditions. There are also marked variations in religious affiliations, within each of these nations (e.g. Buddhism, Shintoism, and Catholic and Protestant Christianity).

Moreover, in any given religion there are many differences as well as similarities. Among Muslims, for example, there are two main branches, Sunni and Shia, that disagree about who should be the leader of the Islamic faith, and within those branches there are sub-sects (e.g. Hanafi and Maliki within Sunni, and Ismaili and Alawite within Shia). So, at best, using broad-brush cultural identifiers in everyday conversation ignores considerable differences and, at worst, they can be used as a basis for national and cultural discrimination. And that has important implications for establishing truly multicultural schools and classrooms.

REFLECTION

Thinking back over your own time in school, can you think of examples when broad-brush essentialist labels were used to identify cultural groups that overlooked the considerable differences within cultures and the differences between individuals within those cultures?

Within the social science literature, the definition of culture has varied over time and the concept of culture has been contested (Heath, 1997). For much of the twentieth century, the discourse around culture employed essentialist categorisations, but that changed in the 1980s and 1990s. Shirley Heath (1997, p. 114) explains that advances in complexity science (dynamic systems) helped researchers see the problems of 'strict divisions between order and disorder, as well as fundamental notions of stability and the linear nature of order' that underpinned the essentialist view of culture, based on racial and ethnic membership. Researchers started to view cultural groupings as much more unstable and dynamic than essentialist models had assumed.

Despite recent advances, **culture** is still variously defined within recent literature, though there are commonalities across the definitions. For example, definitions usually include the notion that culture is a system of *values, conventions and artefacts* that constitute *daily social realities* (Kitayama & Park, 2010). Culture also includes *symbolic resources*, such as icons and scripts that are accumulated and transmitted across generations (D'Andrade, 1995).

Nowadays, there is usually strong recognition that culture is *dynamic* and that it changes over time, both within and across generations (Kitayama et al., 2007). This is because 'cultural ideas and practices have multiple meanings that are constantly in flux, negotiated, manipulated, and arbitrated for a variety of reasons by all individuals who participate in a cultural community' (Kitayama et al., 2007, p. 138). From a dynamic systems perspective, cultures – like all societies – are always highly complex systems (Sawyer, 2005), in which there is a

culture: highly complex systems of values, conventions and artefacts that constitute daily social realities. Culture includes symbolic resources, such as icons and scripts that are accumulated and transmitted across generations.

constant process of ongoing recursive modification (constant back and forth change) between what constitutes a given culture and the individuals within that cultural community.

Culture as embrained

The brains of human individuals (human persons) are socially and culturally embedded. There is a continuous process of mutual, recursive modification, whereby brains create and recreate cultures and cultures physically shape and reshape brains. In addressing how cultures impact brains, Shinobu Kitayama and Ayse Uskul (2011, p. 219) refer to '**embrained culture**', meaning that 'recurrent, active, and long-term engagement' in cultural practices can 'powerfully shape and modify brain pathways' (2011, p. 421). They also point out that this occurs as a result of the 'Hebbian principle of long-term potentiation' (2011, p. 424).

Figure 7.2 attempts to illustrate the mutual, recursive modification process. The arrows pointing from culture to brain are operating in the way the brain is developed. Although any given culture has diverse beliefs and practices, there are some common elements that transcend this variability. For example, core values such as filial piety (Bedford & Yeh, 2019) and independence are still reflected in diverse practices of groups within one culture. 'As the common elements of culture are engaged through everyday practice and consistent feedback, they will contribute to a cumulative change of the neural networks' (Kitayama & Salvador, 2017, p. 844) (Figure 7.2).

embrained culture: a process whereby recurrent, active and long-term engagement in cultural practices shapes and modifies brain pathways

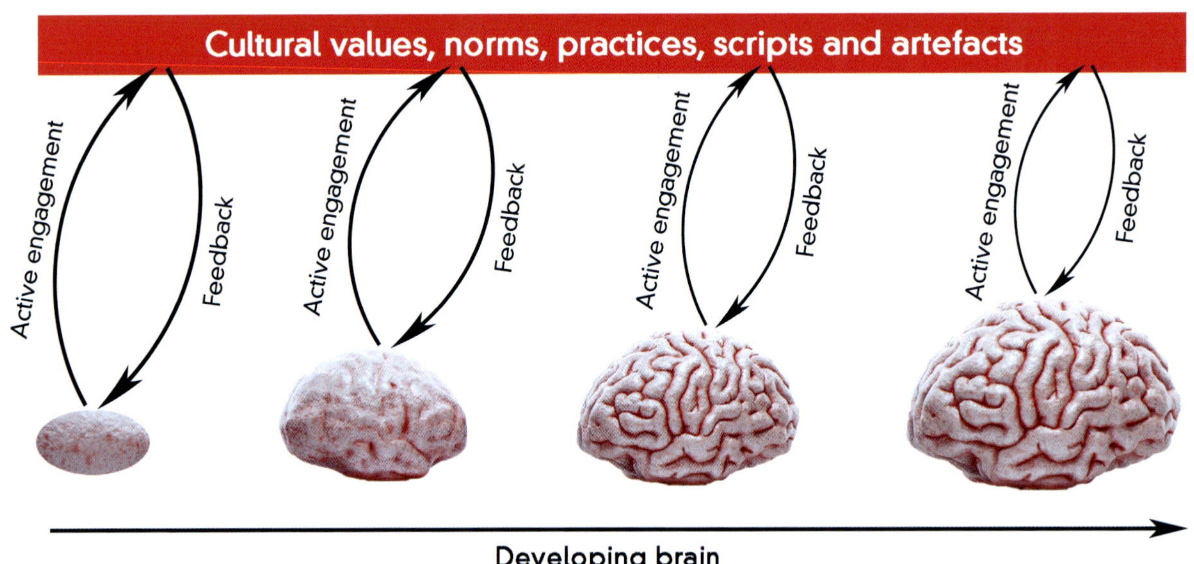

Figure 7.2 Feedback-loop cycles between culture and developing brains
Source: Adapted from Kitayama & Salvador (2017).

Neurodiversity in the classrooms

neurodiversity: the range of variation across all human brains

Armed with the idea that culture is embrained, the notion of **neurodiversity** can become a powerful tool in understanding cultural diversity across individual learners within classroom settings. The key point is that, just as culture is embrained, so also is cultural diversity. Each child's brain is neurobiologically diverse, moulded differently, largely through Hebbian plasticity. Within educational settings, those differences can have a direct impact on their thinking and learning, moment by moment.

The term 'neurodiversity' was first used to describe the range of variation across all human brains, especially in order to avoid deficit views of people with neurological differences (Baron-Cohen, 2017). Gerald Edelman, who you met in Chapter 2, said, 'each individual's brain is more like a unique rainforest, teeming with growth, decay, competition, diversity, and selection' (Edelman, cited in Armstrong, 2012, p. 12). If each brain can be viewed by analogy as a unique rainforest, perhaps a culturally diverse classroom or other learning environment can be viewed as a large ecosystem in which there are many unique rainforests, each of them having developed in distinctive climates and soils. We return to this idea in Chapter 9.

The use of the term 'neurodiversity' to describe cultural diversity within a classroom (Figure 7.3) provides teachers with a helpful tool that goes beyond the normal references to differences in home language, ethnicity and so on. Being neurodiverse, students have brains moulded by their culture, and may therefore have different perceptions and reactions to personal and interpersonal issues. In neurodiverse classrooms, students from diverse backgrounds may assign different values and take different approaches to the same learning task.

Figure 7.3 Cultural neurodiversity within a classroom

Culture and emotions

We can see how neurodiversity operates by noticing how people from different cultural backgrounds interpret and regulate emotional states. While avoiding the simplicities of essentialism, there are some generalisable differences between cultural groups, including different ideals and norms regarding how much one should display or regulate one's emotions (Matsumoto et al., 2008). These ideals and norms may operate as a form of folk theory, transmitted by wisdom dictums or proverbs, as shown below (Schouten et al., 2020).

> Unexpressed emotions will never die. They are buried alive and will come for the later in uglier ways. (Sigmund Freud)
>
> 秘すれば花（世阿弥・風姿花伝）'If hidden, it's elegant.'
>
> 顔で笑って心で泣く（背中で泣く）'Smiling face, crying heart' (Schouten et al., 2020)

From what might be called a 'Western' perspective, Sigmund Freud (1856–1939), the founder of psychoanalysis (which became very popular in the United States in the twentieth century), says that suppressing emotion is potentially harmful and that negative emotions should be *openly expressed*. By contrast, the two Japanese expressions represent a common sentiment found within East Asian countries that *expressing negative emotional states is not to be encouraged*. Such expression of emotions is often viewed as a sign of immaturity, naivety or a lack of social skills. From early in their development, in many East Asian settings, children are taught not to openly express negative feelings.

Neuroimaging studies help to show the extent to which these cultural norms are 'embrained'. Members of different cultural groups exhibit different brain imaging outcomes when asked to supress their emotions while engaging with an emotionally evocative stimulus. When European American adults have to suppress their negative emotions, controlling 'facial motor muscles in the presence of emotionally evocative film clips, [this] suppression [leads to] elevated responses in the amygdala and insula' (Goldin et al., 2008, p. 584) in fMRI images. By contrast, Japanese participants are capable of deactivating the amygdala when asked to suppress their emotions (Ohira et al., 2006), showing greater efficiency in emotion regulation.

Another study (Murata et al., 2013), using EEG recordings, found that people who were born in East Asian countries (China, Japan, Singapore, South Korea) and who spent their early childhood (minimum seven years) in the country of their birth, are much more efficient at suppressing their emotions than European Americans. When having to suppress their negative emotions while watching extremely negative pictures, EEG recordings of the two groups showed significant differences in what are called late positive potentials (LPP), which is an indication of sustained regulation of emotion. The Asian-born participants were much more efficient at suppressing their emotions.

Furthermore, people from different cultural groups tend to construe or appraise similar situations differently, leading to different emotional meanings. For example, the same negative experience, such as 'social exclusion', may bring about very different emotions (e.g. anger vs sadness) depending on the cultural groups to which they belong (Kimel et al., 2017).

The recognition that culture is embrained and that neurodiversity can be a powerful concept in understanding diversity across individual learners and within classroom settings can provide teachers with new ways to understand their culturally diverse students. It helps to explain why students from certain cultural backgrounds may not always want to express their feelings. It is not just a matter of social preference; it may well be deeply ingrained in their brain. Moreover, teachers need to be aware that different cultural norms regarding emotional expression can result in misunderstandings or miscommunications between members within a classroom. In addressing these kinds of cultural issues, teachers might find that introducing their students to how culture is embrained and to the notion of neurodiversity within the classroom will open up new ideas that their students will find helpful in working with their classmates.

Multilingualism and neurodiversity in the maths classroom

Van Rinsveld, A. V., Dricot, L., Guillaume, M., Rossion, B. & Schiltz, C. (2017). Mental arithmetic in the bilingual brain: Language matters. *Neuropsychologia*, 101, 17–29.

In Chapter 5, when addressing the relationship between literacy and numeracy, you were introduced to neuro-imaging studies showing that 'neural circuits underlying numerical and complex mathematical processing are largely distinct from neural circuits engaged during language' use (Ansari, 2016, p. 4887). The use of the word 'largely' implies that there are *some* aspects of mathematical processing that still remain under the influence of language, apart from comprehending the instructions given in maths classes. So which aspect of mathematical problem solving is influenced by one's language?

It is now widely acknowledged that multiplication tables that early primary students should master are actually a 'syntactically organized sequence(s) of words' (Dehaene & Cohen, 1995), although they are represented in numbers. Imagine how you mastered these tables in your own language – for example, simple multiplication problems such as 2 x 2 = 4 (Figure 7.4). These problems are memorised via repetition and rote verbal rehearsal ('Two

Figure 7.4 A child completing the multiplication table

times two equals four' in English, 二二得四 in Chinese, '*i i nun sa*' in Korean pronunciation); hence, 'fluent use of verbal retrieval strategies' of memorised facts is associated with faster and more accurate performance for these problems (Wicha et al., 2018, p. 145). In fact, converging neurological evidence support the role of language in multiplications. For example, while 'subtractions are possible in case of lesions within the language areas but not in case of parietal lesions, the inverse is observed for multiplications' (Van Rinsveld et al., 2017, p. 18).

With respect to addition, learning and practice of 'simple addition problems that are composed of two one-digit operands' (i.e. addition of numbers smaller than 10 such as 4 + 5 = 9) are stored as verbal memory in the form of 'arithmetic facts' (Ashcraft, 1992; Delazer & Benke, 1997; McCloskey, 1992; Van Rinsveld et al., 2017, p. 18), more complex addition problems 'require the execution of mental computations', rather than mere retrieval (Van Rinsveld et al., 2017, p. 18).

This then raises questions about linguistically diverse students who were introduced into arithmetic in their mother tongue, and then have to study mathematics in their added (or second) language. Will all children successfully transfer mathematical facts from one language to another language? If mathematics problems were presented in verbal or written form of new language ('eight times six equals what?', 'eight × six = ?'), would that slow down these students' performance – perhaps because they have to switch between two languages? In what ways would different proficiencies in two languages affect the speed and accuracy of arithmetic problem-solving? Studies addressing these kinds of questions are still in a 'nascent state' (Wicha et al., 2018, p. 166), although some light is beginning to be shed by studies using fMRI and EEG.

One very recent fMRI experiment (Van Rinsveld et al., 2017) studied adult participants whose first language of instruction was German, and the second language of instruction was French. All were bilingual, with high proficiency levels in both German and French, and were pursuing their degrees in a French-speaking Belgian university. Although they were highly proficient with French, these German bilinguals solved complex addition problems (e.g. fifty-six plus thirty-two equals?) faster and committed slightly fewer errors when solving problems in German than in French.

The fMRI images show that these participants used the same neural network that they relied on for German problem-solving but with additional engagement of visuo-spatial pathways when solving problems in French. According to Dehaene (1992; Dehaene et al., 2003, 2004), bilinguals rely on the visuo-spatial circuit to a greater extent 'when solving arithmetic problems in the language for which the verbal route might be more difficult to use' (Van Rinsveld et al., 2017, p. 28). In sum, the findings indicate that mathematics educators need to be aware of potential burdens that linguistically diverse students may have and devise strategies that might encourage these students to bridge the mathematical learnings from two different languages.

Mirror neurons and understanding the intentions of others

An important aspect of forming productive social relationships with others is the ability to sense what they are *thinking* and make judgements about their *intentions*. But how can we do that – where does that ability come from? It seems that the discovery of **mirror neurons** in the 1990s could provide an answer.

Discovery of mirror neurons

From time to time, science gets lucky. Discoveries are made accidentally when they were not the focus of research. For example, in medicine the 'accidental' discovery of x-rays by the German physicist Wilhelm Roentgen in 1895 immediately impacted medical diagnosis and therapy. The 'accidental' discovery of penicillin in 1928 by Scottish researcher Alexander Fleming, at St Mary's Hospital in London, revolutionised the treatment of infection. The discovery of mirror neurons in monkeys in 1992, at the University of Parma in Italy, was also accidental and it has subsequently had a large impact on brain research, not least because mirror neurons seem to be linked to our capacity for effective socialising.

Dr Giacomo Rizzolatti and his research colleagues at the University of Parma were studying the role of neurons in the brains of macaque monkeys when they were performing motor actions such as reaching for peanuts. Very thin electrodes were inserted into individual neurons in the brains of these monkeys, to record their firing patterns (their action potentials – see Chapter 2). On one occasion, they accidentally left the recording running in one monkey when feeding another. To their great surprise, even though that monkey wasn't reaching for food, his neurons fired as though he were, simply as a result of observing the other monkey being fed. His neurons were acting as a 'mirror', responding in the same way as if he were actually being fed himself – hence the term 'mirror' neurons. Subsequently it was discovered that mirroring only occurred when the other monkey was *intending* to grasp food, not when he was moving his arms randomly. The mirror neurons were not just mirroring actions, but *intentions*.

The researchers wrote up their findings and submitted their paper to the journal called *Nature*. However, it was rejected because the reviewers said it lacked interest. Given the enormous interested it received when published in a less prestigious journal, the verdict of *Nature* was clearly wrong. This is not the only time that peer review has rejected important novel findings. Another clear example was the ground-breaking work of Boris Belousov, recounted in Chapter 3. A further one was the seminal paper on epigenetics by Marcus Pembrey (introduced below), which 'had to be published in an obscure Italian journal because the major journals rejected it' (Sankey & Kim, 2013, p. 185).

Does mirror neuron research on monkeys have educational relevance?

Ferrari and Rizzolatti (2014, p. 1) claim there is now 'overwhelming evidence for the existence of mirror neurons in humans from hundreds of experiments using a variety of techniques'.

mirror neurons: neurons that fire when a person performs an action but also fire as a result of watching (mirroring) another person perform the same action. They are thought to play a role in imitation, speech and language, and are also associated with emotion and empathy.

Evidence also suggests that, like monkeys, mirror neurons in humans are also concerned with *intentions*. For example, some researchers have claimed that mirror neurons can discriminate between whether, in picking up a cup of tea, your intention is to drink it, or you are simply clearing the table (Iacoboni, 2005). If correct, this would suggest that mirror neurons may be of particular relevance to teachers and their daily experience of maintaining good classroom order. Perhaps this explains how experienced teachers are able to anticipate the behaviour of students through 'reading' their intentions. For example, when seeing a 15-year-old student pick up a pencil, they can discern whether the student intends to use it for drawing or to throw it at his mate across the classroom!

For newly qualified teachers, 'reading' the intentions of students may not come 'naturally'; they have to be acquired through practical classroom experience, which may suggest that mirroring is not genetically built into neurons. On the other hand, counter-evidence that it might be genetic comes from the ability of newborn babies to imitate and 'mirror' the funny faces pulled by adults, something that is well known to parents in their early bonding with babies. However, it could be that babies are actually training the neurons to *mirror what is perceived* when imitating adults. Some research has suggested that the sight of a protruding tongue is enough to trigger an imitative response in neonates and that this one gesture is enough to explain most if not all accounts of facial mimicry in babies (Anisfeld, 1996).

Ferrari and Rizzolatti (2014) initially found that the part of the brain in humans that corresponds to where the neurons were found in monkeys is Broca's area – which, as you will know from Chapter 5, is implicated in producing speech. This suggests that mirror neurons play a role in our social linguistic interactions with others. Subsequent research has shown, however, that mirroring capacities are found elsewhere in the brain, including areas associated with learning and social behaviours that are of interest to teachers. For example, they have been found to be implicated not only in imitation by infants, but also in older children and adults. Imitation does not just play an important role in social learning; it is also a key part of **socialised learning** approaches such as apprenticeships. Mirroring is also found in brain areas implicated in empathy, and the ability of humans, from around the age of five onwards, to 'read' what others may be thinking and feeling, which is called having a *theory of mind*.

Autism and the 'broken mirror' hypothesis

Based on the claim that the mirror neuron system is responsible for understanding the thoughts, feeling and actions of other people, and also imitating them (Yates & Hobson, 2020), some research has claimed that malfunctioning mirror neuron systems could explain the social deficits experienced by people diagnosed with autism spectrum condition (ASC). This claim has often been referred to as the **broken mirror hypothesis**, although that is misleading because, if mirror neuron systems are *a* cause, it is probably more to do with 'underdeveloped mirror neuron systems' (Geake, 2009, p. 110), rather than their being developed and then broken.

Within the literature, there certainly is research supporting the view that mirror neurons are implicated in autism, though not as a result of the broken mirror hypothesis, which is generally viewed as too simplistic. For example, a paper published in 2020 reviewed the

socialised learning: learning in social contexts. In opposition to the view that learning is 'socialisation' or 'internalisation', the learner is an active generator of theory arising out of their practice.

broken mirror hypothesis: a belief that malfunctioning mirror neuron systems could explain the social deficits experienced by people with autism spectrum condition

existing evidence and concluded that 'there is insufficient support for the broken mirror hypothesis' in its original form (Yates & Hobson, 2020), although it identified better, more sophisticated ways to understand the relationship between mirror neurons and autism.

In 2006, a team of researchers published a study conducted with a group of high-functioning autism spectrum disorder (ASD) adults and a carefully matched control group without ASD. This research found 'local decreases of grey matter in the ASD group in areas belonging to the mirror neuron system (MNS), argued to be the basis of empathic behaviour. Cortical thinning of the MNS was correlated with ASD symptom severity. Cortical thinning was also observed in areas involved in emotion recognition and social cognition' (Hadjikhani et al., 2006, p. 1276). In short, this research found that all the areas of the brain associated with mirror neurons were thinner in the ASD group than in the control group; moreover, the ASD participants with the most severe symptoms showed the greatest thinning.

Although this research had a small sample, it does suggest a link between autism and the mirror neuron system, which is marked by disabilities in social cognition, social skills, theory of mind and empathy. The thinning of the relevant brain areas suggests reduced capacity. There are implications in this and similar research for special education. It may well be that children who experience a range of social difficulties at school have a *physical* impairment in their mirror neuron system, which means it's not just a result of social factors as it is often assumed to be.

There is an important point here for teachers to note: displaying social symptoms does not necessarily imply social causes – the causes could be physical or at least partly physical. Indeed, as this chapter has argued, there is no need to drive a dichotomist wedge between the physical and the social because each continually impacts the other through dynamic feedback.

The social, cultural, political context of intelligence testing

Experience suggest that simply mentioning the notion of intelligence raises alarm bells for many teachers and parents, perhaps for very good reasons. The concept of intelligence is associated with **intelligence quotient (IQ)** testing and the *labelling of children* – some being accorded high IQ, some average IQ, while others have low IQ, allegedly indicating that they 'lack' intelligence. When coupled with the idea that IQ is largely inherited – that it is what you are born with and won't change significantly as a result of learning – it becomes depressing. The culprit is said to be our genes: we are born with them, they are inherited from our parents and, we have been told, they are our destiny – they make us who we are. But is this correct?

intelligence quotient (IQ): a total score derived from a set of standardised tests designed to assess intelligence

Educational and social dangers of intelligence testing

The belief that in any given society there are those who are 'bright' and those who are 'dim' (employing a light-bulb metaphor) because of their genetic inheritance has been used inhumanely to demarcate people from one another. It has also been used in education to demarcate children on the basis of alleged 'innate ability', as we will see shortly. Claims for

Figure 7.5 Francis Galton (1822–1911), the father of eugenics

eugenics: the 'science' of selective mating of those deemed to possess desirable genes, in order to advance humanity

genetic, or biological, determinism: the doctrine that what we inherit genetically plays a significant, if not total, role in determining human nature, personality and intelligence

the validity of IQ measures continues to be heard in the context of medicine, where they are used as a baseline for assessment – 'the patient's sight was impaired by the accident, but his IQ was preserved'.

Some teachers and parents will be very wary of the concept of intelligence, because they know it is deeply stained by some appalling episodes in history, including the **eugenics** movement in the United States, which was based on the views of Francis Galton (1822–1911), a cousin of Charles Darwin (Figure 7.5). Galton coined the term 'eugenics' (literally meaning 'well born') to advocate the selection of those who were deemed to possess high intelligence in order to advance humanity. This provided what was claimed to be 'scientific' justification for conducting forced sterilisation programs of those deemed to be of low intelligence – the so-called 'feeble-minded' – in the United States, some parts of Europe and ultimately Nazi Germany. In Australia in the 1920s and 1930s, Aboriginal and Torres Strait Islander peoples were subjected to medical experimentation, supposedly focused on their experience of pain but in fact underpinned by racist assumptions about intelligence. So what is going on here, all in the name of science?

Genetic or biological determinism

The short answer is that, one way or another, these abuses have been underpinned by **genetic, or biological, determinism**, which stands at the opposite end of the spectrum to social determinism, discussed earlier in this chapter. According to the American Psychological Association's *Dictionary of Psychology* (APA, 2020), genetic determinism is 'the doctrine that human and non-human animal behaviour and mental activity are largely (or completely) controlled by the genetic constitution of the individual and that responses to environmental influences are for the most part innately determined'. A closely related notion is hereditarianism, the belief that heredity (what we inherit genetically) plays a significant if not total role in determining human nature, personality and intelligence. Hereditarianism often goes further in claiming that this belief should be used to solve social and political problems.

From the perspective of dynamic systems theory, although genes play an important role in development, including acting as constraints, they are *not in control* as determinists insist: 'The scale and complexity of the mature animal far exceeds the scale of the DNA code … Thus, while guided by the genes, complexity of form must arise during development in self-organising fashion' (Thelen & Smith, 1994, p. 153). Rather than being rule-driven, as advocated by genetic determinism, each individual child's learning and development pathways are messy, fluid and context sensitive. However, please note that dynamic systems theorists are not advocating the other extreme of the nature/nurture dichotomy, 'where the organism is viewed as containing none of this information for its final destiny, but as absorbing structure and complexity from the order in the environment, through experience' (1994, p. xv). Nowadays, it is fair to say that most 'developmentalists at least pay lip service to the view that development is a function of the *interaction* between genetically determined processes and input from the environment' (1994, p. xv). However, from a dynamic systems perspective, even 'the commonly accepted interactionist position is inadequate to explain the grand sweep of developmental progress' (1994, p. xv). Development is emergent, self-organising and much more dynamic and multi-causal than interactionist theories generally envisage. Moreover, as a correction to the excesses of genetic determinism, there is also the new science of epigenetics.

Epigenetics: Social, cultural and environmental impacts on gene transmission

In November 2009, scientists celebrated 150 years since the publication of Darwin's *On the Origin of Species* (Darwin, 1965). Just two months later, in January 2010, *Time* magazine carried the banner headline 'Why Your DNA Isn't Your Destiny'.

The carefully written article said that the 'new field of epigenetics is showing how your environment and your choices can influence your genetic code – and that of your kids' (Cloud, 2010, p. 27). From a strictly Darwinian position, this begins to sound like heresy. It implies that parents can hand on genetic characteristics acquired from the physical and social environment. But that is Lamarckian theory, which Darwin's theory strongly opposes. The *Time* article came out a few months before the publication of David Shenk's (2010) book *The Genius in All of Us: Why Everything You Have Been Told About Genetics, Talent and IQ is Wrong*, in which he claims the new paradigm of epigenetics shows how bankrupt the notion of nature versus nurture really is.

The article in *Time* told the story of the ground-breaking discovery by health specialist Lars Olov Bygren that adults in his hometown of Norrbotten, in the cold far north of Sweden, had been *handing on genetic traits* to offspring, acquired in response to times of plenty when the harvest was good, and of times of crop failure and famine when children would experience starvation. When there was an over-abundant harvest, on the other hand, children could go from a normal diet to gluttony in a single season. By studying social data from the nineteenth century, he found that children who enjoyed over-abundant winters produced sons and grandsons who experienced significantly shorter lives. Somehow, he realised, these powerful environmental conditions were leaving an imprint on the genetic material in the ova and sperm of these children – but how?

In 2000, by chance, Bygren came across a paper published by Marcus Pembrey (University College, London) in an obscure Italian journal. In the paper, Pembrey had postulated that industrialisation might be shortening the excessively long course of evolution described by Darwin, where change occurs over countless generations and millions of years. What if it is occurring over a few generations? That wouldn't be enough to change the genes, but it might change the cellular material that sits on top of the genome and is thus referred to as epigenetic (the suffix *epi* is Greek and means 'above').

Pembrey had no evidence to support his thesis until, totally unexpectedly, he received a letter from Bygren (who he did not know) providing his data. The two became friends and set out to find additional data to support the hypothesis. Pembrey knew where to go. His friend Jean Golding at the University of Bristol in England was running a massive longitudinal study involving over 14,000 women who were pregnant over the 20 months recruiting period for the study. Working together, Bygren, Pembrey and Golding found that the data from that study were able to confirm the hypothesis. Their results were released in a joint paper published in the *European Journal of Genetics* in 2006 (Pembrey et al., 2006).

This story is interesting from the perspective of the history of science, but it is also important and relevant for teachers. In schools, *learning is a constructive epigenetic process* (genes responding to social and physical environmental switches) for each individual brain, working in unison with Hebbian repetition

Intelligence testing and scientific fraud in education

Cases of fraud in science are rare, but they do occur from time to time. Sir Cyril Burt (1883–1971) was Professor of Psychology at University College, London, and was the first psychologist to be knighted in Britain. He was also the first British psychologist to receive the distinguished Thorndike prize, awarded by the American Psychological Society. Burt was a genetic determinist and hereditarian, convinced that intelligence largely result from the child's genes and that environmental factors have only a relatively small influence. He also believed that the innate potential of a child could be reliably measured at age 11. Moreover, he believed that if you knew a child's innate ability, you could predict their future academic success. Furthermore, he believed this could be used to design an education system, tailoring it to children's 'innate potential' (Burt, 1921).

Burt claimed that, 'It has been the signal merit of the English school of psychology, from Sir Francis Galton onwards, that it has, by this very device of mathematical analysis, transformed the mental test from a discredited dodge of the charlatan into a recognized instrument of scientific precision' (Burt, 1921, p. 130). However, within two years of his death, it was becoming clear there was something terribly wrong. Burt had a mountain of belief, but his evidence was fraudulent – he was a cheat. Those like his former doctoral student Hans Eysenck, who wanted to believe that his research was valid, jumped to his defence (and some still do), suggesting that he didn't cheat – perhaps he was just careless in his methodology. But the general consensus is that his 'scientific' data was simply fabricated to prove what he believed.

Of course, none of this would matter to teachers and teacher educators if it were just a problem in the field of psychology. However, it wasn't: it massively impacted education. In England, Burt's writings on the heritability of intelligence contributed greatly to the establishment of the so-called 11+ system, in which children were graded by taking the 11+ examination. As the name implies, the exam was taken by all 11-year-old children: those who passed the exam (about 30 per cent) went to prestigious grammar schools to study academic subjects, those who failed (70 per cent) were sent to secondary modern schools to study supposedly practical subjects. Some of those who suffered this social and educational injustice are still waiting for psychology to issue 'an apology' (Sankey, 2011, p. 419) for stealing their educational opportunities, although without much hope.

REFLECTION

From the point of view of teachers and schooling, do we need a concept of general intelligence in addition to language, literacy, numeracy and all the academic subjects? Isn't it redundant and, if anything additional is needed, isn't it enhancing children's wisdom?

The reification of intelligence

In 1981, 10 years after the death of Burt, distinguished Harvard scholar Stephen Jay Gould (1941–2002) (Figure 7.6) provided a very detailed analysis of the Burt dilemma in his book *The Mismeasure of Man* (Gould, 1981). Gould, who we met in Chapter 5 when introducing his term 'exaptation', was especially known for his notion of punctuated equilibrium, the idea that evolutionary development occurs through sudden leaps and long intervening periods of stability. This challenges the *gradualist* (steady and even change) pillar of Darwinian theory, much to the annoyance of 'evangelical' Darwinists such as Richard Dawkins, who seem to believe Darwin is beyond criticism and, as Mary Midgley (1987) claims, treat Darwin's theory of 'evolution as a religion'.

Figure 7.6 Stephen Jay Gould

Gould (1981) deals with the accusations of fraud and their link to Burt's chosen methodology, called factor analysis. However, he also says that Burt's most important error was not the fraud, but the **reification of intelligence** – turning the abstract concept of intelligence into a concrete *thing*. There is no such *thing* as intelligence, as it isn't a thing. Intelligence is one of those words that works very well as an adjective, but not as a noun. There is no such thing as beauty, but many things can be described as beautiful. Similarly, it is entirely possible to describe people as intelligent and perhaps machines might even be thought to be intelligent, but intelligence is an abstract concept and doesn't exist as an independent entity, somewhere in the world or in the head, waiting to be tested.

Gould says the principal error of IQ is 'the notion that such a nebulous, socially defined concept as intelligence might be identified as a "thing" with a locus in the brain and a definite degree of heritability – and that it might be measured as a single number, thus permitting a unilinear ranking of people according to the amount of it they possess' (Gould, 1981, p. 239).

Reading those words, you certainly get the impression that Gould thinks there is something wrong with the concept of intelligence and testing it using IQ. You will not be surprised to learn that what he said upset a lot of people who had invested their careers in heritability studies. In some cases, the criticism also had a decidedly political flavour because of the long association in America of heritability and eugenics with right-wing causes.

Hans Eysenck, who had jumped to the defence of Burt, was especially dismissive of Gould, saying that he was a palaeontologist who didn't understand psychology. Eysenck had reason to dislike Gould. Gould (1981, p. 235) notes that 'Burt's supporters tended at first to view the charges of fraud as a thinly veiled *leftist plot* to undo the hereditarian position'. As an example, he reports that Eysenck had written to Burt's sister, saying, 'I think the whole affair

reification of intelligence: turning the abstract concept of intelligence into a concrete thing

is just a determined effort on the part of some very left-wing environmentalist determined to play a political game with scientific facts' (1981, p. 235).

For teachers, this unhappy episode in the history of science is a reminder that *science is a thoroughly human endeavour*, even when its focus is understanding the workings of the physical world, including the brain. Scientists are human beings and, in their passion to defend their cherished ideas (and their careers), there is always the possibility – even if remote – that they will cross acceptable social and ethical boundaries, and politics can certainly play a part.

As a footnote to this sorry story, many of Eysenck's publications are now also considered to be scientifically problematic. In 2019, a review conducted on behalf of King's College, London, where he had been Professor of Psychology, found that 26 of the papers that he co-authored with distinguished German medical sociologist Ronald Grossarth-Meticek were scientifically 'unsafe' (Boseley, 2019). Inquiries are currently continuing into the scientific reliability of Eysenck's other publications.

REFLECTION

Why do you think that heritability studies, including the study of intelligence and IQ testing, can have such strong political associations? Given what you have read in this chapter about 'epigenetics', do you think this might help resolve the politics of heritability?

The Flynn effect

In 1996, Gould produced a second edition of *The Mismeasure of Man* and, as expected, that too met with criticism. One of the critics of that edition was James Flynn (1934–2020), an American political philosopher who settled in New Zealand and taught at Otago University. Flynn argued that Gould had not really addressed some of the important issues around racial differences revealed by IQ testing and the so-called black–white IQ gap. However, unlike those critics who used IQ scores to defend hereditarian beliefs, Flynn's thesis was entirely different. While generally accepting the cross-cultural figures for average IQ scores that are often condemned as inherently racist, he nevertheless turned the whole argument on its head.

Whereas the hereditarians were using IQ scores to claim that they could be used to demarcate the intelligence of different racial groups, and that these differences were genetic, Flynn argued that they were not genetic. Moreover, in all countries that administered IQ tests throughout the twentieth century, average IQ scores had been *increasing* at a rate of about three points each decade. This is known as the **Flynn effect** (Flynn, 1984), but what is going on? Learning can increase, through education, but if intelligence is meant to be innate and therefore largely fixed or stable, how can it be increasing all around the world? Over evolutionary time, perhaps, but surely not over a short period of just a few decades. Could it be that IQ tests are actually measuring something, but it's not a thing called intelligence and it's NOT a measure of general intellectual ability as Hereditarians (and also Piaget) claimed?

Flynn effect: the tendency of IQ scores to change over time

Very briefly, as background, over the years there have been many studies of global differences in IQ, to which Flynn was responding. Many of these studies were brought together into a meta-analysis conducted by Richard Lynn (2006), published in *Race, Differences in Intelligence: An Evolutionary Perspective*. In his table of differences, East Asians are ranked at the top with an average IQ score of 105, then Europeans with 99 then all the way down to the Australian Aboriginal peoples with 62 and finally Sub-Saharan Kalahari Bushmen and Pygmies with 54.

This book caused a considerable stir, but Flynn had already provided an alternative to the hereditarian interpretation of such findings. Unlike many critics, Flynn accepts that there are differences between different social groupings – the scores represent something, but what, precisely? The answer to this enigma came when he focused on the kinds of questions asked in IQ tests. First, many are highly language-dependent; second, many questions focus on abstract pattern recognition, which is where the largest gains in IQ scores are found. So what has improved is not general intelligence, but rather an *abstract problem-solving ability* (Flynn, 2012) – just the kind of ability associated with the ever-accelerating growth of science and technology over the past century and more.

It is therefore little wonder that traditional peoples with their traditional languages and ways of communicating and problem-solving will score lowest on such tests. Significantly, it is also the reason why the *greatest increases in IQ scores are found in traditional peoples*, as they become increasingly familiar with the outputs of science-based technology. In short, IQ scores are a measure of *social evolution* – or, to put it slightly differently, IQ tests are indeed measuring something, but it is not intelligence: it is a measure of *adaptation to modernity*.

Aboriginal children's numerical thought without number words

RESEARCH LINK 7.3

Butterworth, B., Reeve, R., Reynolds, F. & Lloyd, D. (2008). Numerical thought with and without words: Evidence from Indigenous Australian children. *The Proceedings of National Academy of Sciences of the USA*, 105(35), 13179–84.

In Research link 7.2, we were introduced to the findings that, in cultural societies whose languages have vocabulary for numbers, learning and practice of simple addition (4 + 5 = 9) and multiplications (2 × 2 = 4) are stored as verbal memory. So what is happening to the children who are raised in societies whose languages possess very restricted number vocabularies? Can we conclude that children who do not possess the words for numbers lack numerical concepts? Or do these children understand numerical concepts and perform arithmetic without involving words?

A team of developmental scientists (Butterworth et al., 2008) from University College, London and the University of Melbourne attempted to answer these questions by studying Aboriginal children aged from four to seven years from two distinct communities in the Northern Territory (Figure 7.7). One group was from the edge of the Tanami Desert and they belonged to the Warlpiri language group. The Warlpiri language has a very limited range of words for numbers – for example, only those for 1, 2, few and many.

The other group of children was from the Anindilyakwa language group on Groote Eylandt in the Gulf of Carpentaria. Anindilyakwa language has numerals up to just twenty. The Warlpiri language is spoken by about 3000 people in Australia's Northern Territory,

Figure 7.7 The Northern Territory, where the two Aboriginal communities are located

while Anindilyakwa is spoken by about 1500 people. Participating Aboriginal children did not learn languages other than their own indigenous language.

The researchers assessed the performance of the Aboriginal children at four tasks. First, in order to assess 'memory for counters', the researcher placed objects (2, 3, 4, 5, 6, 8, or 9 objects randomly) on her mat and covered them with a cloth, then asked the children to make their mat like hers. Second, in order to assess 'cross-modal matching', children had to decide whether the number of tapping shown by the experiment was equal to the number of objects on the mat by saying 'yes' or 'no'. Third, in order to assess 'non-verbal addition', the researcher placed objects on her mat and covered them after four seconds, then showed another set of objects. After sliding the additional objects under the cloth, children were asked to make their mat like hers. In this manner, children had to provide answer to 2 + 1, 3 + 1, 4 + 1, 5 + 3, etc.

Finally, the 'sharing' task assessed children's ability to divide 6, 9, 7 and 10 items into three sets. The performance of the Aboriginal children was compared with that of English-speaking children from Melbourne aged from four to seven years. In all four tasks, the performance of the Aboriginal children was comparable with the numerical abilities of English-speaking children, and the Aboriginal children solved more addition problems correctly (see Figure 7.8).

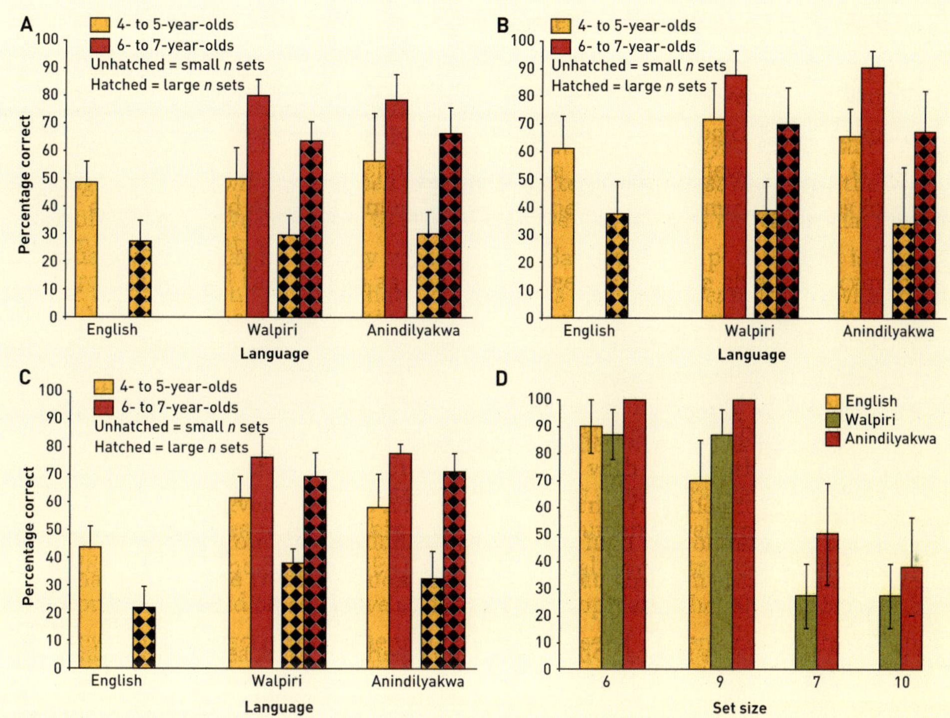

Figure 7.8 Comparison of performance in addition problems between Aboriginal children who speak Warlpiri or Anindilyakwa and English-speaking children
Source: Butterworth et al. (2008).

The team interpreted these results to argue that, whereas English-speaking children remember the word denoting the number of objects, Aboriginal children use an alternative strategy to remember the number of objects in a display by employing a visuo-spatial enumeration strategy (Butterworth et al., 2011). However, a later study conducted by the same team (Reeve et al., 2018) discovered that this visuo-spatial strategy is an important prerequisite for early arithmetic development in both English-speaking and Indigenous language-speaking children. The only difference they found was that, compared with English-speaking children, Anindilyakwa-speaking children needed slightly more time (a few seconds) to decide when comparing the numbers of two sets of objects.

STANDARDS   APST and ACECQA curriculum specifications

APST Standard 1.3 Students with diverse linguistic, cultural, religious and socioeconomic backgrounds
Demonstrate knowledge of teaching strategies that are responsive to the learning strengths and needs of students from diverse linguistic, cultural, religious and socioeconomic backgrounds

APST Standard 1.4 Strategies for teaching Aboriginal and Torres Strait Islander students
Demonstrate broad knowledge and understanding of the impact of culture, cultural identity and linguistic background on the education of students from Aboriginal and Torres Strait Islander backgrounds

APST Standard 2.5 Literacy and numeracy strategies
Know and understand literacy and numeracy teaching strategies and their application in teaching areas

ACECQA curriculum apecifications
1.3 Social and emotional development
1.6 Diversity, difference and inclusivity
2.2 Multicultural education
2.3 Aboriginal and Torres Strait Islander perspectives
2.4 Socially inclusive practice
2.5 Culture, diversity and inclusion
3.6 Working with children who speak language other than, or in addition to, English

This chapter has provided a frontline overview of recent research into the dynamic interaction of culture, biology and context in the process of learning and development. This is highly relevant in advancing newly qualifying teachers' knowledge and understanding of how to respond to learning strengths and needs of students from diverse linguistic and cultural backgrounds. Given the focus on the 'social' and 'cultural' context of learning and development, this chapter critically evaluated some of the classical tenets of socio-cultural theories of learning, including that of Lev Vygotsky. It then went beyond what is generally available to newly qualifying teachers with regard to cultural influences on learning by introducing the notions such as 'embrained culture' and 'neurodiversity'. In doing so, it adds considerably to their understanding of how students' cultural and linguistic backgrounds have implications for the teaching of specific subjects – for example, mathematics. It also provides them with an understanding of emotional diversity, drawn from neuroscientific research on emotion regulation and perception.

The chapter has provided a solid grounding for newly qualifying teachers to address two long-standing contentious issues: intelligence testing and biological determinism. It also introduced findings on mirror neurons, which are implicated in disabilities in social cognition, social skills, theory of mind and empathy. Overall, the messages in this chapter encourage teachers to embrace human diversities in the classroom and to provide carefully designed teaching strategies that are socially and culturally inclusive.

SUMMARY

- From the moment each child is born and even in the womb, children's biological brains are already embedded in dynamic and constantly shifting, physical, social and cultural contexts.
- From a science of learning and development perspective, the idea that we somehow grow out of our biology and into culture, out of the physical, natural world into the world of society, simply establishes a false and totally unnecessary dichotomy.
- Advocating a false dichotomy between our biological and social/cultural selves can lead to a form of social determinism in scholarship that systematically excludes biological explanations from its account of human behaviour. Arguably, what is needed in education is a notion of socialised learning.
- The term 'culture' is often used as an essentialist categorisation of entire groups with labels based on racial and ethnic membership. However, this ignores a vast array of national, social and cultural differences within these putative essentialist groupings.
- Brains create and recreate cultures, and cultures physically shape and reshape brains. Culture may therefore be said to be *embrained*, and in the context of education, schools, classrooms and other learning environments are shaped by *neurodiversity*.
- Recent and ongoing research suggests that socialisation is aided in the brain by mirror neurons that enable us to read the intentions of others. It is suggested that an underdevelopment of mirror neurons may be implicated in autism, although it is highly unlikely to be the single cause of autism.
- Genes are turned on or off by environmental factors, so 'nature' is not the opposite of 'nurture' as the traditional nature/nurture dichotomy insists. Teachers need to be aware that although genes are certainly implicated in learning and development, they are not fixed and are not a child's destiny.
- Genetic or biological determinism, which stands at the opposite end of the spectrum from social determinism and its closely related notion hereditarianism, asserts that what we inherit genetically plays a significant if not total role in determining human nature, personality and intelligence.
- Within education, the concept of intelligence has been viewed quite widely as a misapplication of science in understanding the social and cultural brain, partly because it has been used to falsely label children on the basis of assumed genetic differences in ability.
- Hereditarians have used IQ scores to claim that they can be used to demarcate the intelligence of different racial groups, and that these differences are genetic. James Flynn, however, argued they were not genetic. Moreover, in all countries that administered IQ tests throughout the twentieth century, average IQ scores increased at a rate of about three points each decade. This is called the Flynn effect.

KEY POINTS FOR TEACHERS

- Research shows that one very important aspect of the socially and culturally embedded brain is that children are *relational* (needing relationships for their survival and wellbeing). From the moment each child is born, and even when still in the womb, their brain is *relationally* embedded in dynamic and constantly shifting physical, social and cultural contexts.

- Dynamic systems theorist Alan Fogel points out that, when we say human beings are fundamentally relational, we are also admitting that, as individuals, we are inherently incomplete. We find ourselves in others (family, friends, teachers) and become who we are through others.

- Teachers should be aware that the idea of the child as *moulded through socialisation* invokes a very passive view of childhood, in which the rules and conventions of society are passed down to them as they grow older from above, by adults. On the contrary, Ingold insists that children are agents with purposes and perspectives of their own, and every child is immersed in the society and culture into which they are born, from the very beginning.

- From the perspective of brain science, our brains and those of many other animals are both inherently social and also inherently biological. Within cultural psychology, it is now realised that it is not possible to demarcate the domain of culture from biology. However, this is contrary to the core claims of social determinism, sometimes encountered in educational theory and research.

- Teachers should be careful about using 'essentialist' labels when thinking about the children they teach, basing their categories on some basic *essence* to define and demarcate students, one from the other – perhaps geographical (e.g. East Asian, Western), or based on religion or ethnicity.

- Cultures are *embrained*, meaning that recurrent, active and long-term engagement in cultural practices can powerfully shape and modify brain pathways. Armed with the idea that culture is embrained, the notion of *neurodiversity* can become a powerful tool in understanding cultural diversity across individual learners within classroom settings.

- Recognition of neurodiversity in the classroom and the idea that culture is embrained provide teachers with new ways to understand their culturally diverse students – for example, why students from certain cultural backgrounds may not always want to express their feelings.

- From the perspective of dynamic systems theory, rather than rule-driven genetic determinism, each individual child's learning and development pathways are messy, fluid and context sensitive. Although genes play an important role in development, including acting as constraints, they are not in *control*, as determinists insist.

- Discovery of the epigenetic process, whereby genes are turned on or off by environmental factors, is important and relevant for teachers. In schools, for each individual brain, *learning is a constructive epigenetic process* (genes responding to social and physical environmental switches), working in unison with Hebbian repetition.

- Teachers need to be aware that while it is entirely possible to describe people as intelligent and perhaps machines might even be thought to be intelligent, intelligence is an abstract concept and doesn't exist as an independent entity, somewhere in the world or in the head, waiting to be tested.

- Research conducted by James Flynn in the 1980s accepted that the differences in IQ scores between different social groupings represent something, but it is not intelligence;

rather, it is abstract problem-solving ability, or what he called an adaptation to modernity. That is why people in traditional cultures have the lowest IQ scores, and it is also why, as they increasingly engage with modernity and technology, their IQ scores are increasing the fastest.

REVIEW QUESTIONS

1. What are the main tenets of the classical view of 'socialisation' in human development? Why are they problematic?
2. Why is it problematic to use the term 'culture' as an 'essentialist categorisation'?
3. What does it mean by saying that 'culture is embrained'? Can you give one example that illustrates such a process?
4. What is the broken mirror hypothesis of autism and why is it considered scientifically suspect?
5. What problem does Stephen Jay Gould refer to in using the term 'the reification of intelligence'?

Guided responses

FOOD FOR THOUGHT

1. Within education, one very important implication of the socially and culturally embedded brain is that children are *relational* (needing relationships for their survival and wellbeing). However, would you agree that in much of teacher education and in many schools, issues of student relationships and wellbeing are addressed at a social level of analysis and very little attention is given to what might be occurring in children's brains? If you agree, are there ways in which this might be improved?
2. Culture is not a fixed entity; rather, cultures are constantly in flux, and cultural ideas and practices are constantly negotiated, manipulated and arbitrated by all those who are members of any given cultural community. Do you think this has any implications for how teachers understand culture in the context of classroom practice, and do you find the notion that culture is *embrained* and the *neurodiversity* of classrooms helpful in this regard?
3. What messages do you think arise from the Flynn effect, for teachers, schools and education policy-makers? For example, in arguing that IQ scores are a measure of social evolution and the adaptation to modernity, it is also suggesting that education plays a key role in any given society, and without that important factor, the gap between classes, races and genders will not be closed.

RESEARCH ACTIVITY: SUPPORTING STUDENTS FROM CULTURALLY AND LINGUISTICALLY DIVERSE BACKGROUNDS

Figure 7.9 provides comparisons of academic performance in science at school between immigrant and non-immigrant students across various countries that participated in the

Program for International Student Assessment (PISA) in Year 2015. PISA, coordinated by the Organisation for Economic Cooperation and Development (OECD), assesses 15-year-old students' academic performance (i.e. reading, mathematics, and science literacy) every three years. The PISA data from Year 2015 show that

> in most countries, both first- and second-generation immigrant students tend to perform worse than students without an immigrant background. The average science performance of foreign-born students whose parents were also born outside the host country is 447 score points, about half a standard deviation below the mean performance of non-immigrant students (500 score points), on average across OECD countries. Second-generation immigrant students perform between the two, with an average science score of 469 points. (OECD, 2016, p. 248)

Of course, there are exceptions. In some countries, such as Singapore, Macao, the United Arab Emirates and Qatar, immigrant students outperform their non-immigrant peers.

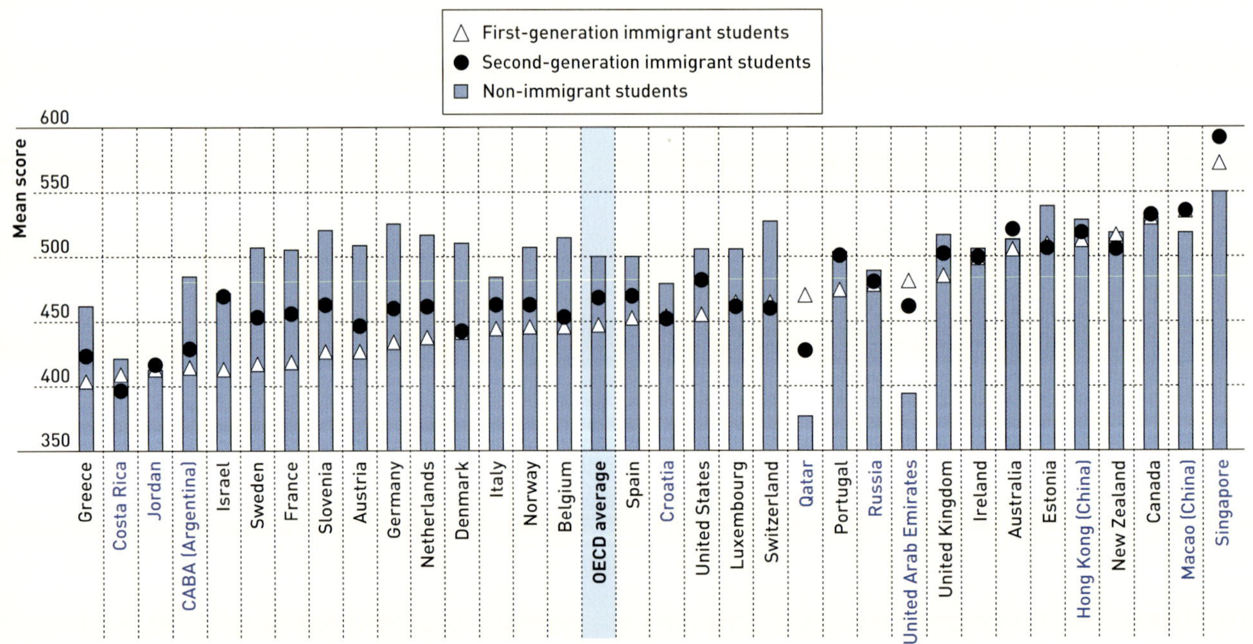

Figure 7.9 Student performance in science, by immigrant background
Source: OECD (2016, Figure 1.7.4).

Suggested resources

Moore, E., Evnitskaya, N. & Ramos-de Robles, S. L. (2018). Teaching and learning science in linguistically diverse classrooms. *Cultural Studies of Science Education*, 13(2), 341–52.

Ryu, M. (2015). Positionings of racial, ethnic, and linguistic minority students in high school biology class: Implications for science education in diverse classrooms. *Journal of Research in Science Teaching*, 52(3), 347–70.

Saravia-Shore, M. (2008). Diverse teaching strategies for diverse learners. In R. W. Cole (ed.), *Educating everybody's children: Diverse teaching strategies for diverse learners* (2nd ed.). Alexandria, VA: Association for Supervision and Curriculum Development, pp. 41–97.

Research activity links

Questions

1. Why do you think immigrant students in some countries perform *worse* in science than students without an immigrant background?
2. Why do you think in some countries students with an immigrant background perform *better* in science than students without an immigrant background?
3. What do you think schools and teachers should do in order to provide students from culturally and linguistically diverse (CALD) backgrounds with an inclusive science learning experience?

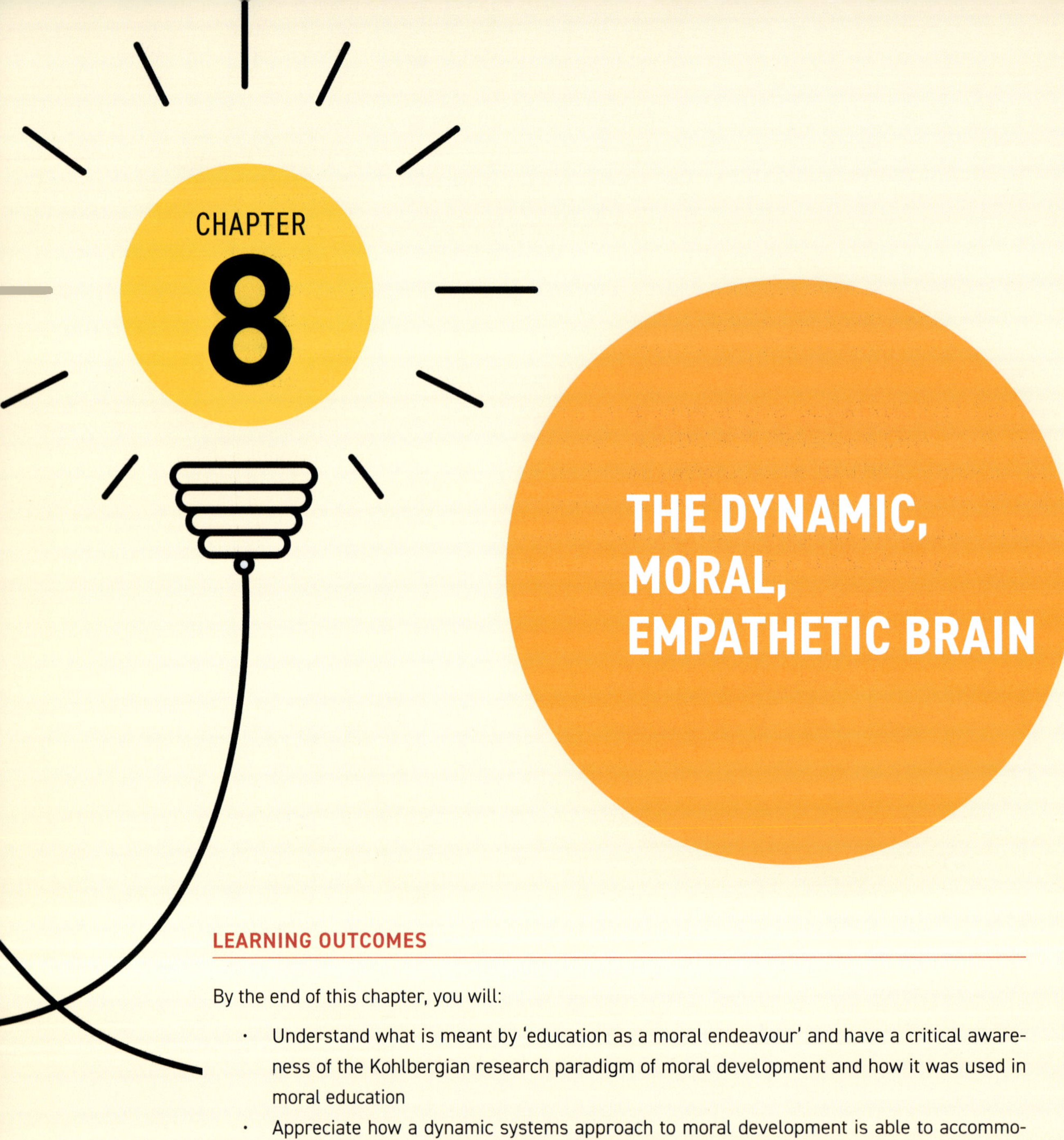

LEARNING OUTCOMES

By the end of this chapter, you will:

- Understand what is meant by 'education as a moral endeavour' and have a critical awareness of the Kohlbergian research paradigm of moral development and how it was used in moral education
- Appreciate how a dynamic systems approach to moral development is able to accommodate, but also go beyond, the Kohlbergian developmental paradigm
- Understand the possible interrelationship between empathy and moral thought and action, and what is meant by the three gradations of empathy
- Possess a basic knowledge of the neurobiology of trust and care and the neural processes involved in moral judgement and moral decision-making

Moral reasoning and moral development in education

There would seem to be a very close relationship between education and **moral values**. For example, most schools in Australia are keen to stress the *values* that they uphold; they post them up on notice boards and they publish them on their school website. In Asian countries, moral education is a compulsory subject on the school curriculum, as it is in some other countries, although it may go under other titles such as character education or personal and social education. In China, it is closely aligned with patriotic education, although it may also be concerned with 'instilling' moral values (Figure 8.1). In many 'Western' countries, moral education is not taught *explicitly*, as a discrete subject on the curriculum; instead, it is said to be taught *implicitly*, through the values the school upholds and role-modelling by teachers.

moral values: 'guiding principles, beliefs, sensitivities, held and displayed by individuals or groups in respect to how they relate to and deal with others and the world' (Sankey & Kim, 2016)

Figure 8.1 Children in a prestigious school in Shanghai, China having a moral education lesson, in English, on care

Education as a moral endeavour

It is also frequently claimed that education is itself a moral endeavour, or, alternatively, 'a moral practice' (Pring, 2001). The aims and purposes of education are said to be fundamentally moral, concerned with the enhancement of each individual child as a member of society. Education, by its very nature, seeks what is *good,* and *intrinsically worthwhile* in attending to the development of children's minds (brains) and their mental and physical wellbeing. Indeed, there 'is a strong consensus in the educational literature that teaching is an inherently moral endeavour, and that the moral work of teachers is of central importance to education' (Sanger, 2008, p. 169).

Figure 8.2 A school's outdoor board presenting the school's values

This suggests that teacher education should help beginning teachers to carefully and systematically appreciate the moral and ethical dimensions of education in preparation for their own school and classroom practice (Figure 8.2). Certainly, some teacher education courses do address these issues, and some incorporate it into students' assignments – perhaps as part of a child and adolescent development course; however, evidence suggests that at best it is patchy. A major reason is that the accreditation criteria for teacher education in many countries often do not require it. For instance, in Australia, a high-quality, innovative government-funded Australian Values Education Program (AVEP) that ran between 2003 and 2010 produced a National Framework for Values Education in Australian Schools (DEST, 2005). Yet currently there is no recognition in the Australian Professional Standards for Teachers that teaching is a moral practice. In fact, the word 'moral' does not get a mention and the only mention of 'ethics' is in regard to the use of information technology in classrooms. To say the least, 'There is a troubling disconnect between contemporary scholarship on the moral nature of teaching practice, and the practice of teacher education' (Sanger, 2008, p. 169). So what is the problem?

Thinking back over your own school experience, were you aware that your school espoused a set of values? Did values implicitly or explicitly influenced how the school was run, how teachers responded to students and how the school viewed education?

Moral values are embrained

Arguably, part of the problem is that moral values are thought to be mainly, if not entirely, social and cultural. That can easily translate into **moral relativism**, the belief that moral values are personal. There is no touchstone to ground morality, it is just personal opinion – your truth versus my truth. In education, this can lead to the view that moral learning and development can safely be left to the preferences of schools and governments. Moral values are indeed manifested socially and culturally but, like everything social and cultural, including our emotions and our ability to empathise with others, moral values are also *embrained* (Kim et al., 2021), the product of neuronal processes in the brain (Sankey, 2006; Sankey & Kim, 2016). Moreover, like all human development, their development is the product of emergent processes occurring in the dynamic, self-organising brain (Kim & Sankey, 2009, 2010).

moral relativism: the belief that moral beliefs are personal and there is no touchstone to ground morality – it's just personal opinion, your truth versus my truth

However, that is not how moral development has conventionally been understood in education. As with so much other development theory, moral development has been conceived as progressing though distinct *stages*. This is especially the case in the **Kohlbergian paradigm of moral development,** built on the pioneering work of Lawrence Kohlberg (1927–1987) and his colleagues. Kohlberg was strongly influenced by Piaget, and his theory of moral development can be viewed as an extension of Piaget's approach. Kohlberg was also opposed to moral relativism. We begin this chapter by considering the Kohlbergian paradigm of moral learning and development because it strongly influenced research and the teaching of values in schools during the second half of the twentieth century and the first decade of this century.

Kohlbergian paradigm of moral development: cognitive developmental perspective of moral development that comprises six distinctive, invariant stages

Moral development and the problem of moral relativism

Kohlberg 'was profoundly moved by the horrors of the Holocaust ... and viewed his own theory as a response to the Nazi ethos' (Lapsley, 1996, p. 41). How could such a thing happen, how could it ever be defended, given that it seemed to be so utterly immoral? However, when he looked to academia, he could find few resources by which to condemn the Nazi persecution; instead, he found social and moral relativism. Kohlberg became convinced that he needed to find a strong argument against moral relativism. For that, he believed, he had to focus on **moral reasoning** and how it develops over time. He also needed a methodology, and for that he used moral dilemma scenarios. For example, he told a story about Heinz known as the Heinz dilemma (Kohlberg, 1981):

moral reasoning: the process by which individuals try to determine the difference between what is right and what is wrong

> In Europe, a woman was near death from a rare form of cancer. There was one drug that the doctors thought might save her, a form of radium that a chemist in the same town had recently discovered. However, he was charging ten times what the drug cost him to make.
>
> The sick woman's husband, Heinz, went to everyone he knew to borrow the money, but he could only get together about half of what the drug cost. He told the druggist that his wife was dying and asked him to sell it cheaper or let him pay later. But the druggist said no. So Heinz got desperate and broke into the man's store to steal the drug for his wife. (Kohlberg, 1981)

REFLECTION

So here is Heinz's dilemma. Before reading any further, pause to consider whether you think Heinz is right to steal the drug and why. The reasons you give are what Kohlberg is really interested in. If possible, try to put yourself empathetically into the position of Heinz, then discuss his dilemma with three or four classmates.

Kohlberg's developmental stages and levels

When Kohlberg (1981) exposed children, adolescents and adults to this kind of dilemma, he found a clear pattern emerging. This formed the basis of his six stages of moral reasoning development, which he subdivided into three distinct levels that he called *pre-conventional,*

conventional and *post-conventional*. Examine Table 8.1 and look particularly at the right-hand column, which provides the *reasoning a*ssociated with each 'stage' – that is Kohlberg's focus. Whether the respondent believes Heinz was right or wrong is not the main issue; it is the reasons *why* that matter to Kohlberg.

Table 8.1 Kohlberg's (1981) six stages of development, operating at three distinct levels

Pre-conventional reasoning level		
Stage 1	Heteronomous, egocentric morality	What is right or wrong is defined by those who have power. Moral transgressions are punished because they are immoral, and they are immoral because they are punished.
Stage 2	Individualism and tit-for-tat reciprocation	The child now sees it's important to get on with other individuals, who have their own needs. It is good to reciprocate – if you scratch my back; I will scratch yours.
Conventional reasoning level		
Stage 3	Interpersonal and mutual expectations morality	The limited tit-for-tat reciprocity of stage 2 gives way to an interpersonal, relational duty of mutual trust and loyalty, to live up to the expectations of others close to me.
Stage 4	Social system and society-maintained morality	Interpersonal duties give way to the duty we owe to society at large. The social system provides laws that are applied impartially and regulate what is morally right.
Post-conventional reasoning level		
Stage 5	Human rights and social welfare morality	What is legal may not be what is moral. The moral individual identifies liberty and human dignity as universal values that transcend any given society or culture and permit cross-cultural moral discourse.
Stage 6	Universal ethical principles	There are a number of noble ethical ideals or principles that one should always take towards all others.

Although all six stages and three levels are concerned with how people reason when faced with moral issues, such as the Heinz dilemma, Kohlberg believed that only *post-conventional reasoning* actually constitutes *moral reasoning*; the pre-conventional and conventional levels fall short of what is meant (or should be meant) by *moral* reasoning.

This distinction basically comprised Kohlberg's response to the Nazi ethos. Hitler and his Nazi associates were acting and thinking legally (stage 4) – after all, they made the laws! But Kohlberg believed they were morally bankrupt. Stage 4 reasoning is simply conventional, following what is generally, socially, politically accepted or decreed. Action may well be legal, but that doesn't make it moral. To be truly moral, Kohlberg believed, human reasoning has to make the leap into *post-conventional* (stage 5) reasoning, where liberty, human dignity, universal values and rights, and a concern for the welfare of others provide the touchstone of moral reasoning.

The stark difference between the legal and the moral, identified by Kohlberg, is explored in *Les Misérables*, the nineteenth-century novel by French author Victor Hugo (1802–1885). It is the story of a convict, Jean Valjean, and a police inspector called Javert, who is determined to ensure that Jean Valjean follows *the rule of law*. Valjean has suffered a brutal life as a convict, including the brutality of Javert, but after 19 years of penal servitude he has been granted parole. He is homeless and hungry when he is shown great kindness and understanding. This experience transforms him into a man of high moral sensitivity. He becomes rich, owning a factory that he strives to run on moral principles. He is so respected that he is elected mayor of his city. But all the time, Javert is trying to track him down. Valjean has not reported to the authorities, so has *broken the law* a second time. You may have read the book, or seen the movie or the highly successful stage musical. If not, please do. You will see what Kohlberg means by the conventional/post-conventional divide.

In the context of education, teachers often establish classroom rules, but perhaps they might spend a little time negotiating a set of classroom *values* – respect for others, care, honesty and so on. They then post them up on the classroom wall. Instead of *violating an imposed rule*, the child who offends will thus have *disregarded an agreed value* – 'we've agreed to respect one another, haven't we? Let's keep to our values and be respectful'. In adopting such a strategy, the teacher is starting to role-model post-conventional thinking – going *beyond* the rules.

REFLECTION

Do you think there are social/political contexts today where the Kohlbergian distinction between acting morally as opposed to acting in obedience to the law is especially relevant and, if so, might this have political implications nationally and/or globally?

The under-determination of theory by evidence

Kohlberg's account of moral development produced a research paradigm that lasted more than half a century. Time and again, it was shown to be correct – the evidence was overwhelming. As teachers, you could actually try it out in your own school, taking a sample across various age-groups, especially if your school encompasses K–12. If you abide by the Kohlbergian methodology, you will very likely find it works. But, despite adding even more evidence in its favour, this 'does not show … that Kohlberg's theory provides a correct and comprehensive account of moral development' (Moshman, 2011, p. 75). The same point was made in Chapter 6 regarding the extensive research evidence supporting the notion that emotions can be universally recognised in facial expressions. So much evidence, yet still problematic.

The core problem is what philosophers of science call the **under-determination of theory by evidence**, which means that the observational evidence used to support any hypothesis or theory will often (some say always) be less than required and, moreover, it will always be compatible with more than just that theory.

A lot depends on how the research providing the evidence was conducted. If, as in the case of the Kohlbergian research paradigm, research is done a certain kind of way, using certain kinds of methodological rules or algorithms, you will produce a particular kind of data. In other words, if you conduct research the way Kohlberg conducted it, you will get stages; however, if you conduct research differently, you may well not.

under-determination of theory by evidence: the observational evidence used to support any hypothesis or theory will often (some say always) be less than required and, moreover, it will always be compatible with more than just that theory

This does not mean that evidence simply can't be trusted. The point of questioning evidence in science is to demand the best evidence available. When presented with 'evidence', we should always consider whether it is sufficiently reliable and robust, and whether it could be interpreted differently. Now let us put this advice into practice in Research link 8.1.

RESEARCH LINK 8.1

People's perceptions of whether theism (belief in God) has an important relationship to morality – thinking and acting morally

Tamir, C., Connaughton, A. & Salazar, A. M. (2020). *The global God divide: People's thoughts on whether belief in God is necessary to be moral vary by economic development, education and age*. Washington, DC: Pew Research Center.

It is often claimed that there is a close link between belief in God and morality. The core idea is that religious belief often has a strong moral content – for example, Judaism and the Ten Commandments.

In the West, however, there has been a strong philosophical argument that if you act in order to obey God's will or follow God's commandment, you are not acting morally, but religiously. To act morally, one must have a moral reason, not a religious reason. Moreover, historically and currently, many clearly immoral acts are committed in the name of God.

So the research question is: Do you think there is a *connection* between belief in God and morality (moral thought and action)? Of course, whether or not you have a religious faith may well influence your opinion of whether religion and morality are closely interconnected. However, in 2020 the Pew Research Centre in the United States published findings from a major survey, across 34 countries on six continents, totalling 38,426 respondents, to see what people thought. Participants were asked to answer the following question: 'Which of the following statements comes closest to your opinion? It is NOT necessary to believe in God in order to be moral and have good values OR It is necessary to believe in God in order to be moral and have good values?'

A summary shows that 'a median of 45% say it is necessary to believe in God to be moral and have good moral values' (Tamir et al., 2020, p. 4) (Figure 8.3). Survey participants from emerging and developing economies were more likely to have a positive view of religious belief and to consider it important in their lives. They were also 'more likely than people in this survey who live in advanced economies to say that belief in God is necessary to be moral' (Pew Research Center, 2020, p. 4). Moreover, most countries surveyed displayed a 'generational gap' (Tamir et al., 2020, p. 10), with those aged 50 years or older significantly more likely to think that belief in God was necessary for morality than those aged from 18 to 29.

Now, following the advice given about how to view research 'evidence', check the data supplied and carefully consider whether you think the research question is sufficiently robust to yield data that is clearly informative. This research question asks people to state whether or not it is *necessary* to believe in God in order to be moral and have good values? However, notice that the responses are then used to state whether there is a *perceived link*

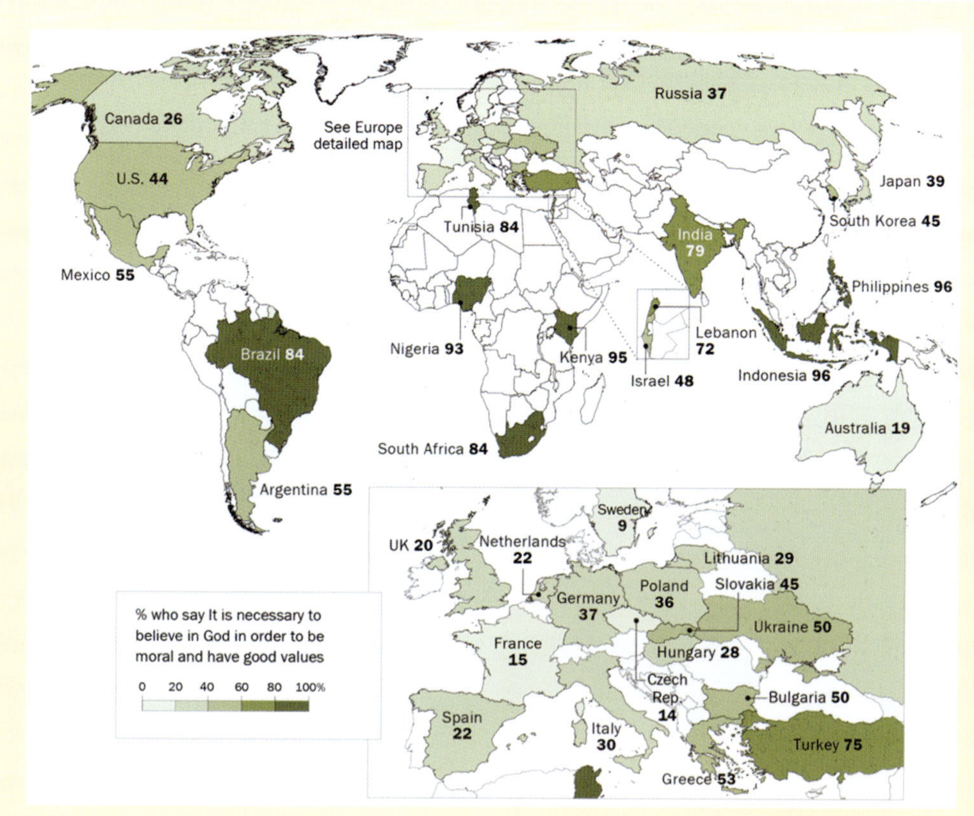

Figure 8.3 Percentage of people who say it is necessary to believe in God to be moral
Source: Tamir et al. (2020).

between religion and morality – but surely that is a different issue. For example, it might not be logically *necessary* to believe in God in order to be moral and have values, but it might nevertheless make a very big difference to how people think and act.

In other words, to answer the earlier research question you were asked, there is a *connection* between belief in God and morality (moral thought and action), although it is not a *necessary* connection.

What do you think?

For this research link, you are invited to look carefully at the results of this research by the Pew Research Center (2020) named *The Global God Divide*.

Research link 8.1

Towards a dynamic systems approach to moral development

Over the years, many refinements were made to the original Kohlbergian paradigm. A major methodological change was made by James Rest, one of Kohlberg's students. Kohlberg's own research was based on interview. Rest and his colleagues at the University of Minnesota, who referred to themselves as Neo-Kohlbergians (Rest, 1974; Rest et al., 1999), devised what they called the *Defining Issues Test* (DIT), which aimed to provide what they believed was a more objective measure of moral reasoning. They also attempted to extend the theoretical foundations of the Kohlbergian paradigm by designing what they called the *Four Components Model* of moral development – comprising moral *awareness*, moral *judgement*, moral *intention* and moral *action*.

However, the tide was beginning to turn. By 2008, it was becoming increasingly clear that the Kohlbergian model 'now looks a bit shop worn' and 'there is increasing recognition that the field of moral development is now at an important crossroads' (Lapsley & Hill, 2008, p. 284). The paper just cited was published in the *Journal of Moral Education* (*JME*), which, under the editorship of Monica Taylor, had made a major contribution to the field of moral education, for some 30-plus years (Figure 8.4). The journal was strongly aligned with the *Association of Moral Education* (AME), and many of the AME members were closely associated with the Kohlbergian tradition. In September 2008, the *JME* published a special edition in recognition of half a century of Kohlberg's academic work. One year later, in September 2009, a number of papers were published in response to the 2008 special edition.

Figure 8.4 The September 2008 and September 2009 editions of the *Journal of Moral Education*

One argued that moral development (as well as moral education) requires a major paradigm shift, from the Kohlbergian paradigm to a dynamic systems paradigm (Kim & Sankey, 2009). Changing paradigms does not necessarily mean completely obliterating what has gone before: elements of the old paradigm may well be retained in the new. As the paper notes, 'a previous paradigm may retain validity within certain parameters. Though replaced by relativity theory, Newtonian mechanics was nevertheless used to successfully navigate the stricken Apollo 13 back to earth' (2009, p. 285). Also, a dynamic systems paradigm is able to incorporate Kohlberg's identification of different kinds of moral reasoning while abandoning the idea that they occur in progressive stages, from one to the other, onwards and upwards.

REFLECTION

Given that the kind of moral reasoning Kohlberg identifies as Stage 5 requires viewing liberty and human dignity as universal values that transcend any given society or culture, might it be discouraged or even opposed in some cultural or national settings? If so, why?

From a dynamic systems perspective, 'any account of moral development should at least be consistent with our status as biological and neurobiological organisms' (Kim & Sankey, 2009, p. 289); moreover, every account of human development necessarily incorporates brain development. However, Kohlberg and the majority of the Neo-Kohlbergians ignored the biology of development, including brain biology, though there were a few exceptions. There are other important changes of perspective when adopting a dynamic systems approach:

- *Bringing moral thought and action together.* Kohlberg's research methodology placed an enormous emphasis on moral reasoning, but not enough focus on moral behaviour or moral action. From a dynamic systems perspective, however, thought and action are intimately entwined and not easily pulled apart.

- *Recognising that moral decision-making and action are not always a direct result of rational judgement, as Kohlberg assumed.* Reasoning often occurs *post-hoc*, after the action. It is often a story we construct to justify what we did, but it is not the actual reasons or causes, which may be entirely subconscious (Sankey, 2011, p. 419). This is important for teachers. Don't assume that a child always knows the *reasons* why what they did was wrong. And don't be surprised when, if you ask for reasons, you are given a made-up story. We all do it, all the time.

- *Emphasising the role of emotion and **empathy** in moral thought and action, which Kohlberg largely overlooked.* We will come to these four issues later in this chapter.

empathy: the ability to understand and share the feelings of another person

A key assumption of a dynamic systems approach is that 'moral development shares the same dynamic processes found within the whole of human development' (Kim & Sankey, 2009, p. 284). Thelen and Smith (1994), in their seminal work introducing dynamic systems to development, focused on cognitive and motor development. The 2009 *JME* paper by Kim and Sankey is using exactly the same principles as those of Thelen and Smith, but applying them to moral development. Thus, just as cognitive and motor development is inherently 'messy', 'exploratory, opportunistic, syncretic' and 'context sensitive' (Thelen & Smith, 1994, p. xvi), so too is moral development.

Likewise, human development, including moral development, is not pre-programmed, but instead is self-organising and the developmental pathways each individual child follows emerge from multiple interacting factors and causes. This means that each child's moral developmental pathway is different because each child is a product of the *multiple interacting factors and causes* that have physically crafted their brain's neuronal connections and pathways.

The rudiments of morality – a predilection to value

In presenting a dynamics systems account of moral development, Kim and Sankey (2009) argue that the rudiments of morality reside in what it called a **predilection to value** common to all species. In making this claim, a distinction is being drawn between moral norms (rules) that are clearly a human, social, cultural construct and moral values that have ancient evolutionary origins. The argument is that we are born with a *value bias*, as are all other

predilection to value: an in-born value bias within organisms, a fundamental ability to discriminate what is better and what is worse, what is beneficial and what is harmful

creatures – the ability to discriminate what is better and what is worse, what is beneficial and what is harmful, what is helpful and not helpful. A predilection to value is found even within bacteria:

> Bacteria possess highly developed sensory systems for the detection of nutrients, energy sources, and toxins, and the capacity to store and evaluate the manifold information provided by these diverse receptors. The final outcome of this sensory integration is the decision to continue swimming in the same direction or tumble into a different course. Thus, some of the most fundamental features of brains, such as sensory integration, memory, decision-making, and the control of behaviour, can all be found in these simple organisms. (Allman, 2000, pp. 5–6)

> Bacteria discriminate between nutrients and toxins and take avoiding action when necessary. Animals discriminate between the odour of food and the odour of a predator. Without this ability, no organism could survive. (Kim & Sankey, 2009, p. 294)

Elsewhere, in clarifying what is meant by a 'predilection to value' in the context of *moral development*, Sankey and Kim (2013) emphasise that it is not being claimed that human beings are genetically 'pre-programmed for morality, which would be contrary to the notion of emergent self-organisation', central to a dynamic systems account. Rather, it is 'simply a recognition of a value bias that seems to be built into the very fabric of life' (2013, p. 186). Genes are clearly part of the overall developmental process, but a predilection to value is saying something about how the world works. Thus, as noted in Chapter 3, even a developing mountain stream will 'seek' the optimum (best) route, given the available terrain – it self-organises without being pre-programmed (Thelen & Smith, 2006).

The idea that *human beings are born with a predilection to value*, an in-built value bias, whether genetically endowed or otherwise, was presented without evidence by Kim and Sankey (2009). However, it is was subsequently found that babies as young as three months old 'show a negativity bias in their social evaluations' (Hamlin et al. 2010, p. 923). They are able to discriminate between *helping* (pro-social) and *hindering* (anti-social) behaviours, and show a preference against what is hindering. In an elegant experiment conducted by Kiley Hamlin and colleagues, babies as young as three months old were presented with the image of a red circular block trying to climb a hill.

The basic experimental design is shown in Figure 8.5. In one scenario, the round circular block is being prevented by a blue square block that keeps bumping it back as soon as it makes any progress. In a contrasting scenario, the red block received *help* from a yellow triangular block that stays behind it, supporting it and bumping it up the hill. Having viewed these different scenarios a number of times, to get accustomed (habituated) to the scenarios, the experimenters tested for preferences. An independent experimenter who did not know the identities of the blocks (what role they had played) held them up in front of each baby, about 50 centimetres from the infant's face, while the parent closed their eyes. A second independent experimenter subsequently coded (from videotape) the amount of time each infant spent looking at each character.

This experiment found that 'three-month-olds have an aversion to antisocial actors – they recognise the negative valence of actions that impede the goals of others'. The researchers concluded that the ability 'to evaluate others on the basis of their behaviour is a fundamental

Figure 8.5 Two experimental conditions (hinderer vs helper) used in the study
Source: Adapted from Hamlin et al. (2010).

aspect of social cognition' (Hamlin et al., 2010, p. 928). Although this research was not arguing for an in-born *predilection to value* as such, empirical evidence of a 'negativity bias in very young babies' social evaluations' is precisely the kind of evidence this notion requires. Social evaluations exist before the babies have been taught to make such discriminations, and certainly well before the onset of Kohlberg's stages of moral development.

This suggests that teachers do not need to instil a sense of value, as children already have it. Rather, teachers need to *nurture* what newborn babies already possess. There is another thought, too: could it be that these very young babies are already *empathising* with the plight of the red block in striving to climb the hill – are they born with the ability to empathise too?

What is empathy, and is it needed when thinking and acting morally?

There is a widespread view that empathy and morality are very closely related. In order to think and act morally, it is claimed, one has to possess the ability to sense other people's thoughts, feelings and emotions. Those who lack this ability – for example, psychopaths – are prone to act immorally, without care and compassion for others, because their brains cannot process how those they offend are feeling. And remember Phineas Gage – the way he behaved after his traumatic accident that destroyed parts of his brain, described in Chapter 6. It seems that empathy is needed in order to act morally. But what about sympathy, is that the same as empathy or are they different?

In everyday conversation, the words 'empathy' and 'sympathy' are often used interchangeably, but in academic debate they are usually separated. Empathy requires an emotional act of *feeling into* what another person (or perhaps other animal) is feeling. The English word 'empathy', though based on Greek, was introduced into English by the German psychologist Edward Titchener (1867–1927) as a translation of the German word *Einfühlung*, which means *feeling into*. Sympathy, on the other hand, is an awareness and *understanding* of another persons' *suffering,* and it may also include taking action to help to alleviate it. This distinction seems to imply that empathy is primarily emotional and sympathy mainly cognitive; however, as argued in Chapter 6, this is a false dichotomy. Another problem is that the literature in moral philosophy and also psychology on empathy, sympathy and morality is vast, and there is also considerable disagreement – it is a highly contested area of research.

Empathy in the context of education science and practice

To be honest, some of the philosophical debate separating empathy and sympathy seems to be 'splitting hairs' – surely there is a lot of overlap between these two concepts, uniting emotion, thought and moral decision-making and action. Aren't there occasions when we need empathy in order to be truly sympathetic, to really put ourselves empathetically into the shoes of the person who is suffering in order to take action to help alleviate their pain? Aren't empathy and sympathy both concerned with compassion and consolation, especially in educational contexts when teaching children to behave morally, one to another?

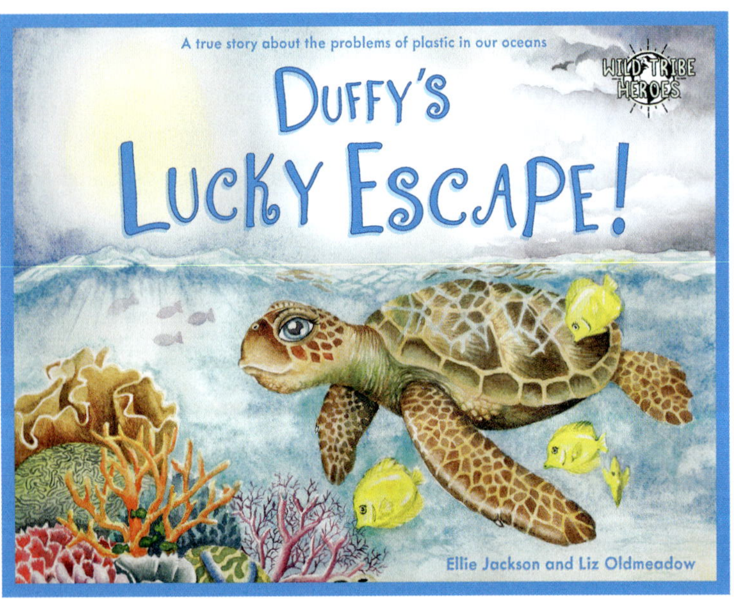

Figure 8.6 A story book that encourages children's empathy on animals affected by environmental issues

Teachers often observe that children and young people's moral engagement is initiated by their feelings and concerns for people who suffer from injustice or those who are in need. In educational discourse, empathy is often introduced as a precursor of morality. For example, pre-schoolers and kindergarten children may be introduced to the stories of animals who have lost their habitat due to climate disasters (e.g. melting ice in the Antarctica or bushfires in Australia) (Figure 8.6). Their empathetic and sympathetic concern for the animals' plight can lead to a deep understanding of, and moral concern for, climate justice in the later years of schooling.

Conversely, a lack of empathy is seen as a major cause of aggression, violence (Marshall & Marshall, 2011) and other anti-social behaviours (Van Langen et al., 2014; De Wied et al., 2009), which are matters of major concern, as schools aim to create safe learning environments for all students. In Denmark since 1993, empathy lessons have been a part of Danish education under the mandatory national *Step by Step* program. These empathy lessons, taken for one hour each week by all students aged from six to 16 years, are called *Klassens Tid*. Danes believe that this program can help children build relationships while practising empathy and compassion. More recently, in 2019, a UNESCO project called Learning for Empathy: A Teacher Exchange and Support Programme was launched, with generous sponsorship from the Japanese Government. The project aims to highlight appreciation of cultural diversity and the promotion of a culture of peace and non-violence (UNESCO, 2020).

In search of more recent perspectives

Although the concept of empathy is discussed and used in the context of education, researchers such as Jean Decety note that there is still 'much misunderstanding and confusion over the concept of empathy and its relations to moral cognition' (Decety & Cowell, 2014, p. 526). For example, these authors note that empathy does not necessarily pave its way straight to moral behaviour. Indeed, at times empathy can interfere with moral decision-making by sensitising a person's self-concern. It may also favour concern for close kin and in-group members over the needs of others. Generally, humans are better at understanding the state of other people when they are familiar with the other person and their circumstances. Some children might therefore show more empathy-related behaviours towards familiar people rather than others who are unknown (Masten et al., 2010). This may inhibit them from making fair and impartial decisions when different social groups are involved.

Research has also shown that children aged from three to nine years may only feel obliged to obey moral rules or norms (i.e. do not harm others!) when within a known group of people. In addition, children at that age may only perceive that harming another who is not in their group is wrong only when presented with the rule (Rhodes & Chalik, 2013). If this is indeed correct, it suggests that the teaching of moral rules should be part of the explicit curriculum in schools – it has to be *taught* and can't be left to be caught implicitly. What needs to be taught is the importance of extending empathy to all, as exemplified in the Danish curriculum.

Arguably, one of the main reasons for what Decety and Cowell (2014) call 'misunderstanding and confusion' in the literature on human empathy is that much of it has little or nothing to say about *where and how empathy arises in the brain*, which we will turn to shortly, first with regard to empathy and later when considering the neurobiology of moral thought and action. Moreover, primatologist Frans de Waal (2009), in his provocative book *The Age of Empathy: Nature's Lessons for a Kinder Society*, says that if we are to truly understand the nature of human empathy, we would do well to start by carefully studying other primates, then we would discover that much of what we are told about our own biological origins is wrong. Referring particularly to American society, he says, 'Every debate about society and government makes huge assumptions about human nature, which are presented as if they come straight out of biology. But they almost never do' (de Waal, 2009, p. 4).

De Waal is especially critical of *social Darwinism* and its influence in America. This ideology was first promulgated by British political philosopher Herbert Spencer (1820–1903), who applied Darwin's laws of natural selection to justify political conservativism and racism, and to oppose social reform. In America, it was closely associated with the rise of eugenics, discussed in the previous chapter. Spencer coined the phrase 'survival of the fittest' (often incorrectly attributed to Darwin), which was quickly used to justify raw competition in business, calling it 'a law of biology' (de Waal, 2009, p. 28). For John D. Rockefeller, unbridled competition and survival of the fittest in business was even married to religion, with him calling it 'a law of God' (de Waal, 2009, p. 28). Given that 'the book found in most American homes and every hotel room urges us on almost every page to show compassion', de Waal wonders how some Christians 'can embrace such a harsh ideology without a massive case of cognitive dissonance, but many do' (de Waal, 2009, pp. 28–9).

Spencer's concept of the 'survival of the fittest', and the allied concept of 'nature red in tooth and claw' (a nineteenth-century idea that pre-dated Spencer's writing), provides a very biased view of nature, ignoring the prevalence of symbiosis (where two types of plants or animal mutually cooperate to the benefit of the other). It is also contrary to what we are now learning about primate behaviour from primatologists such as de Waal, especially regarding empathy and what he calls **consolation** in primate behaviour. He says, 'I have seen literally thousands of consolations – that's how common the behaviour is.' He then adds, 'A victim of aggression, who not long ago had run for her life, or screamed to recruit support, now sits alone, pouting, licking an injury or looking dejected. She perks up when a bystander comes over to give her a hug, groom her, or carefully inspect her injury' (de Waal, 2009, p. 90).

In what follows, we will largely follow Decety and de Waal, to gain some recent consensus on the definition of empathy and the relationship between empathy and morality.

Empathy: Three gradations and their neurobiology

Converging evidence from both animals and humans, encompassing primatology, developmental science and neuroscience, now shows that human empathy is a process with three gradations, as shown in Figure 8.7. De Waal uses the notion of three layers because he pictures it like three layers of a Russian doll, but the use of the term 'layer' can mistakenly align it with the notion of a 'triune brain' (see Chapter 6), which is not what he means.

At its core is what is called **emotion contagion**, which refers to a tendency to share another's emotional state. 'When an observer attends to another's state, he or she spontaneously accesses information about the other (the "target"), their feelings, the situation and other related concepts through a distributed associative process' (de Waal & Preston, 2017, p. 499). This is an '"automated process" shared with a multitude of species' (de Waal, 2009, pp. 208–9). Two outer gradations that surround this core are **empathic concern** and **perspective taking**. Each layer takes a different developmental time course, some showing maturity earlier than the others, depending on the extent of brain maturation (Decety & Michalska, 2010). These outer gradations enable the primitive automatic emotion contagion to become fine-tuned, thereby making us more considerate of others. We will now consider each of these in more detail.

consolation: 'reassurance behaviour directed at a distressed party, such as a victim of aggression' (de Waal & Preston, 2017)

emotion contagion: an automatic tendency to share another's emotional state

empathic concern: 'concern about another's state and attempts to ameliorate this state' (de Waal & Preston, 2017)

perspective taking: perceiving or understanding a situation from an alternative perspective, including that of another person

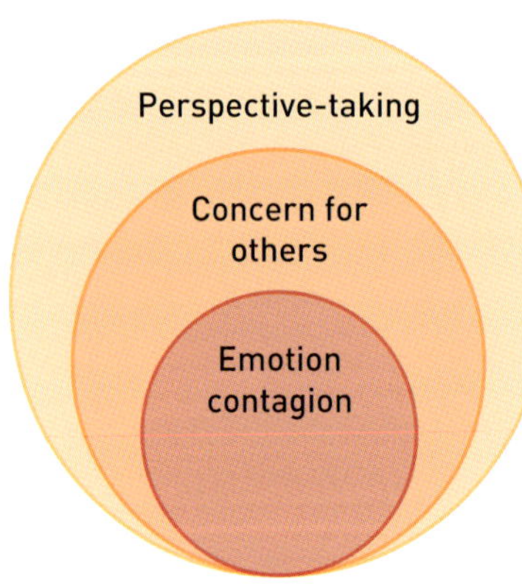

Figure 8.7 Three gradations of empathy
Source: Based on de Waal (2009); de Waal & Preston (2017); Decety & Cowell (2014).

Emotion contagion

The term 'emotion contagion' is used because one literally *catches* the emotion of another when experiencing it. This idea goes back to at least the middle of the twentieth century, so predates the discovery of mirror neurons, though that may give it empirical support (but see below). Emotion contagion is something with which we are born. For instance, within 18 hours of birth, newborns become very distressed if they hear another infant starting to cry (Martin & Clark, 1987). This distress is not observable when newborns hear recordings of their own crying or the crying of other species (i.e. an infant chimp), showing that the newborn's reaction is *other-oriented* and *species specific*. Moreover, as mentioned in Chapter 7, within 36 hours of birth, babies are capable of imitating facial expressions of fear, sadness and surprise by changing their mouth movements: widening of the lips as in a happy face, somewhat protruded lips as in a sad face, wide opening of the mouth as in a surprised face (Field et al., 1982).

Emotion contagion relies on what is called a **perception-action mechanism (PAM)** with which babies are born (Preston & de Waal, 2002). From our earliest years, we have an empathetic sense of others' feelings (including pain) because our nervous system evolved to resonate between other and self. Imagine you are watching a picture of someone else's hand being hit by a hammer. Our body immediately reacts as if our own hand is hit! When we attend to another person's state (pain), we have access to their state using *our own* individual somato-sensory experiences. Degrees of precision of 'understanding other's state gained from this resonation range from a basic sense that the other feels "bad" to a "highly tailored, accurate understanding of that specific state and its entailments"' (de Waal & Preston, 2017, p. 504).

Whether or not this empathetic ability to 'catch' another's emotional states results from mirror neurons is contested (Blair, 2011; Decety & Michalska, 2012; de Waal & Preston, 2017). If mirror neurons are involved, they are likely not the whole story. Emotion contagion involves a widely distributed network related to relevant feelings, memories and associated situations, encompassing the amygdala, hypothalamus, hippocampus and orbitofrontal cortex (OFC) (Decety & Michalska, 2012).

perception-action mechanism (PAM): spontaneous activation of an individual's own personal representations for the other person, their state and their situation when perceiving the other's state (de Waal & Preston, 2017)

Concern for others

Emotion contagion can often result in empathic 'concern about another's state and attempts to ameliorate this state' (de Waal & Preston, 2017, p. 500). However, obversely, it can also result instead in *aversion*; in some cases, feelings gained from observing others' distress induce concerns about oneself and feelings that make us want to avoid the situation.

Until quite recently, it was believed that humans could not exhibit empathic concern until about two years of age, due to the lack of self-awareness (Hoffman, 1975). However, accumulating empirical findings now show that a rudimentary form of other-oriented concern is observable at as early as six months. Six-month-old infants reach out to the distressed peers (e.g. leaning or gesturing toward, touching, looking at a peer's mother) while (unlike newborns) they rarely becoming distressed themselves (Hay et al., 1981). Also, between eight and 16 months, infants show gestures reflecting concern in response to their mother's distress (Roth-Hanania et al., 2011). This suggests that empathic concern is observable well before the second year of life (Davidov et al., 2013).

Whether the emotion contagion leads to self-concern or concern for others seems to depend on emotion regulation. In Chapter 6, two processes of emotion regulation were introduced: *reactive* and *deliberate* regulation (Woltering & Lewis, 2009), and you learnt that vertical integration across brain stem, limbic structures and cerebral cortex enables efficient emotion regulation. Empathic concern is strongly related to deliberate regulation (Rothbart et al., 1994). In addition, those who are good at regulation and control over their ability to focus and shift attention are more likely to show sympathetic concern to others (Decety & Michalska, 2012).

'Deliberate emotion regulation is most consistently associated with frontal cortical regions such as the *dorsolateral prefrontal cortex* (DLPFC)' (Woltering & Lewis, 2009), approximately Brodmann area 46; *the dorsomedial prefrontal cortex* (DMPFC), mainly Brodmann areas 8 and 9; the *ventrolateral prefrontal cortex* (VLPFC), Brodmann areas 44, 45 and 47; and the paralimbic regions such as the *dorsal anterior cingulate cortex* (ACC), Brodmann areas 24, 32 and 33 (Woltering & Lewis, 2009). See Figure 8.8 to identify the location of these areas.

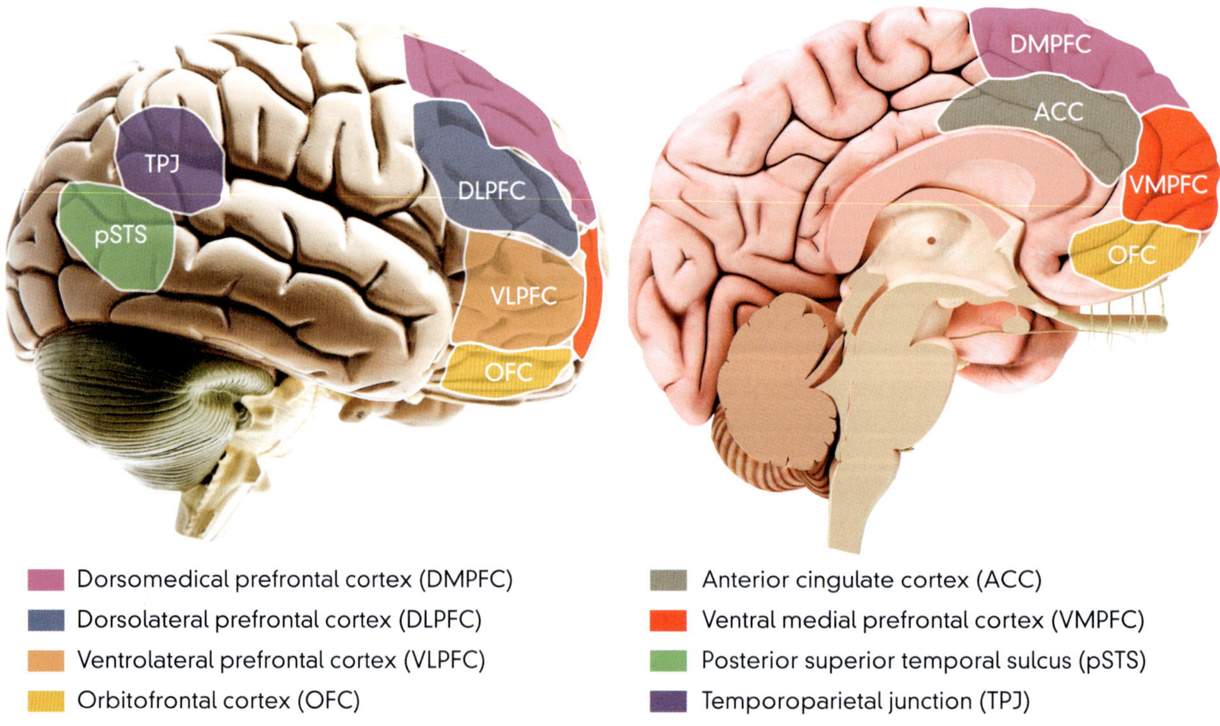

Dorsomedical prefrontal cortex (DMPFC)
Dorsolateral prefrontal cortex (DLPFC)
Ventrolateral prefrontal cortex (VLPFC)
Orbitofrontal cortex (OFC)
Anterior cingulate cortex (ACC)
Ventral medial prefrontal cortex (VMPFC)
Posterior superior temporal sulcus (pSTS)
Temporoparietal junction (TPJ)

Figure 8.8 Cortical regions that are associated with the three gradations of empathy

Perspective taking

Whereas the two previous gradations (emotion contagion and concern for others) are common in many species, perspective taking is observable from only 'intensely social animals such as humans, apes, elephants and dolphins' (de Waal & Preston, 2017, p. 499). Humans and these other highly social animals have an increased ability to differentiate self and other, and hence adopt another's perspective. Already at the age of 18 months, infants not only appreciate that another person can hold a *desire* that differs from their own, they can also appreciate that the other person's *desires* are related to their felt emotions (Repacholi & Gopnik, 1997). By four or five years of age, 'this desire-based concept of emotion develops to included beliefs and expectations' (Decety & Michalska, 2014, p. 174).

Taking another's perspective involves having a *theory of mind*, reading what others may be thinking and feeling (see Chapter 7) and working memory which you learnt about in Chapter 4 (Decety & Michalska, 2012). Several empirical studies confirm that the prefrontal cortex plays a particularly important role in taking the perspectives of others. In a study with four- to six-year-old children (Liu et al., 2009), children took what is called the **false-belief test**. Children's EEGs were recorded during the experiment, so that researchers could compare EEG/ERP wave patterns (see Chapter 4) between those who were able to take another's perspective and those who could not.

In these experiments, children were given scenarios, such as the following:

> There are two boxes (yellow and green). A boy puts a dog in a yellow box, and he goes away without looking at the boxes. When he is absent, the dog moves from the yellow box to a green box. The boy returns.

false-belief test: a critical theory-of-mind task that assesses whether an observer knows what another knows, even if the reality is inconsistent with what the observed person knows

After watching the scenario on the screen, children were asked which box they think the boy will open. Children who were able to take the perspective of the boy pointed to the yellow box, and thus passed the test. However, those who could not make distinctions between the 'belief' held by the boy and the 'reality' pointed to the green box, and so failed the test. The two groups of children showed notably different patterns in their frontal late slow wave that are observable between 775 and 850 ms.

Children who passed the false-belief task showed pronounced a frontal late slow wave as in adults' waveforms. Other studies show that neural circuits that are activated when reading what others are thinking and feeling include the dorsomedial prefrontal cortex (DMPFC) and posterior STS (pSTS) at the junction of the parietal cortex (TPJ) (Moriguchi et al., 2007; Sabbagh et al., 2009). See Figure 8.8 (above) to identify these cortical areas.

For teachers, the key point is that the ability of children and adolescents to empathise with one another is rooted in evolution and starts to develop within hours of being born. In helping children to become more empathetic, from early childhood education through to high school, teachers are developing what has been developing from the moment they entered their physical, social and cultural world. Moreover, it has all been happening in their embodied brain as a result of their physical, social and cultural experience of the world, which has crafted each individual brain.

One important task for teachers is to help children and adolescents extend their ability to empathises *beyond their immediate group* – from 'us' to 'them'. Given what we have just been discussing, this will involve building on their in-born ability to 'catch' what others outside our group are feeling (emotion contagion), to deepen their concern for these 'others' and progressively hone their perspective-taking abilities. Remember, as emphasised over and over again in this text, in doing this you are physically shaping and changing children's brains.

RESEARCH LINK 8.2

Teacher's empathy in K–12 education

Although the concept of empathy is widely used, and there are many studies on students' empathy, research into K–12 teachers' empathy remains very limited and fragmented. Recently, Izhak Berkovich (2020) ran a systematic review of peer-reviewed empirical research that addressed K–12 teachers' empathy as a key concept. Only 28 publications met the inclusion criteria. These studies alert us to the critical value of teachers' empathy in students' learning and perceptions of the school environment. Here are three important findings on teachers' empathy and students' learning and development:

1. *In some countries, teacher empathy levels are higher than for those who are not teachers.* Data collected from Serbia (Stojiljković et al., 2012) and China (Huang et al., 2012) indicate that teachers' scores on perspective-taking and empathic concern are higher than the scores of other groups. This indicates that, in these countries, the teacher training system might be successful at attracting into the profession and retaining those who have high empathy. Additionally, a Polish study (Klis & Kossewska, 1996) found that special education teachers' empathy levels are higher than those of teachers in regular secondary education. At present, there is no systematic international investigation into whether prospective teachers are screened for empathy.

2. *A teacher's empathy influences the teacher's own sense of self-confidence.* Studies from Serbia (Stojiljković et al., 2014) and Israel (Goroshit & Hen, 2016) show that teachers with higher level of empathic perspective taking have more positive and healthier perceptions of themselves. A study in the United States (Barr, 2011) indicates that teachers with higher levels of empathic perspective taking tend to believe that students in their school have more respectful, friendly and helpful peer relationships. The same study found that a teacher's personal distress correlated with negative perceptions of student–peer relationships. While these results are based entirely on teachers' self-reporting, nevertheless they suggest a need to focus on empathy training programs for teachers.

3. *A teacher's empathy skills can be improved by carefully designed training programs.* Findings from various studies indicate that training programs can enhance teachers' 'empathic skills, including awareness of the importance of empathy, improving emotion recognition ability, expressing empathy, and managing empathy in a professional context' (Berkovich, 2020, p. 561). Successful elements incorporated into the training programs include regular meeting with mentors with high empathic communication skills; cultivating critical cultural awareness and reflection; and self-evaluation of one's own empathic skills.

The neurobiology of moral thought and action

Although, as noted in the previous section, empathy does not necessarily pave its way *straight* to moral behaviour, our discussion nevertheless suggests there is often a close relationship between having empathetic feelings, and thinking and acting morally. For example, we noted that children and young people's moral engagement is often initiated by their feelings and concerns for people who suffer from injustice or those who are in need. Part of the reason is the role played by *emotion* in how we *feel*, what we *think* and what we *do* – in the moral judgements we make, in our choices and moral decision-making. Here is an example of moral judgement and decision making. It is called the trolley dilemma. What will you do?

The trolley dilemma

Here is the trolley dilemma:

> A runaway trolley [tram] is headed for five people who will be killed if it proceeds on its present course. The only way to save them is to hit a switch that will turn the trolley onto an alternate set of tracks where it will kill one person instead of five. Ought you to turn the trolley in order to save five people at the expense of one? (Greene et al., 2001, p. 2105)

This dilemma (Figure 8.9) was first posed by the English moral philosopher Phillipa Foot in 1967. In 2001, a research team led by Joshua Greene used it to study the relationship between reason and emotion, using MRI. This study found that 'moral dilemmas vary systematically in the extent to which they engage emotional processing and that these variations in emotional engagement influence moral judgement' (Greene et al., 2001, p. 2105).

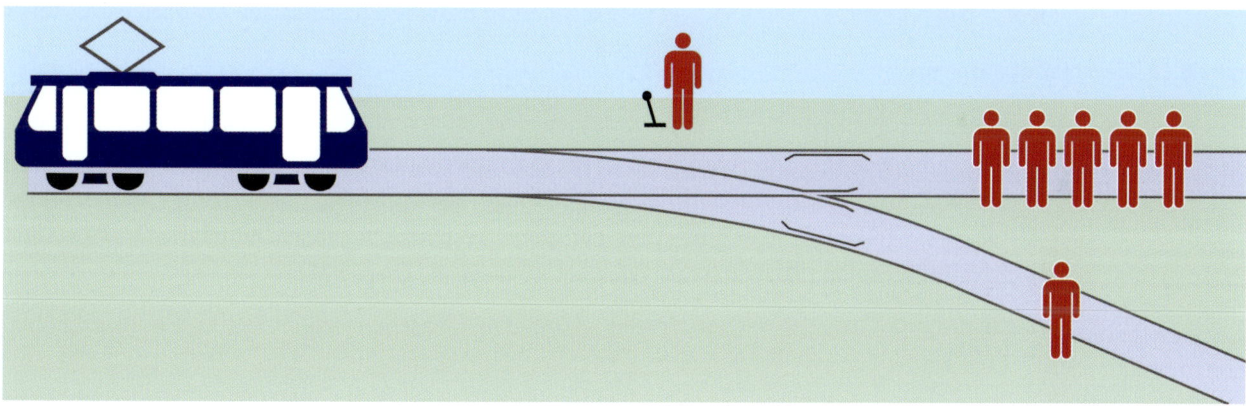

Figure 8.9 The trolley dilemma

So what are you going to do, and why? Will you sacrifice one person to save five, which means you have to act – change the track? Or will you do nothing and let five die? Think about it carefully and, if possible, discuss it with your classmates.

The footbridge dilemma

You will be interested to know that most people say they will sacrifice the one individual to save five in the trolley dilemma. That seems to be a safe **utilitarian choice** – the greatest benefit

utilitarian choice:
a choice based on the belief that the consequences of any action are the only standard of right and wrong

Figure 8.10 The footbridge dilemma

for the greatest number. But now let us tweak the scenario slightly (Figure 8.10):

> As before, a trolley threatens to kill five people. You are standing next to a large stranger on a footbridge that spans the tracks, in between the oncoming trolley and the five people. In this scenario, the only way to save the five people is to push this stranger off the bridge, onto the tracks below. He will die if you do this, but his body will stop the trolley from reaching the others. Ought you to save the five others by pushing this stranger to his death? (Greene et al., 2001, p. 2105)

What the researchers found is that most people say no, they will not push the stranger to save the five. In this case, their judgement not to act is consistent with what is called a **deontological choice**, which focuses on the rights of the individual rather than trying to satisfy the utilitarian concern for the greater good. But this is still rather odd. Most people say they will sacrifice one to save five in one scenario but not in the other. This poses a major problem for moral philosophers. It is very difficult to provide a justification for both cases, taken together.

Green and colleagues concluded that the key difference is the degree to which *emotion* is playing a part in the decision-making and, moreover, this difference is occurring in competing neural systems in the brain. The more personal and 'harmful action in the footbridge case, pushing the man off the footbridge, triggers a relatively strong negative emotional response, whereas the relatively impersonal harmful action in the switch case does not' (Greene, 2015, p. 197).

REFLECTION

What is your reaction to the footbridge dilemma? Can you think of occasions (perhaps even in schools and classrooms) where a clash could arise between trying to serve the needs of the greatest number and a concern for the rights of the individual?

Although referring to differences of moral judgement occurring in competing neural systems in the brain, Greene nevertheless makes the very important point that 'morality has few, if any, neural mechanisms of its own' (Greene, 2015, p. 197). There was a time, not so long ago, when some researchers believed that it should be possible to find dedicated pathways in the brain associated with moral judgement and decision-making; that hope has largely, if not entirely, been abandoned. Greene (2015, p. 198) very importantly concludes that, 'To truly understand the neuroscience of morality, we must understand the many neural systems that shape moral thinking, none of which so far, appears to be specifically moral' – though, as we have been noticing, many are related, one way or another, to emotion.

The chemistry of care and trust

So far, not much has been said about the very important role of hormones and peptides in the brain, although we have considered the role of neurotransmitters in synaptic transmission. In 2011, neurophilosopher Patricia Churchland wrote a book called *Braintrust: What Neuroscience Tells Us About Morality*, which discusses the brain chemistry of moral values. Churchland points out that 'the values rooted in the circuitry for caring – for wellbeing of self, offspring, mates, kin and others – shape social reasoning' (2011, p. 8). These circuits for *care and trust in the brain*, which are central in addressing social and political concerns and other aspects of social life, are driven by the neuropeptide **oxytocin**.

Oxytocin, which is a 'very ancient peptide (chain of amino acids), is at the hub of the intricate network of mammalian adaptations for caring for others' and 'is found in all vertebrates' (Churchland, 2011, p. 14). It also plays a core role in what Churchland refers to as 'the mechanisms that "maternalize" the female mammalian brain' (2011, p. 31), raising the threshold of tolerance for others, and in reducing fear and avoidance responses that are essential ingredients of trust. In primates, the widespread practice of grooming is associated with increased levels of oxytocin.

Churchland's core thesis is that '*caring* is a ground floor function of nervous systems. Brains are organised to seek well-being and to seek relief from ill-being. Thus, in a perfectly straightforward way, the circuitry for self-maintenance and the avoidance of pain is the source of the most basic values – the values of being alive and of well-being' (Churchland, 2011, p. 30, emphasis in original).

Starting with a ground-floor concern for self-maintenance and the avoidance of pain, shared by all organisms as a necessary condition for survival, over evolutionary time human beings and other social creatures extended that to care for their offspring (Sankey, 2016). The neural processes that care for self were recruited to perceive a sense of danger to offspring, as a danger to one's self (Sankey, 2016). Eventually, the imperative to care for self and offspring 'has in some mammalian species extended further, to encompass kin or mates or friends or even strangers, as the circle widens. This widening of other-caring in social behaviour marks the emergence of what eventually flowers into morality' (Churchland, 2011, p. 14).

Notice, however, that this claim is concerned with *moral values*, not moral norms or rules. Churchland believes that values are 'more fundamental than rules' (2011, p. 9). She adds that, 'Moral values need not involve rules, though they sometimes do; they need not be explicitly stated but may be implicitly picked up by children learning to get along in their social world' (2011, p. 10).

oxytocin: a hormone produced by the hypothalamus and secreted by the pituitary gland

Trust and the community of school

In 2016, a paper titled 'The Neurobiology of Trust and Schooling' (Sankey, 2016) provided a critical evaluation of Churchland's thesis and applied it to education. Although there were many points of agreement with Churchland, there were also some important and some minor differences. One minor difference was the suggestion that while care is a foundational moral value in the context of our social lives, trust is not. Rather, it should be viewed even more fundamentally than care because it constitutes 'the precondition for a social life. Without trust, a caring, sharing, considerate, compassionate, empathetic social life would be impossible' (Sankey, 2016, p. 184). Trust forms the bedrock of our social lives, but it requires considerable social skill in learning who we should trust.

The paper argues that 'trust and the moral values that are built on trust are foundational for the social life of each and every school community' (Sankey, 2016, p. 189). Moreover, they are the product of each and every individual's brain within that community. Despite what some in education may believe, they 'are not simply socially constructed or socially conditioned nor are they simply the product of culture, though social and cultural experience continue to shape each individual brain throughout life' (2016, p. 189). All human beings, across all regions and cultures, possess brains that have been 'shaped by evolution to trust and care and otherwise ground our social lives' (2016, p. 189).

Care and trust should therefore not be viewed as some kind of bolted-on addition to education, in the way that schools are run and what is taught at school. They are part and parcel of what it means to be human and what it means to be educated. We have just noted, above that oxytocin in the brain results in 'increased tolerance of others and a corresponding decrease in levels of anxiety, fear and avoidance' (Sankey, 2016, p. 190). Schools and classrooms can sometimes be places of fear and avoidance; they can be places where there is a lack of tolerance and where some children are deliberately excluded by peers and even teachers. These are not just social phenomena – they are also profoundly neurobiological.

In the context of Australia, the concept of 'values pedagogy' underpinned the National Framework for Values Education in Australian Schools (DEST, 2005). In discussing the National Framework, Toomey et al. (2010) explore aspects of what they call 'student well-being pedagogy'. In Chapter 2 of that book, Neville Clement states that teacher 'beliefs and values emerge as a significant ingredient in the learning environment, especially in the form of caring, support and competence, irrespective of subject domain' (Clement, 2010, p. 28). Values matter in education, and that is hardly surprising: trust and core values such as caring are deeply embedded in our brains (Sankey, 2016) (Figure 8.11).

Figure 8.11 A caring and cooperative classroom as a core ingredient in student wellbeing, in contrast to bullying in a school

Nurturing the moral, empathetic brain should not be considered an option for education; arguably, it should be front and centre of teaching and learning in every school. To repeat what was said at the start of this chapter, there 'is a strong consensus in the educational literature that teaching is an inherently moral endeavour, and that *the moral work of teachers is of central importance to education*' (Sanger, 2008, p. 169, emphasis added).

APST and ACECQA curriculum specifications   STANDARDS

APST Standard 1.1 Physical, social and intellectual development and characteristics of students
Demonstrate knowledge and understanding of physical, social and intellectual development and characteristics of students and how these may affect learning

APST Standard 1.3 Students with diverse linguistic, cultural, religious and socioeconomic backgrounds
Demonstrate knowledge of teaching strategies that are responsive to the learning strengths and needs of students from diverse linguistic, cultural, religious and socioeconomic backgrounds

APST Standard 4.3 Manage challenging behaviour
Demonstrate knowledge of practical approaches to manage challenging behaviour

APST Standard 6.2 Engage in professional learning and improve practice
Understand the relevant and appropriate sources of professional learning for teachers

ACECQA curriculum specifications
1.3 Social and emotional development
2.3 Guiding behaviour/engaging young learners

This chapter has encouraged readers to reflect on the unique responsibility of teachers by understanding what is meant by 'education as a moral endeavour' and the importance of moral education. This chapter has gone beyond what is generally available (e.g. the cognitive approach to moral development) to newly qualifying teachers by providing a frontline overview of recent research into the development of the moral and empathic brain. It has added knowledge on the neurobiology of trust and care, and the neural processes involved in moral judgement and moral decision-making. It has also provided readers with an understanding of interrelationship between empathy and moral thought and action, and what is meant by the three gradations of empathy.

The chapter has provided sufficient grounding for newly qualifying teachers to address some of the challenging and antisocial behaviour of students, with informed perspectives about how such behaviour emerges, drawing on relevant neurobiological evidence. Understandings gained from this chapter should inform newly qualifying teachers of how they might be able to create socio-emotionally safe classrooms and support students' development into caring and responsible citizens.

SUMMARY

- Education is often said to be a moral endeavour, meaning that the aims and purposes of education are fundamentally moral, concerned with the enhancement of each individual child as a member of society. Education, by its very nature, seeks what is *good* and *intrinsically worthwhile* in attending to the development of children's minds (brains) and their mental and physical wellbeing.

- Kohlberg's theory of moral development played a central role in education. For more than half a century, his six stages of moral development provided the basis for understanding moral development and the design of moral education. However, from a dynamic systems perspective, his 'stage theory' is now seen to provide only a rough approximation of moral development.

- Kohlberg's account of moral development incorporated a valid distinction between acting conventionally and legally, abiding by the accepted rules of society, and acting morally. To be truly moral, one must reason post-conventionally, where the emphasis is on liberty and human rights, which transcend any given culture and permit cross-cultural discourse.

- A dynamic systems approach to moral development, proposed in 2009, attempted to retain what was valuable in Kohlberg's work, but also move beyond it by applying the same principles used in a dynamic systems account of cognitive and motor development to moral development. It abandoned the notion of progressive developmental stages while retaining the Kohlbergian idea that there are different forms of moral reasoning.

- This dynamic systems proposal also incorporated the idea of a 'predilection to value', a *value bias* common to all other creatures, including bacteria. Humans are born with the ability to discriminate what is better and what is worse, what is beneficial and what is harmful, what is helpful and not helpful. It has since been discovered that babies as young as three months exhibit such an ability: they can identify what is helpful and not helpful, and show a preference for what is helpful or pro-social.

- Recent research suggests that empathy is a process with three gradations: emotion contagion (you can catch if by observing it in others); empathetic concern; and empathetic perspective taking. These are instantiated in the brain.

- Research comparing responses to the trolley dilemma and the footbridge dilemma has revealed identifiable differences between people's utilitarian and deontological choices. A key difference is in the degree to which emotion plays a part in the decision-making; moreover, this difference is occurring in competing neural systems in the brain.

- According to Sankey (2016), 'trust and the moral values that are built on trust are foundational for the social life of each and every school community'. Moreover, they are the product of each and every individual's brain within that community. Nurturing the moral, empathetic brain should not be considered an option for education; arguably, it should be front and centre of teaching and learning in every school.

KEY POINTS FOR TEACHERS

- Given that teaching is an inherently moral endeavour, and that the moral work of teachers is of central importance to education, teacher education should systematically help beginning teachers to carefully and systematically appreciate the moral and ethical dimensions of education in preparation for their own school and classroom practice.
- Moral values are manifested socially and culturally but, like everything social and cultural – including our emotions and our ability to empathise with others – moral values are also embrained, the product of neuronal processes in the brain. Their development is the product of emergent processes occurring in the dynamic, self-organising brain.
- From a dynamic systems perspective, any account of moral development embraced by teachers should at least be consistent with our status as biological and neurobiological organisms. Moreover, teachers should appreciate that moral development is not a special or different kind of development; it shares the same dynamic processes found in all human development.
- Research has found that babies as young as three months are able to discriminate between helping (pro-social) and hindering (antisocial) behaviours and show a preference against what is hindering. This supports the claim that human beings are born with a predilection to value, or value bias. Teachers do not need to instil a sense of value; they already have it. Rather, teachers should nurture what newborn babies already possess.
- Research shows that children's ability to empathise with one another is rooted in evolution and starts to develop within hours of being born. When helping children to become more empathetic, from early childhood through to high school, teachers are developing what has been there from the moment they entered their physical, social and cultural world.
- One main task for teachers is to help children and adolescents extend their ability to empathise beyond their immediate group. This will involve building on their inborn ability to 'catch' what others outside the group are feeling (emotion contagion), to deepen their concern for these 'others' and progressively hone their perspective-taking abilities.
- Although researchers have previously believed that it should be possible to find dedicated pathways in the brain associated with moral judgement and decision-making, that hope has largely been abandoned. Teachers should be wary of claims that there are specific 'moral neurons'; in fact, many neural systems are involved in moral thinking and decision-making.
- Despite what some in education believe, moral values are not purely a social construct, nor are they simply a product of culture, though social and cultural experience are key factors shaping each individual brain throughout life. All human brains, across all geographical regions and cultures, have been shaped over evolutionary time to trust and care, and otherwise ground our social lives.
- Brains are organised to seek wellbeing and relief from ill-being. Oxytocin, a very ancient and also very basic peptide, with only nine amino acids, plays a central role. Care and

trust should therefore not be viewed as some kind of bolted-on addition to education, in the way that schools are run and what is taught at school; instead, they are part and parcel of what it means to be human and what it means to be educated.

Guided responses

REVIEW QUESTIONS

1. What is meant by the claim that education is a moral endeavour?
2. In Kohlberg's developmental account of morality, what are some of the main differences between the conventional and post-conventional levels of moral reasoning?
3. Compared with Kohlbergian moral development theory, what important changes of perspective are brought when adopting a dynamic systems approach to moral development?
4. What are the three gradations of empathy?
5. From a neurobiological perspective, what is the difference between utilitarian and deontological decision-making processes?

FOOD FOR THOUGHT

1. Do you think the distinction Kohlberg made between conventional (law and order) and post-conventional moral reasoning (with its emphasis on human rights) provides a sound criterion by which to condemn the Nazi persecution? Does it avoid the trap of social relativism (while respecting social pluralism) as he hoped, and do you think this kind of criterion remains useful nowadays in response to political tyranny?
2. Do you find the suggestion helpful, educationally, that teachers might abandon the idea of setting classroom *rules* to focus instead on classroom *values*? Would that change the dynamic of classroom control, or do you think it wouldn't make much difference?
3. Given what you have read in this chapter about empathy, what teaching strategies and curriculum initiatives might teachers use to help children and adolescents enhance their empathy, particularly empathy for those who are not perceived as part of their own group?

RESEARCH ACTIVITY: SCHOOL-BASED INTERVENTIONS TO PROMOTE EMPATHY

The promotion of empathy learning and development has often been discussed in the context of school-based social-emotional learning (SEL) programs. SEL curricula aim to enhance the development of five interrelated abilities in the areas of emotional and social development: responsible decision-making; self-management; self-awareness; relationship skills; and social awareness (CASEL, 2020).

Self-awareness and social awareness are particularly related to empathy development. Self-awareness is defined as 'the ability to *understand one's own emotions, thoughts, and values and how they influence behavior across contexts*', whereas social awareness is defined as 'the ability to *understand the perspectives of and empathize with others*, including those from diverse backgrounds, cultures, and contexts' (CASEL, 2020, p. 2).

Below is a list of school-based SEL programs that specifically emphasise the promotion of empathy-related skills:

- Collaborative Classroom: 'Caring School Community: A CASEL SELect program for grades K–8'
- MindUP
- Open Circle
- Roots of Empathy
- CASEL Program Guide: 'RULER Approach'.

The effectiveness and quality of these chosen programs was carefully investigated with experimental studies (Malti et al., 2016). After checking each program's website, please think about the following questions.

Research activity links

Questions

1. In each program, which aspect of empathy (i.e. emotion contagion, empathic concern, perspective taking) is addressed?
2. In what ways does each program offer opportunities to practise empathic skills?
3. What has been improved as a result of implementing each school-based program?

PART **3**

LEARNING, WELLBEING AND THE ECOLOGY OF LEARNING ENVIRONMENTS

In the final three chapters of this textbook, the many strands explored in the earlier chapters are brought together in exploring the *ecology of embodied learning and development*. The use of the term 'ecology' is both literal and metaphorical. Ecology is concerned with the dynamic, holistic relationship of living things to their environment and to each other. In education, this applies directly, as the relationship of students to each other and to their learning environment. When used as a metaphor, ecology denotes the *holism* of education, where all the many aspects of learning and development, including the values that underpin education and the role of assessment and feedback in learning, combine to form the practice of education and, as a product, the emerging, educated child, adolescent and adult.

Chapter 9 opens with a discussion of the central role of metaphor and analogy in everyday language and also their important heuristic role in science before introducing the analogy of *classrooms as ecosystems*. The similarity relations between ecosystems and classrooms as learning environments are provided for readers to fully understand what this analogy entails. The chapter then explores the ecological concept of *affordance*. The word 'affordance' was coined by James Gibson, who worked with his wife Eleanor in developing the notion of ecological perception and learning as part of what they called *ecological psychology*. The pioneering work of the Gibsons was at odds with much classical cognitivism. It also contributed directly to the introduction of dynamic systems theory in human development, discussed in Chapter 3. James Gibson argued that in order to understand the behaviour of animals, it was necessary to view them as embedded in an environment; they could not be successfully studied in isolation from their environment. Within education, this speaks to the necessity of always viewing *students as embedded in learning environments*. Gibson's notion of affordance is widely used outside education, particularly in design science. This chapter brings the notion firmly within the fold of education by applying the notion of affordance to the design of learning environments in school.

Chapter 10 focuses on the science of learning, assessment and feedback. Teaching, learning and assessment are closely interrelated at all levels of education. This chapter extends what we learnt in Chapter 4 regarding working memory and metacognition, plus notions of 'error-related negativity' and 'feedback-related negativity' in brain science. In taking a science of learning perspective on assessment, readers are reminded of what they learnt in Chapters 1 and 2 about *Hebbian learning and brain plasticity*, and how these Hebbian processes

Concept map 3 The ecology of embodied learning and development

are occurring in the brain when students are preparing for assessment. Assessment can, however, lead to failure in education, and this chapter emphasises that failure has a positive role to play in learning. Readers are introduced Karl Popper's important work on the role of *learning by mistakes* in science – what he called learning through conjectures and refutations. Rethinking the role of failure in learning leads to the notion that *classrooms should be safe places to be wrong*, not only for students but also for teachers. The chapter ends by introducing the very recent notion of the predictive brain and how the brain functions from the top down by making predictions and through the employment of millions of predictive loops sensing prediction errors. Assessment and the detection of error are thus central to the ecology of the brain.

Chapter 11 takes these many ideas one ecological step further, holistically uniting the science of learning and development with the science of student wellbeing. The extent to which learning and development are intimately related to student wellbeing in school is initially explored through the findings of recent large-scale surveys. A key issue is *whether schools are safe places to be,* both physically and mentally. The statistics provide some comfort, but also come with cautionary messages, including that one in seven children in Australia does not feel safe at school *a lot of the time*. The chapter takes that message and considers student wellbeing in relation to the affordance of *value-based learning environments*, building on the discussion in Chapter 9, then relating the science of student wellbeing to the notion of the interoceptive, predictive brain, introduced in Chapter 10. The chapter ends with the hope – which has permeated this book as a whole – that in the context of education, the science of learning and development will always be viewed *educationally*, as a values-directed endeavour, concerned with student wellbeing and flourishing, where students are viewed as sentient, emotional and relational beings.

THE LEARNER, EMBODIED COGNITION AND THE AFFORDANCE OF LEARNING ENVIRONMENTS

LEARNING OUTCOMES

By the end of this chapter, you will:

- Understand the role that analogy and metaphor play in all human thinking, including their role in science
- Know what is meant by the analogy 'classrooms are ecosystems' and understand the similarity relations involved
- Understand what is meant by the ecology of 'affordance', 'embodied cognition' and 'perception–action loops'
- Understand how the formation of concepts, including abstract concepts, is grounded in perception–action experiences, and appreciate how the embodied learner learns through doing
- Know that physical learning space has a significant influence in shaping students' learning experiences and understand how the notion of affordance can be applied when designing innovative physical learning environments in schools

Analogy, metaphor and classrooms as ecosystems

Learning never occurs in a vacuum; it is always highly contextualised. When studying the science of learning, it is therefore important to keep the multiple contexts in which learning occurs in view. Learning begins when we are in the womb and continues until the day we die. We are learning creatures. As you have heard throughout this text, all learning occurs in the brain: it is a core activity of our brain, necessary for our survival and wellbeing. Embodied brains are also inherently embedded, physically, socially and culturally. Learning is therefore also embedded – physically, socially and culturally – in multiple learning environments.

To accommodate the kind of learning that occurs in schools, we create learning environments that we call classrooms – although of course classrooms can take many different forms within buildings and in open spaces. In this chapter, we will apply ideas drawn from science in order to deepen our understanding of classrooms (using that word generically to represent the many different kinds of leaning environment found in schools). In this first section, we will explore the analogy of *classrooms as ecosystems*; however, by way of introduction we need to say something about the use of *analogy* and *metaphor* in everyday language and the important role it plays in science, in terms of generating and presenting new ideas.

REFLECTION

Think back to your own experience of the many different learning environments that can be found in schools. In what ways did these differences impact students' learning and the teaching methods and strategies that were used by the teachers working in those environments?

The role of analogy and metaphor in science

The language of science can often seem very opaque and not easy to understand, which can be very off-putting to those who do not have much of a background in science. It may therefore come as a surprise to learn that scientists frequently use language in much the same way that we all use language, employing analogies and metaphors to compare things that are already known to describe things we are trying to understand – or understand more deeply. Metaphors 'are crucial in the production of knowledge in that they allow us to make concrete connections between abstract concepts and everyday experiences' and 'the language of science is largely metaphorical' (Taylor & Dewsbury, 2018, p. 1). It often comes as a surprise to scientists to be reminded how much they rely on analogy when generating new hypotheses, as well as the extensive role of metaphor in communicating their research findings.

In Chapter 4, we saw Steven Mithen argue that 'the human passion for analogy and metaphor (Mithen, 1996, p. 52) is a product of *cognitive fluidity*, which he believes laid the foundation for the modern human achievement of art, religion and science. By 'cognitive fluidity', Mithen means the ability to interweave and integrate the specialist knowledge our ancestors gained from interacting with their world and solving the many problems they faced, such as social skills, communicating, tool-making and hunting. Although these abilities can be found in other animals and our earliest ancestors, they are not integrated. The achievement of cognitive fluidity by modern humans (Homo sapiens), opened up 'the possibility for the use of powerful metaphors and analogy, without which science could not exist' (1996, p. 246). We also saw, John Geake use the term *fluid analogising* (Geake, 2009, p. 95) when discussing creativity and imagination. He argues that 'the essence of intelligent behaviour lies in making insightful metaphors and analogies' (2009, p. 95).

An analogy is a comparison between two things that have similarities, but also may have important differences. Within classrooms, teachers frequently use analogies to help explain difficult ideas. Geake goes so far as to say that 'a characteristic of good teachers is their ability to create appropriate analogies for explanation and clarification' (Geake, 2009, p. 95). Young children often ask searching questions: 'Why do we have to plug in the television to make it work?' The teacher replies, 'Because it uses electricity and electricity comes through the wires.' But the child is not satisfied, asking, 'What's electricity and how does it come through wires?' Respecting the child's inquisitiveness, the teacher might explain, 'Well, electricity is a bit like water and wires are a bit like a garden hose – the electricity *flows* through the wires, from the plug in the wall to the television, to make it work.' The child thinks for a while and then asks, 'Is electricity wet?' The teacher has used an analogy with valid similarities, but also differences that are potentially misleading.

Within science, analogy is very frequently used to assist explanation, but it also plays a heuristic (discovery or problem-solving) role when trying to discover and understand difficult or puzzling phenomena. In the seventeenth century, Christian Huygens (1629–95) based his theory of light on the analogy of *waves* in water, and the same analogy can be used to explain properties of sound. Go to the beach and watch the waves, and you will quickly notice that they have different heights (amplitude) and different frequencies with which they arrive at the shore; they also appear to bounce (reflect) off objects such as rock, and the wave forms are obviously being propagated in the water. All very familiar stuff – especially to those who love the sea. Can these same properties be applied, by analogy, to understanding sound and light?

In Figure 9.1, you can see the similarity relations of this analogy. This is not trivial; in the twentieth century, the property of reflection in water waves was exploited in the development of RADAR – without which present-day air traffic control would be impossible. However, in the nineteenth century, the validity of these analogical relations was hotly disputed by scientists.

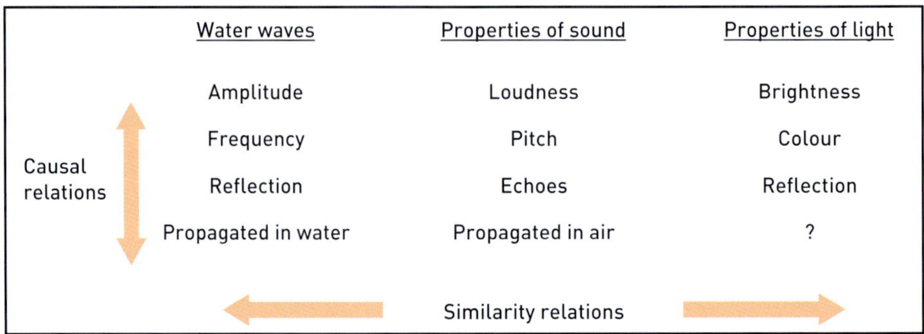

Figure 9.1 Similarity relations in applying the metaphor of 'waves' to sound and light

Notice the question mark in regard to the propagation of light. Does light require a medium for its propagation, as water waves need water and sound waves need air? Some scientists said yes, and they postulated the existence of what they called the 'luminiferous aether'. At the same time, however, other scientists following Isaac Newton (1642–1727) believed that light didn't travel like waves at all. They maintained that the analogy was wrong: rather, light travels as particles, what they called corpuscles. As one historian of science notes:

> A contentious issue between the two schools was the luminiferous aether, which was considered to be an essential part of the wave theory. Although no direct, independent evidence was available to confirm the aether, Airy accepted its existence as almost certain and supported his belief by the analogy between the propagation of light in the aether and the propagation of sound in air. (Cantor, 1975, p. 122)

Metaphors also make comparisons between different phenomena, and often contain a core analogy; however, unlike analogies, metaphors imaginatively structure language by asserting that one thing *is* another. In everyday language, for example, affection *is* warmth (e.g. I was greeted very *warmly*); difficulties *are* burdens (e.g. she is *weighed down* with responsibilities).

George Lakoff and Mark Johnson (1999) made an important contribution to our understanding of how central metaphors are to every aspect of our thinking in claiming that all human knowledge *emerges* from our *embodied physical and social experiences*. Metaphors aren't just poetic embellishments within language, as often described; they are deeply embodied and shape the very concepts we use. They say, 'concepts are created as a result of the way the brain and body are structured and the way they function in interpersonal relations and in the physical world' (Lakoff & Johnson, 1999, p. 37). Does this really matter? Yes, enormously – for example, consider the *age-old perception/conception dichotomy* that you will continue to find in some textbooks, the idea that we use our body, our physical senses in perception, but conception is purely mental (independent and entirely separate from our ability to perceive and act. The claim that *concepts are metaphorical and also embodied* undercuts this perennial perception/conception dichotomy and the associated philosophical notion that there can be such a thing as 'pure reason'.

In biology, we have all heard the word 'cell' used over and over again. In this book, we have had a lot to say about brain cells (neurons), but where did that word come from? The person who introduced the word 'cell' into science was the English natural philosopher Robert Hooke (1635–1703), who was one of the first to use a microscope to examine the construction of plants and other natural objects. It is said that the structure of plants reminded him of the little rooms used by monks in monasteries that are called cells, hence the term 'cell' in biology, used as a metaphor.

Also in biology, we hear of genetic *blueprints*, ecological *footprints*, food *chains* and so on. All these are metaphors, but it is very important to note that some may be very misleading, even to the point of reinforcing outdated ideas and 'contributing to public misunderstanding about complex scientific issues' (Taylor & Dewsbury, 2018). This is particularly the case in regard to the metaphors of genes as 'blueprints', or 'codes' or 'instructions', which can reinforce genetic determinism and ignore the lessons of epigenetics, which you studied in Chapter 7.

Electricity is not wet, light waves do not require the medium of a luminiferous aether and genes are not blueprints. As Cynthia Taylor and Bryan Dewsbury (2018, p. 2) note:

> If genes really do function as blueprints, we should expect a one-to-one correspondence between particular genetic 'instructions' and phenotypic outcomes in organisms, with limited input from the environment in structuring variation between individuals. Yet this is not the case. Often, single genes can, and do, direct multiple phenotypic outcomes through epigenetic processes that are responsive to the environment.

The key point is that *we cannot escape the use of analogies and metaphors* in all of our thinking, and in science, but we also need to ensure we are not over-stretching the positive and helpful comparisons they engender. Teachers need to be particularly aware, when using analogies and metaphors to clarify difficult ideas, that they are not also unintentionally misleading their students – electricity may be said to *flow* through wires, like water through a hose, but it is not wet. With that important caution in mind, let us use the scientific term 'ecosystem' as an analogy to try to deepen our understanding of classrooms as learning environments.

The analogy of classrooms as ecosystems

To set the scene, let us provide a bit of background context. In 1994, in the United Kingdom, pre-service teacher education was about to undergo a massive change, initiated by government legislation. Initial teacher training had hitherto been conducted mainly in universities, largely under the control of each university (though also under the watchful eye of an accreditation body). But it was about to become much more *school-based*. Briefly, this meant that students would spend a lot more time in schools and the practical training of teachers would be a *partnership* between universities and schools, with schools in the lead. Instead of student teachers undertaking two or three short periods of teaching practice or practicum in one or more schools during their teacher education course, they would spend much more time in carefully selected schools, in the role of mentees or interns, where a designated experienced teacher would act as a mentor and the school would have a major say in assessing the student's professional performance.

Clearly this had major resourcing implication that were not only financial but also intellectual. For example, in forging such a partnership, school teachers (mentors) and university lecturers would presumably need some kind of *shared framework* for observing and assessing student teachers when in schools. Observation is not a neutral activity, there is much more to it than the careful and attentive use of our senses. What any given observer 'sees' is strongly influenced by *past experience, current knowledge, expectations, beliefs* and much more. That is what is meant by the *theory-dependent nature of observation* (Mulkay, 1979) and it applies to everyday observation, observation in science and when observing teachers.

When applied in education, this means that any two people – for example, the mentor and lecturer – observing the same lesson will likely interpret and assess what they see differently, including whether the lesson was educationally worthwhile or not. But more than that, what they actually *see* (what they actually notice, when observing) might be different, even though the images reaching the retinas in their eyes are identical. Observation is *highly selective*, as noted in Chapter 4, where the gorilla experiment was cited as an example.

Moreover, the theory dependence and selectivity of observation is compounded in education because classrooms are highly *dynamic*, *complex*, *multifaceted* and *shifting* environments, so understanding what is going on is not at all self-evident. Furthermore, a major disruption can result from a quite small event and external influences, such as weather, time of day, day of the week and what occurred just prior to any lesson, can strongly influence what actually happens in the lesson being observed.

Given these many complicating factors when observing lessons and analysing what is actually going on, it was argued that using the analogy of *classrooms as ecosystems* could provide a common perceptual and conceptual framework (Lambert & Sankey, 1994) (Figure 9.2). Although the notion of classrooms as multifaced environments had previously been used in research (Doyle, 1977), Lambert and Sankey took the idea forward by suggesting six similarity relations. The basic analogy is set out in Table 9.1. The similarity relations are those employed in the original paper, though the descriptions have been updated to include the language of complexity (dynamic systems) theory.

Table 9.1 Classrooms as ecosystems analogy

Ecosystems	Similarity relations	Classrooms
Ecosystems are complex and self-organising; they teem with activity that is highly dynamic and variable, resulting in constant change.	Dynamic and variable	Classrooms are complex and self-organising; they teem with activity that is highly dynamic and variable, resulting in constant change.
Ecosystems are emergent and have a history, where previous events feed into and influence current events.	A given history	Classrooms are emergent and have a history, where previous events feed into and influence current events.
Ecosystems are fragile; they are easily impacted and disrupted by events large and small.	Fragile	Classrooms are fragile; they are easily impacted and disrupted by events large and small.
Ecosystems comprise many and varied parts (non-living and living, multiple species and processes), working together to produce the whole.	Part/whole relations	Classrooms comprise many and varied parts (students, teacher, furniture, resources, pedagogy), working together to produce the whole.
Ecosystems thrive when functioning with a high degree of balance. When out of balance, they quickly become dysfunctional and chaotic.	Striving for balance	Classrooms thrive when functioning with a high degree of balance. When out of balance, they quickly become dysfunctional and chaotic.
The dynamic interrelational processes that constitute an ecosystem are not self-evident; they only become apparent when it is realised that what is observed is operating as a systematic whole.	Not self-evident	The dynamic interrelational processes that constitute a classroom are not self-evident; they only become apparent when it is realised that what is observed is operating as a systematic whole.

Source: Based on Lambert & Sankey (1994).

Figure 9.2 Two ecosystems: a rainforest, and a classroom

Take a close and careful look at the way classrooms are described when using the analogy of classrooms as ecosystems. Check the right-hand column – can you think of examples to illustrate those descriptions, drawn from your experience of classrooms?

Affordance, embodied cognition and perception–action loops

embodied cognition: the belief that human thinking and reasoning are not an autonomous, disembodied mental process, but rather deeply dependent on features of the physical body

In the quest to deepen our understanding of classrooms, in this section we will stay with the science of ecology and explore the notion of *affordance* with regard to learning environments. We will also draw on what has previously been said about observation as we focus on the *nature of perception* and we will pick up on the previous discussion of *analogy and metaphor* in science, looking at how they can both help and mislead. We will also incorporate the idea that *concepts are embodied*, thus rejecting the perception/conception dichotomy. All of this will take us to the notion of **embodied cognition** and how it is challenging **cognitivism or representationalism**, which has been the standard model of cognition, hugely influential across may disciplines over the past half-century, including education. And it generally still remains so, not least because of its strong metaphorical alliances with computer learning.

cognitivism or representationalism: the belief that 'thinking can be understood in terms of the representational structures in the mind and computational procedures that operate on those structures' (Thagard, 2014, para. 11)

The affordance of the learning environment

affordance: a term coined by James J. Gibson to draw attention to the necessary relationship between an animal and its environment and what that environment 'offers the animal, what it provides or furnishes either for good or ill' (Gibson, 1979, p. 127)

The term **affordance** was first coined by James J. Gibson (1904–79) (Figure 9.3). Although the verb 'afford' already existed in English, Gibson turned it into a noun to draw attention to the necessary relationship between an animal and its environment, and what that environment 'offers the animal, what it provides or furnishes either for good or ill' (Gibson, 1999, p. 127). It should however be noted that James Gibson did not work alone, he often worked closely his wife Eleanor Gibson (1910–2002), also an eminent scientist, especially in jointly developing what they called *ecological perception and leaning*.

Figure 9.3 James J. Gibson (1904–1979) and Eleanor Gibson (1910–2002)

Gibson argued that if we are to understand the behaviour of animals, they have to be viewed as embedded in an environment – the animal and its environment are not separate items; they can only be understood in relation to one another. Richardson (2000, p. 106) says that the **organism-in-environment** 'is the only unit of analysis when affordances are the language of psychology'. When applied in the context of education, the unit of analysis is *learner-in-learning environment*. Despite what happens in much research focused on learning, the learner should not 'be abstracted and studied in isolation from the dynamics of the learning environment in which she is nested' (Duncan et al., 2021, p. 9).

organism-in-environment: an alternative approach to unite an organism and the environment in which it is situated, in opposition to pure 'organicism' or 'environmentalism' in developmental science

However, it is easy to see how this holistic insight put Gibson in conflict with the dominant research method used in psychology when studying human behaviour and learning, which is to extract the learner from the learning environment and conduct laboratory-based controlled experimentation. Within the context of education research, Gibson's notion of affordance and the notion of the *learner-in-learning environment* strongly argues for school-based research. Whether it employs conventional methods such as questionnaires or the use of brain technology such as EEG, it should be conducted in the *natural setting* of the school. In Chapter 10, you will be introduced to one such school-based EEG study.

Viewing the animal as embedded in a highly dynamic environment or ecosystem is, of course, fundamental to ecology, and Gibson (1999) described his approach as 'ecological psychology'. The core analogies in his ecological approach are inherently organismic, unlike the many mechanistic analogies found in cognitivism. Indeed, his account of human thinking and learning was in stark contrast to both behaviourism and cognitivism. He 'opposed any thinking which was not premised on the assumption that to understand the nature of the organism one had to understand it through its relationship with its ecology. This put him at odds with existing theorists' (Richardson, 2000, p. 94). For the past 50 years or so, cognitivism, or representationalism, has largely dominated psychological theories of thinking and learning. It is closely aligned with computing and frequently uses computer-based analogies – for example, the term 'hard-wired' to refer to what is innate or known without learning. However, that analogy can so easily reinforce belief in genetic determinism.

Classic cognitivism and the 'brain is a computer' metaphor

The core metaphor, the *brain is a computer*, has taken hold over the past half-century, to the point where it is often no longer presented as a metaphor, but rather taken to be a literal account of how the brain works – precisely like an advanced machine.

The main assumption of classical cognitivism (or representationalism) has been that 'thinking can be understood in terms of the representational structures in the mind and computational procedures that operate on those structures' (Thagard, 2014, para. 11). Figure 9.4 provides a summary of how, in cognitivism, the brain is conceived as a disembodied central processing unit (a computer). According to cognitivism, the brain receives information from the world via the five bodily sense (sight, smell, sound, touch, taste) and categorises it using representations. These representations are processed using disembodied, abstract rules, schemas and logical operations of one kind or another, resulting in thought and action, in the same way that computers process algorithms.

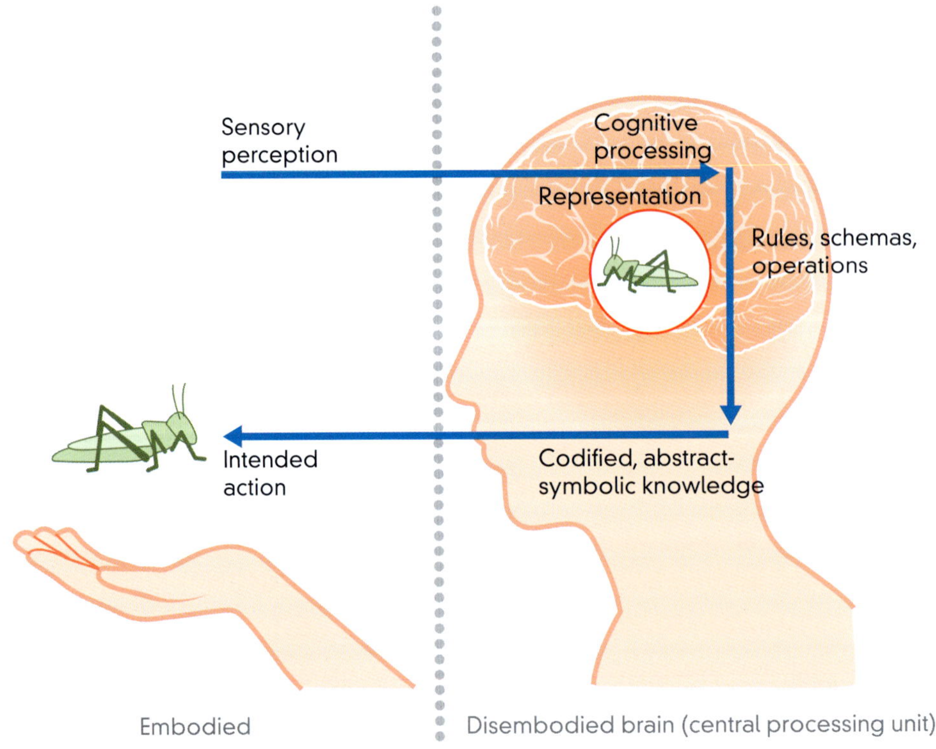

Figure 9.4 A diagrammatic representation of disembodied cognition

For example (see Figure 9.4), when a learner gains a conceptual understanding of 'insect', sensory or motor features of objects and events (e.g. sounds of insects, shapes, colours) are said to be transformed into a common symbolic representation format (i.e. animal with six legs), and the original sensory information is lost (Kiefer & Pulvermüller, 2012; Sheya & Smith, 2019). Piaget viewed these putative symbolic representations as distinctive feature of what he called the 'formal operation' stage of thinking, which he claimed follows the stage of concrete operations at around the age of 11 or 12 years.

REFLECTION

Thinking back over your own development, you will be aware of many changes occurring around the age of 11 and 12, but do you remember the seismic transition from concrete reasoning to abstract (formal operational) reasoning that Piaget postulated?

According to the classic cognitivist account, cognition is conceived as 'fundamentally distinct and separate from sensory-motor process' and it is 'this symbolic character that accounts for the varied phenomena of cognition and that explains the structure of the cognitive system' (Sheya & Smith, 2019, p. 1266). However, this 'places cognition between the perceptual processes that provide the input and the motor processes that execute its output' (Pezzulo & Cisek, 2016, p. 414). Essentially, this is a 'serial sense–think–act model of behaviour' (2016, p. 414). James and Eleanor Gibson's ecological approach is opposed to the cognitivist separation of perception (sensing), cognition (thinking) and acting. It is similarly opposed to the behaviourist's input/output separation of stimulus and response, which also inserts cognition as the activity of the brain (conceived as a largely unknowable black box) between perception and action.

Gibson argued that just as we should not separate the animal from the environment when trying to understand the animal and its behaviour, we should also not attempt to separate *perception, thought and action* into separate mental faculties. Nor, for that matter, should we separate out other parts of the organism, such as genes or brains, and give them priority. Thus, there is no place in Gibson's thinking for the notion of 'a complex mind divorced from a simple body or simple perception' (Richardson, 2000, p. 106). Given his ecological perspective, Gibson realised that what was required was a way of conceptualising the parts working together as a dynamic, functioning whole. In 1979, he wrote, 'What psychology needs is the kind of thinking that is beginning to be attempted in what is loosely called systems theory' (Gibson, 1979, p. 2). Indeed, his important insights into the ecology of perception, thought and action contributed to the then-emerging dynamic systems account of human development, which you studied in Chapter 3.

However, that was over a quarter of a century ago. Much more recently, Gibson's work on affordance and perception, and the unity of perception, cognition and action, has contributed to what is now called the new science of embodied cognition or embodied cognitive science. This encompasses aspects of dynamic systems thinking, the seminal work of Lakoff and Johnson on metaphor (mentioned above) and the notion of the embodied brain. The core claim of embodied cognition is that human thinking and reasoning are not an autonomous, disembodied mental process, but rather 'deeply dependent on features of the physical body'

(Wilson & Lucia, 2017, p. 1) and the body plays 'a significant causal or physically constitutive role in cognitive processing' (2017, p. 28). In other words, the brain is not disembodied, nor is it alone in producing thought and action. Rather, our bodies are intimately involved in the process of cognition, to the point where perception and cognition (abstraction) overlap and it is often difficult to know where one ends and the other begins.

The brain as a feedback control system with perception–action feedback loops

Figure 9.5 attempts to capture the main features of this holistic model. The brain is conceived not as an information processing system, where its main purpose is to understand the world, but rather as a **feedback control system** (Pezzulo & Cisek, 2016), the main purpose of which is to guide *interaction* with the world through the lived experience of the perceiving organism, the perceiving child. The system incorporates perception and action loops that continuously *sample* the environment, make *predictions* on the basis of that sampling and *act* in response. However, this is not a one-way process: each element impacts the others in the overall dynamic of perception, thought and action.

feedback control system: a system that is able to keep itself controlled within a given range via monitoring congruence and incongruence between the reference signal and perceived signal (feedback). The brain is proposed to be a feedback control system.

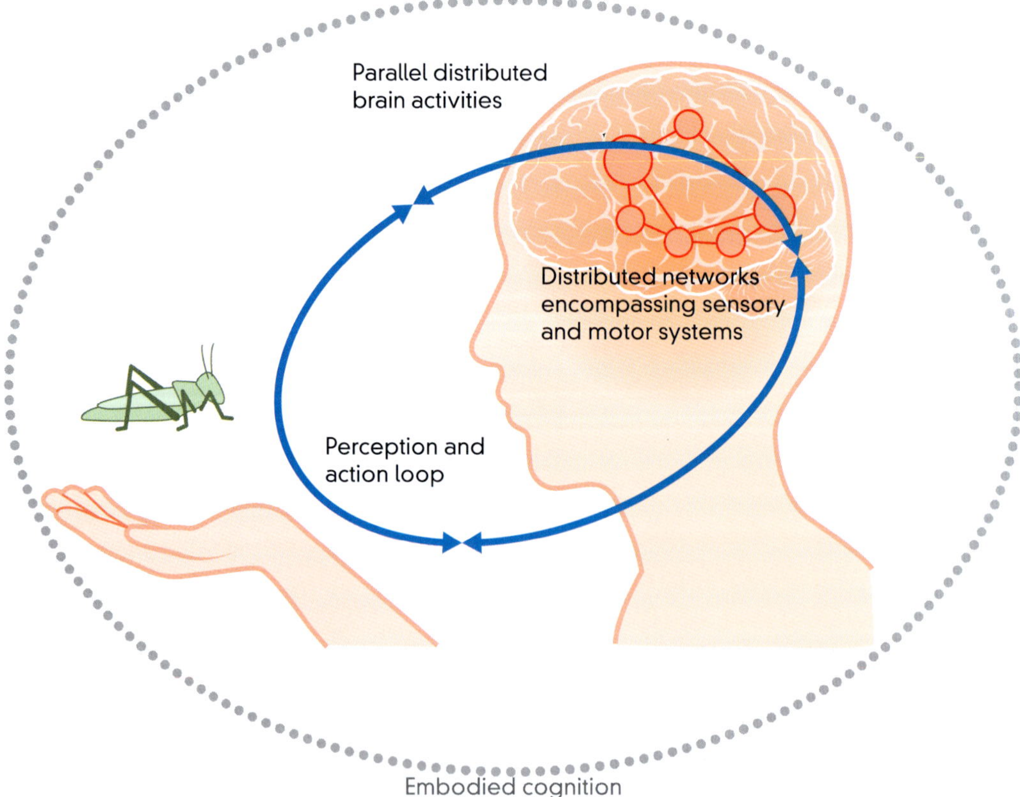

Figure 9.5 A diagrammatic representation of embodied cognition

In the embodied cognition model, the divide erected by classic cognitivism between 'embodied perception' and 'disembodied thought' is gone – perception, thought and action are all embodied. Nor is there any need for disembodied, abstract rules, schemas and logical operations. The embodied cognition model questions the need for representations or representational structures in the brain, and argues instead for the importance of concepts. Moreover, *abstract concepts* are *not* conceived as a special and higher kind of thinking, but rather as emergent flexible entities that are constituted by distributed networks within the brain, encompassing sensory and motor areas (Kiefer & Pulvermüller, 2012; Sheya & Smith, 2019). In short, 'cognitive processes emerge out of and are dynamically coupled to sensory-motor systems' (Sheya & Smith, 2019, p. 1266) and the 'classically proposed dichotomy between so-called concrete and abstract concepts does not exist' (Kiefer & Trumpp, 2012, p. 19).

Paradigm change: The current position

You will probably not be surprised to know that these ideas meet with considerable opposition from classic cognitivists. Strategically, many such as Steven Weisberg and Nora Newcome try to incorporate embodiment while clinging to the cognitivist notion of cognitive representations in the head, which is contrary to Gibson's anti-representationalist position. While acknowledging that embodied cognition and embodied learning are offering 'special promise for bolstering learning in science, technology, engineering and mathematics' (Weisberg & Newcome, 2017, p. 1), the so-called STEM disciplines, they seem determined to stay true to their cognitivist roots. By contrast, distinguished brain scientist and philosopher Walter Freeman (1927–2016) agreed with Gibson that the brain doesn't use representations. He said, 'representations exist only in the world and have no meaning', whereas, 'meanings exist only in brains without being represented there' (Freeman, 2000, p. 16).

One way to understand what is going on with these many contested positions in science is to invoke the notion of *paradigm change*. This notion goes back to the 1960s and Thomas Kuhn (1922–96), who wrote a quite explosive book called *The Structure of Scientific Revolutions* (1962). It was explosive because it claimed that science does not progress in a cumulative way, with knowledge building on knowledge to produce new knowledge; instead, big changes in science usually result from revolutionary change, when old ideas (theories) are thrown out and replaced by new ones. In the process of revolution, some scientists will be converted to the new ideas although others may offer resistance – which will end when they die.

In describing the process of revolution, Kuhn claimed that rival paradigms are incommensurate (it is not possible to understand one paradigm using the ideas of a rival paradigm). Those in rival camps therefore seem to talk past each other, sharing little common ground. However, this might suggest that scientific change is irrational and relativistic (there is no truth, only your truth and my truth). Although Kuhn strongly denied being a relativist, social theorists were happy to grasp his apparent relativism and use the notion of the **incommensurability of scientific theory** in building an anti-realist account of science (e.g. Mulkay, 1979). This debate certainly had an impact on the study of education and education research, especially in the Australian context (Cole, 1997).

The point for now, however, is that if one looks at the current debate in regard to cognitivism versus embodied cognitivism and the notion of representations (whether they are in the world or in the head, and if so, where), it seems to bear many of the hallmarks of a

incommensurability of scientific theory: the idea that there are no shared, objective standards for comparing or evaluating competing theories

Kuhnian scientific revolution. Classic cognitivism and embodied cognitivism are two alternative paradigms where the proponents of each rival camp seem to talk past each other. Indeed, one aspect of that debate, concerning the nature of concepts (whether they are symbolic representations located in a central hub in the brain or, alternatively, related to the sensory motor areas of the brain), has been described as having reached an 'impasse', with both camps claiming that 'the same empirical evidence is compatible with their views' (Michel, 2021, p. 655).

The view taken in this text is that, in time, the notion of the embodied cognition will likely prevail. Within education, it seems to offer a richer and more comprehensive account of learning, one that is entirely consistent with brain science and dynamic systems theory. It also encompasses the notion of affordance, which is especially relevant in understanding classrooms, as we discuss in final section of this chapter.

The embodied learner and learning through doing

According to embodied cognition, perception and cognition are both closely related to action (doing). Cognition, our thinking and reasoning (including abstract concepts), is grounded in our senses (perception) and in our interaction with the environment (our doings). This means that even our 'most complex thoughts are sense-based and not abstract-symbolic' (Kiefer & Trumpp, 2012, p. 19), as cognitivism has claimed.

Admittedly, at first sight this might appear to be very dubious. From a philosophical, analytical perspective, for example, it seems to involve what is called a **category error**, mistakenly applying things that belong to one distinct category to a different distinct category where they do not apply – in this case, incorrectly applying the word 'abstract' to our physical, perceptual interaction with the world. A quick look at any standard dictionary will reveal that the word 'abstract' is defined as 'having no reference to material objects', such as those perceived and acted on in the world. But dictionary definitions and the meanings of words are inevitably influenced by conventional wisdom and theory, and when that changes, related definitions of words should also change.

category error: mistaken application of things that belong to one distinct category to a different distinct category to which they do not apply

Even so, how can abstract concepts in mathematics and concepts such as 'action potential' in brain science, 'luminiferous aether' in physics or notions such as 'justice' in law or 'liberty' and 'human dignity' in moral theory – all of which you have met in this book – be grounded in the sensory (perception) and motor (action) systems of the brain? And what is the evidence?

Well, as mentioned in the previous section, currently the evidence for embodied cognition remains contested. Nevertheless, the evidence is growing and is expected to continue to grow as more studies are conducted. Shortly, we will take what might be the difficult problem and consider evidence related to abstract mathematical concepts and also one example from physics, concerning sound. However, let's start with a very practical example of cognition being closely related to doing, involving our sense experience and our interaction with the environment (our doings). How do teachers learn to teach and become expert at it?

We are entering controversial territory, the kind that existed in the 1990s in the United Kingdom, mentioned above, when teacher education became school based – a partnership with schools in which student teachers spent a lot of time in schools as mentees or interns as part of their training. Although imposed by government legislation, starting in 1994, two major teacher education institutions (at London and Oxford Universities) had already paved the way by adopting a school-based approach (Sankey, 1996). Part of the reason for this change was the claim that student teachers could not learn to teach by listening to lectures or attending seminars at the university; they needed to be in school, *learning to teach by doing it*. Needless to say, not all teacher educators agreed. Even if they agreed that teachers learnt to teach through doing it, they still insisted that they required lectures and seminars on teaching methods.

REFLECTION

So, what do you think? What are the skills of teaching that need to be learnt? What can only be learnt about the skills of teaching by actually doing it, practically, and what can only be learnt by study, reading and discussion? Discuss this with classmates.

Whatever conclusions you reach regarding what is involved in learning to teach, you will probably find that most experienced, practising teachers will emphasise the importance of learning to teach through doing. One of the main reasons why learning through doing is important is the 'tacit' nature of much classroom teaching.

The term 'tacit knowledge' was coined by Michael Polanyi (1891–1976), a distinguished chemist who turned to the philosophy of science and whose ideas regarding the nature of science and scientific knowledge strongly influenced Thomas Kuhn and his notion of scientific revolutions. The concept of tacit knowledge (Polanyi, 1983) is closely associated with embodied knowledge – it means a kind of 'gut feeling'. It is *implicit* knowledge, as opposed to explicit, codified knowledge. Tacit knowledge is often difficult to put into words and, significantly, it is generated in personal experience and is highly contextualised – related to actual events and circumstances. All of which means that the tacit aspects of teaching can only be learnt through doing, through experience and critical reflection on experience.

Embodied cognition, learning and education

RESEARCH LINK 9.1

Kiefer, M. & Trumpp, N. M. (2012). Embodiment theory and education: The foundations of cognition in perception and action. *Trends in Neuroscience and Education*, 1(1), 15–20.

Over the last a few decades, empirical research on embodied cognition has allowed us to answer important questions about learning and education. By collecting several different measures related to learning – both observable performance in assessments and brain activities recorded from experimental situations – these empirical studies offer important insights into how students learn and how educational experiences should be designed.

Kiefer and Trumpp's (2012) paper provides a comprehensive overview of the recent data on embodied cognition in different areas (e.g. literacy, numeracy) and explores the implications for teaching and learning. Below are their key findings from recent empirical studies:

- Handwriting training, which links sensory-motor areas to the region implicated in letter recognition, improves subsequent reading performance. However, typewriting does not improve reading performance as much as handwriting training. Therefore, young children's learning of handwriting should be prioritised over the learning of typewriting with digital devices.
- Establishing relevant sensory and motor traces (by observing, performing actions) during learning improves subsequent memory performance more than when learning the same content in purely linguistic form.
- 'Abstract number concepts are at least partially rooted in our motor experiences' (Kiefer & Trumpp, 2012, p. 19). Explicitly training children in finger counting (or alternative visual spatial experiences) can accelerate learning numbers and has beneficial effects on later mathematical performance in schools.

The sensorimotor grounding of abstract concepts

Are mathematical reasoning and conceptualisation grounded in perception and action? Well, with regard to the issue of embodiment, how did you learn to count? It is very likely that you used your fingers, and you probably still do on some occasions. This is a very good example of embodied cognition. In finger counting, perception and action, seeing and doing are brought together when doing basic arithmetic. Moreover, it uses the sensorimotor networks in the brain. Indeed, the 'finger sensorimotor system grounds numerical processes' (Soylu et al., 2018, p. 108).

Of course, it could be argued that this is basic maths, and surely advanced mathematics is different. However, as we found in Chapter 5, there is compelling evidence that 'advanced mathematics, basic mathematics, and even the mere viewing of numbers and formulas recruit similar and overlapping cortical sites' (Amalric & Dehaene, 2016, p. 4913). In other words, the brain does not have a separate processing area for advanced maths; it employs the same systems used for basic mathematics.

Moreover, this same study showed that when professional mathematicians reflect on highly abstract maths concepts or statements that do not involve Arabic numbers, their brains still recruit the areas we usually use when recognising numbers and running single digit arithmetic calculations. In particular, two regions that overlap across different levels of abstract mathematical engagement are inferior temporal cortex and intraparietal sulcus (IPS). The inferior temporal cortex (lower portion of the temporal lobe, Brodmann area 20) 'is a key part of the ventral visual pathway implicated in object, face and scene perception' (Conway, 2018, p. 381). Areas of the intraparietal sulcus (located at the base of the superior parietal lobule, Brodmann area 7) are highly involved in sensorimotor functions, including 'planning of eye movements, grasping movements, reaching and head movements' (Kaas, 2012, p. 1098).

As we have seen in this chapter, according to classical cognitivism, abstract conceptual knowledge is said to be represented in a unitary 'conceptual centre or hub'; it is said to be symbolic and amodal, meaning that it is distinct from the sensory and motor brain systems (Trumpp et al., 2013). However, this study shows that the common ground in the brain across different levels of mathematical abstraction is the sensory and motor (doing) areas, which is what the embodied view of cognition claims. Additional evidence that concepts are grounded in the brain's sensorimotor networks and are not distinct from the sensory and motor brain systems, as classical cognitivists claim, is coming from lesion studies (Trumpp et al., 2013).

Elsewhere, we have mentioned the importance of lesion (brain damage) studies in understanding the workings of the human brain. In Chapter 2 you met HM and in Chapter 6 Phineas Gage. Now let us meet 33-year-old JR, a German-speaking male who in 2009 was admitted to hospital in Tubingen, Germany with a large abscess in the left posterior superior/middle temporal gyrus of his brain, which was causing him to have seizures. (The left temporal gyrus is located in the left temporal lobe. The superior temporal gyrus is largely Brodmann areas 41 and 42; the middle temporal gyrus is largely Brodmann area 21.) The location of his lesion is known to be associated with conceptual and perceptual processing of sound.

Briefly, JR was given 200 visually presented stimuli in the form of 'words'. There were 100 proper words and 100 pseudowords (word/pseudoword decision methodology) and JR was asked whether or not each one was a word. (The pseudowords were pronounceable as letter strings but did not have any meanings. They were created from real words by replacing one or two letters.) Among the 100 proper words, 50 words referred to objects with high acoustic features (sound-related concepts such as rooster or bell). The other 50 words had low relevance of acoustic features (e.g. rabbit, rope). Each stimulus (either word or pseudoword) was visually presented on a computer screen in randomised order (Trumpp et al., 2013). JR's performance was compared with healthy young adults' performance (control group). After watching each stimulus on the screen, JR 'had to decide as quickly and accurately as possible whether the target stimulus was a meaningful word or a meaningless pseudoword, and indicating their decision by pressing a button with the index and middle finger of his right hand' (Trumpp et al., 2013, p. 479).

JR made a high number of errors when deciding whether each item was a proper word, *if the words had high acoustic features*. When the words did not have acoustic features, he did not make any errors at all. JR's lesion therefore provides evidence that this auditory region 'is *necessary for both conceptual and perceptual sound processing* of everyday objects' (Trumpp et al., 2013, p. 476), and the concepts (words) are employing sensory brain networks, as embodied cognitivists claim.

A clear message from the JR case study is that sensorimotor (perception/action) experiences are an integral part of learning abstract concepts, such as those introduced in school. Also, embodied sensory experiences not only help us learn new abstract concepts, but also apply them efficiently to real-life problems. If we lose such subordinating sensory correlates in the brain, as exemplified in the case of JR, we become unable to use abstract concepts when categorising names of objects. So, what does this mean for teachers who have to introduce numerous abstract concepts in their everyday classrooms?

For example, can teachers enhance students' learning of abstract concepts by carefully incorporating sensorimotor experiences within the learning experiences?

Learning through doing in education

The importance of 'experience' has been recognised in education for more than a century. In his 1938 book *Experience and Education*, John Dewey (1859–1952) argued that learning occurs through hands-on experiences being transformed into knowledge. This same emphasis on experiential learning was provided by David A. Kolb (1984), however, it is only recently that processes underpinning **experiential learning** have been understood. It is now recognised as a form of embodied cognition, and its neural mechanisms (see Figure 9.6) now help us understand why 'hands-on experiences' are vital in the process of learning abstract concepts.

experiential learning: 'the process whereby knowledge is created through the transformation of experience' (Kolb, 1984, p. 38)

In the context of learning science, it is now appreciated that 'hands-on' learning experiences can yield a more enhanced learning of abstract concepts when compared, for example, with passively observing a teacher demonstrate the concept to the whole class (Kontra et al., 2012). Even though students are learning the same scientific concept, those who engage with manual experiments show greater accuracy in their conceptual understandings because the hands-on experience prompts activation of sensory-motor areas when learning the concept, which are then reactivated when students subsequently engage with the concept (Kontra et al., 2015). Research link 9.2 compares two different learning experiences, *hands-on action* vs *observation*, within the context of a university physics course.

Figure 9.6 Hands-on learning in a science classroom

Embodied cognition also helps to explain why hands-on experience can influence learning in a classroom. In Figure 9.6, objects and gestures enhance the learning of abstract concepts by providing a physical medium in which they can be experienced and expressed (Goldin-Meadow et al., 2010; Goldin-Meadow & Beilock, 2010; Kontra et al., 2012). Evidence comes from a research experiment involving a group of novice students in an organic chemistry class.

Initially, prior to attending a training module, the research team assessed the strategies that each novice student used when solving a number of chemistry problems. Students used both verbal strategies (e.g. a student verbally describes 'if reflecting the molecule over a mirror plane then it wouldn't match it up with the original') and gestural strategies (e.g. a student places one hand over one image of the molecule and then flips their hand from palm down to palm up over the top of the original molecule). Sometimes their strategies were relevant, but they used those strategies inappropriately. The researchers counted the number of times each student used the three strategies: verbal strategies only; gesture only; in both speech and gesture.

Students then learnt the concept of 'stereoisomers of molecules', which was totally new to them, and were then asked to solve problems on stereoisomers. The post-training test found that the number of relevant gestural strategies that participants used predicted high performance. The number of relevant strategies expressed in speech (either alone or with gesture) did not predict high performance at all (Ping et al., 2021). This finding supports the claim that 'action experience in the form of gesture can facilitate the learning of complex math and science concepts' (Kontra et al., 2012, p. 737).

Physical experience enhances science learning

RESEARCH LINK 9.2

Kontra, C., Lyons, D. J., Fischer, S. M. & Beilock, S. L. (2015). Physical experience enhances science learning. *Psychological Science*, 26(6), 737–49.

Some science classrooms seem to be moving from traditional learning to virtual and on-line learning environments where student observe phenomena and experiments rather than physically experiencing them. But can virtual science learning replace traditional laboratory-based science learning?

An experiment carried out in a university physics course shows that the *physical experience leads to better learning* than passive observation through virtual learning. This supports the claim that it is crucial to incorporate hands-on experiences in the design of effective science curricula.

In this study (Kontra et al., 2015), students who were enrolled in a university physics class first read a description of 'angular momentum' and then took a pre-test to assess their 'understanding of the relations between physical properties of spinning objects (e.g., moment of inertia, angular velocity) and amount of torque (resistive force) exerted when spinning objects are tilted through space' (Kontra et al., 2015, p. 739). After the pre-test, students were divided into pairs, to deepen their understanding of the concept of 'angular momentum'. One student in each pair was assigned to the *action group* and the other was assigned to the *observation group*. The two groups' *pre-test* performance did not differ significantly.

The students in the *action group* were given bicycle wheels, allowing them to physically experience the consequences of tilting the wheel in different ways. The students assigned in the *observation group* passively observed their partner's experiment, noticing the 'consequences of the wheels' changing angular momentum via the laser dot on the wall' (Kontra et al., 2015, p. 739) (see the blue line and red laser dot on the wall in Figure 9.7).

During the test that ran immediately after the learning experiences, the students in the *action group showed a significant gain in accuracy* compared to those in the *observation group*. Moreover, the *action group* continued to show enhanced improvement in a test that ran several days after the experiment.

fMRI scanning shows that physical hands-on experience, compared to observation, leads to 'increased activation of sensorimotor systems' (Kontra et al., 2015, p. 742) (i.e. primary motor/somatosensory cortex, supplementary motor area/anterior cingulate cortex) important for dynamic physical concepts. Moreover, the level of this activation showed positive

▶ ▶

Figure 9.7 A student in the *action group* participates in a hands-on experiment on angular momentum
Source: Adapted from Kontra et al. (2015).

correlation with enhanced understanding of torque and angular momentum as assessed via the quiz (Kontra et al., 2015).

The researchers argue that hands-on experience with the physical world enables 'the recruitment of sensorimotor brain systems evolved for computing forces, vectors and trajectories, which aids understanding of complex science concepts involving kinetics' (Kontra et al., 2015, p. 737).

Affordance and the design of learning environments in schools

Although the notion of a learning environment is not restricted to the physical environment – for example, it can also include the teacher and pedagogy – there is an increasing recognition that the actual physical space has an undeniable influence in terms of shaping students' learning experiences (Mahat et al., 2018). Applying the notion of affordance that was introduced in the earlier section of this chapter adds to our understanding of how the physical space can influence the learning and development of students when located in that environment.

Brain processes that navigate competing affordance landscapes in school

The physical environment of the school, including the specifically designed buildings, classrooms and furniture, provides an unfolding landscape of affordances to those who are situated within that environment. At school, students and teachers constantly have to navigate through multiple existing environments while trying to achieve their plan (e.g. transiting to another classroom, teaching a planned lesson in a particular classroom environment with its particular affordances), and often it is necessary to prospectively adapt one's plan, based on what is perceived by the brain as affordant opportunities or hindrances.

Yet this does not mean that our brains are simply passively reacting to already available affordances. Rather, brains are constantly engaged in generating predictions about present affordances and future affordant opportunities (Pezzulo & Cisek, 2016). When interacting with learning environments, the brain simultaneously identifies 'the set of desirable actions currently available in its environment' and selects what to do among the competing available options. In this regard, environmental affordances 'are defined with respect to an agent's individual capabilities (e.g. a tree branch might have a 'walkability' affordance for a monkey, not necessarily for a sedentary man)' (Pezzulo & Cisek, 2016, p. 415). Nevertheless, affordances are not purely subjective; rather, they are 'objective in the sense that they do not depend on whether the agent perceives them, attends to them, or chooses to act on them' (Pezzulo & Cisek, 2016, p. 415).

Young children seem especially adept at identifying affordant opportunities or hindrances. Imagine, for example, a kindergarten school that has steps leading up to the school building. Alongside the steps there are rocks laid decoratively, affording a small garden area. The young children attending the kindergarten will likely perceive two *competing affordances*: walking up the boring steps or climbing up the more challenging rocks. For adults, climbing up the rocks may well be doable but not sensible or desirable, or may be contrary to social norms. However, to these playful kindergarten-age children, *exploring and making fun* takes priority (Figure 9.8).

It goes without saying that in school environments, different affordances provided by often competing physical environments have considerable implications for students' safety and wellbeing (Clark, 2001; Sandseter, 2009).

Independently, without copying others, many young children will seize the opportunity to climb the rocks in preference to the stairs,

Figure 9.8 Child's choice between two competing affordances (walking up the stairs vs climbing the rocks)

and thus unintentionally (or intentionally) strengthen their motor capabilities. When climbing the rocks, an 'affordance navigation' (Pezzulo & Cisek, 2016) emerges from a continuous interaction between climber and rocks. Some mature children may plan the route before staring the climb. This requires the climber to *predict* sequences of affordances that will unfold or be created, and involves situated and 'embodied decision making' (Cisek & Pastor-Bernier, 2014; Lepora & Pezzulo, 2015) – considering limb length, the configuration of the rocks, the level of slope and so forth. But others will start to climb without any pre-plan and will make decisions about their next movement as they perceive the needs of this particular affordance navigation.

When a child engages with 'affordance navigation', different brain areas process various aspects of a decision in parallel. This recursive decision-making process during the affordance navigation reuses existing neuronal networks, including the visual system, and feedback loops from other parts, especially the cerebellum. The visual system comprises two parallel pathways (Cisek, 2007): an occipito-temporal 'ventral stream' (in red in Figure 9.9) and an occipito-parietal 'dorsal stream' (in yellow). Cells in the ventral stream (in red) are sensitive to 'information about the identity of objects' (e.g. what is it – steps or rocks?), and cells in dorsal stream are sensitive to 'spatial information' (e.g. where is it?) (Cisek, 2007, p. 1587). Following the dorsal stream, specific actions become more salient and compete for further processing (yellow circles).

In the context of competing affordances (Figure 9.9), the choices they present are biased by top-down level decisions (i.e. choice of rock climbing for fun rather than steps) from prefrontal cortical regions (bidirectional arrows in orange). In performing the final selected action (climbing), two streams of feedback loops continue to bias the decisions (bottom-up): 'overt feedback' from the environment (dashed arrow outside the brain); and 'internal predictive feedback through the cerebellum' (dashed loops) (Cisek, 2007, p. 1587). This bottom-up feedback constantly reports whether there is any mismatch or prediction error (i.e. is it possible to continue to crawl? Could I get my balance with the next step?) and minimises the error if one is found. If the prediction error cannot be minimised, a child will find that there is no choice other than changing their original plan (e.g. returning to the steps rather than pursuing the rock climbing). We will have a lot more to say about the role of prediction in the brain in Chapters 10 and 11.

Designing learning environments that empower teachers' pedagogical choices

Following Gibson's (1979) introduction of the term 'affordance' in the field of ecological psychology, the term has been widely used in other fields, including architecture, engineering and design in regard to how to design the physical environments to enhance human performances. Also, as seen in the previous section, the brain dynamics involved in affordance navigation is investigated by neuroscientists (Djebbara et al., 2019; Pezzulo & Cisek, 2016). Recently, the notion of affordance has started to be employed in regard to what is called **innovative learning space** design (Mahat et al., 2018) in education, not least to improve students' safety (Sandseter, 2009) and wellbeing (Clark, 2001).

innovative learning space: a specially designed learning environment that aims to provide optimal conditions for teaching and learning

However, there is no one-size-fits-all solution regarding how to incorporate insights from environmental affordances into the design of institutionalised structures. Sometimes

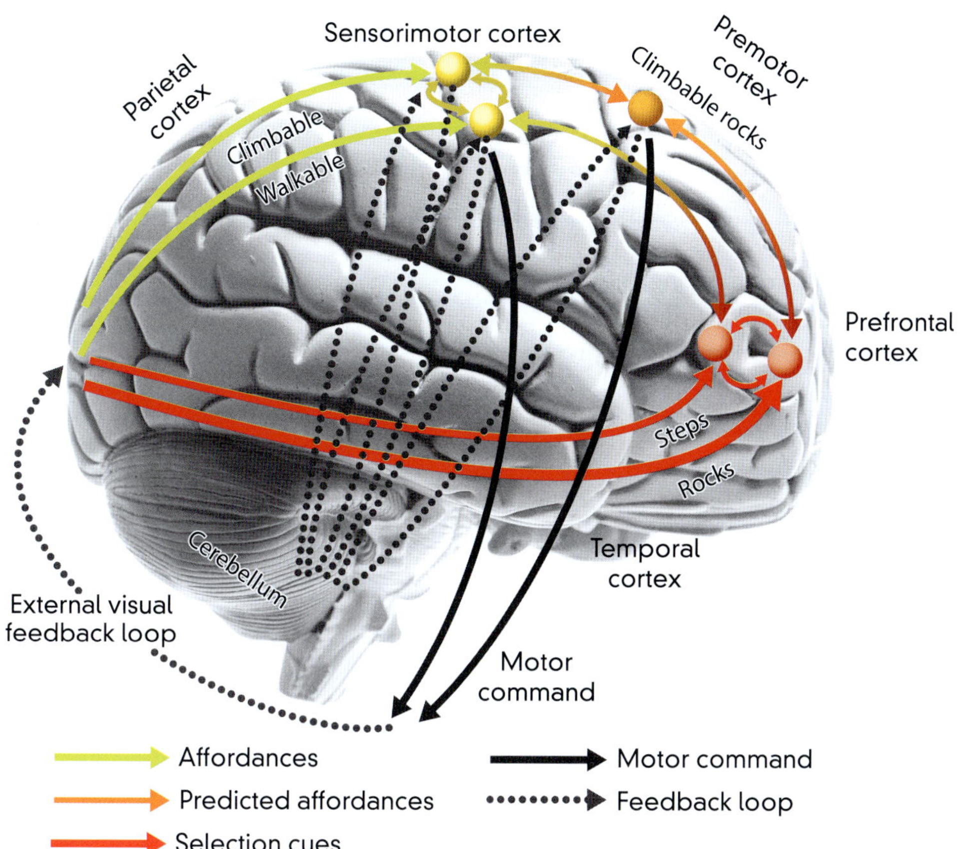

Figure 9.9 Competing affordances in the brain
Source: Adapted from Pezzulo & Cisek (2016) and Cisek (2007).

an ambitious design results in a disconnection between the intent of the design and the outcome, especially 'if the interactive dynamic between the social, psychological and pedagogical influences within the space are not acknowledged' (Mahat et al., 2018, p. 15). The difficulties involved in the translation of the ideas of 'affordance' to the design of a learning space are well illustrated in the contentious debate regarding whether **open-plan classrooms** can better support pedagogical needs.

Open-plan schools first became popular in North America in the late 1960s and 1970s. Advocates of open-plan schools argue that, by removing or reconfiguring walls, open-plan classrooms 'break down existing teacher power structures and increase student empowerment and collaboration' (Mahat et al., 2018, p. 15). Figure 9.10 provides a view of an open-plan classroom at Pratt Community School in Minneapolis, in the United States. With fewer walls, this school runs multi-age classes between different grades. For example, during art and physical education classes, students from fourth and fifth grades are grouped according to their current level and needs, and work in multi-age groups. For mathematics, students go to their class, which has been determined by each student's current level within the grade. For the remainder of the day, student stay in their homeroom class.

open-plan classroom: classrooms that have removed or reconfigured walls within or between classrooms

Figure 9.10 An open-plan classroom at Pratt Community School, Minneapolis, USA

However, there has been debate regarding whether open-plan classrooms actually bring about changes to the learning experiences and support the needs of students (Saltmarsh et al., 2015). Particularly during the open-plan movement in the 1970s, the poor acoustics of newly opened classrooms became an issue as a large number of studies reported that noise in a classroom can affect negatively children's language and socio-emotional development (Elliott, 1979; Johnson, 2000; Klatte et al., 2010; Mealings et al., 2015).

REFLECTION

Look at the open-plan classroom in Figure 9.10. In what ways do you think the concept of affordances has been built into the design to promote collaboration and connectedness across activities and subjects?

Recently, advances in acoustic design and materials, together with critical reflections on earlier open-plan classrooms, have enabled a 'resurgence of research into open-plan classrooms' (Mahat et al., 2018, p. 16). The new generation of open-plan classrooms brings affordances from both traditional closed spaces (i.e. sense of belonging) and those created by openness, such as 'flexible groupings, task dependent space diversification and opportunities for increased interactions, team-teaching, collaboration' (Mahat et al., 2018, p. 16).

Figure 9.11 shows how both 'closedness' and 'openness' are incorporated in this kindergarten space at Central Coast Grammar school, New South Wales, Australia. In essence, this space is an open-plan design, created without walls between classrooms and common areas, but it also provides sliding doors that teachers can use to separate the classroom from the common area when preferred.

The common area is spacious enough to hold a large learning activity involving all kindergarten children (in this case, a total of 80, divided into four classes). Given the total space afforded, varied learning activities can run flexibly across different parts of the open

Figure 9.11 Learning space that affords flexible transition between closedness and openness

space. The yellow decorative material on the ceiling is designed to reduce the noise level by absorbing the sound.

Affordances are not only provided by the structure of the space, but by the design and configurations of furniture placed in a learning space. Figure 9.11 shows how the 'agility' and 'fluidity' (Dovey & Fisher, 2014) are provided by furniture in a classroom. The design shape of the tables placed in this classroom provides considerable flexibility in converting between different arrangements and varied learning activities from pair learning to group learning. Accordingly, children are provided with sense of ownership of their space and individuality in having his or her own desk and chair bag.

A 2012 Space, Design & Use (SDU) survey, compared what are known as teacher 'mind-set frames' (Hattie, 2012), when working in five different kinds of classroom settings, ranging from traditional learning spaces to the totally open-plan learning spaces. Teachers working in the type of space design illustrated in Figure 9.11 reported the strongest positive belief in their teacher role, compared to those working in other types, including complete open-plan spaces without any possibility to separate the classroom space from the common area (Imms et al., 2017). Teachers working in the kind of space featured in Figure 9.11 reported the strongest belief that they were 'change agents', 'learning experts' (2017, p. 23) and developers of 'relationship and trust' (2017, p. 24). They also perceived themselves as more 'engaged in dialogue and challenge, and able to 'see opportunity in error' (Hattie, 2012, p. 159).

STANDARDS   **APST and ACECQA curriculum specifications**

APST Standard 1.1 Physical, social and intellectual development and characteristics of students
Demonstrate knowledge and understanding of physical, social and intellectual development and characteristics of students and how these may affect learning

APST Standard 3.3 Use teaching strategies
Include a range of teaching strategies

APST Standard 3.4 Select and use resources
Describe strategies that support students' wellbeing and safety working within school and/or system, curriculum and legislative requirements

APST Standard 4.4 Maintain student safety
Understand the relevant and appropriate sources of professional learning for teachers

ACECQA curriculum specifications
1.4 Child health, wellbeing and safety
2.4 Teaching methods and strategies

This chapter has elucidated what is meant by the analogy 'classrooms as ecosystems', and has encouraged readers to view classrooms as *dynamic, complex, multifaceted* and *shifting* environments. This understanding will help newly qualifying teachers by providing a common analogical framework for understanding the dynamic nature of classrooms that could enrich their classroom observation. This chapter has gone beyond what is generally available to newly qualifying teachers by providing a frontline overview of recent research into affordance, perception–action loops and embodied cognition. It has also added knowledge on how the formation of abstract concepts is grounded in perception–action experiences and an appreciation of how this can be applied in the planning of teaching strategies and the design of innovative learning spaces.

SUMMARY

- Scientists frequently use language in much the same way that we all use language, employing analogies and metaphors to compare things that are already known in order to describe things we are trying to understand, or understand more deeply.
- Within science, analogy is very frequently used to assist explanation, but it also plays a heuristic (discovery or problem-solving) role when trying to discover and understand difficult or puzzling phenomena.
- Gibson argued that in order to understand the behaviour of animals, they have to be viewed as embedded in an environment – the animal and its environment are not separate items; they can only be understood in relation to one another. When applied to

learning environments, this means viewing the learner as embedded in the learning environment – they should be understood in relation to one another.

- Gibson's work on affordance and perception, and the unity of perception, cognition and action, have contributed to what is now called the new science of embodied cognition, or embodied cognitive science.
- Classic cognitivism and embodied cognitivism are two alternative paradigms, where the proponents of each rival camp seem to talk past each other. Indeed, one aspect of that debate, concerning the nature of concepts (whether they are symbolic representations located in a central hub in the brain or, alternatively, related to the sensory motor areas of the brain) has been described as having reached an impasse.
- In the notion of embodied cognition, the divide erected by classic cognitivism between 'embodied perception' and 'disembodied thought' is gone – perception, thought and action are all embodied. Moreover, abstract concepts are not conceived as a special and higher kind of thinking, but rather as emergent flexible entities that are constituted by distributed networks within the brain, encompassing sensory and motor areas.
- According to embodied cognition, perception and cognition are both closely related to action (doing). Cognition, our thinking and reasoning (including abstract concepts), is grounded in our senses (perception) and in our interaction with the environment (our doings).
- It is very likely that you used your fingers to count – and you probably still do sometimes. This is a very good example of embodied cognition. In finger counting, perception and action, seeing and doing are brought together when doing basic arithmetic.
- Although the notion of a learning *environment* is not restricted to the physical environment, it can also include pedagogy – for example, there is an increasing recognition that the actual physical space has an undeniable influence in shaping students' learning experiences.
- Recently, the notion of affordance has started to be employed with regard to what is called *innovative learning space design* in education, not least to improve students' safety and wellbeing. However, there is no one-size-fits-all solution in regard to how to incorporate insights from environmental affordances into the design of institutionalised structures.
- Recently, advances in acoustic design and materials, plus critical reflections on earlier open-plan classrooms, have enabled a resurgence of research into open-plan classrooms. The new generation of open-plan classrooms brings affordances from both traditional closed spaces (i.e. sense of belonging) and those created by openness, such as 'flexible groupings, task-dependent space diversification and opportunities for increased interactions, team-teaching, collaboration' (Mahat et al., 2018, p. 16).

KEY POINTS FOR TEACHERS

- Learning never occurs in a vacuum; it is always highly contextualised. When studying the science of learning, it is therefore important to keep the multiple contexts in which learning occurs in view.

- Within classrooms, teachers frequently use analogies to help explain difficult ideas, but we also need to ensure we are not over-stretching the positive and helpful comparisons they engender. Teachers need to be particularly aware, when using analogies and metaphors to clarify difficult ideas, that they are not also unintentionally misleading their students.
- Observation, including the observation of teaching in classrooms, is not a neutral activity; there is much more to it than the careful and attentive use of our senses. What any given observer 'sees' is strongly influenced by *past experience, current knowledge, expectations, beliefs* and much more. Any two people (for example, the mentor and lecturer) observing the same lesson will likely interpret and assess what they see differently.
- Observation is *highly selective*. What two experienced observers actually *see* might be different, even though the images reaching the retinas in their eyes are identical. Observation of classroom teaching is compounded because classrooms are highly *dynamic, complex, multifaceted* and *shifting* environments, so understanding what is going on is not self-evident.
- Using the analogy that classrooms are ecosystems provides a way of understanding and assessing classroom activity including teaching, using a number of similarity relations that apply in both natural ecosystems and classrooms.
- Teachers should be aware that recent research suggests that *abstract concepts* are *not* a special and higher kind of thinking; rather, they are emergent flexible entities that are constituted by distributed networks within the brain, encompassing sensory and motor areas. The classical dichotomy between so-called concrete and abstract concepts does not exist.
- Research shows – and most experienced, practising teachers will agree about – the importance of learning to teach through doing. One of the main reasons why learning through doing is important is the 'tacit' nature of much classroom teaching, meaning that teaching requires *implicit* knowledge, in addition to explicit, codified knowledge.
- There is compelling evidence 'that advanced mathematics, basic arithmetic, and even the mere viewing of numbers and formulas recruit similar and overlapping cortical sites' (Amalric & Dehaene, 2016). In other words, the brain does not have a separate processing area for advanced maths; it employs the same systems used for basic mathematics.
- Hands-on learning can yield a more enhanced learning of abstract concepts when compared, for example, with passively observing a teacher demonstrate the concept to the whole class. For example, even though students are learning the same scientific concept, those who engage with manual experiments show greater accuracy in their conceptual understandings because the hands-on experience prompts activation of sensory-motor areas when learning the concept, which are then reactivated when students subsequently engage with the concept.
- There is an increasing recognition that the actual physical space has an undeniable influence in shaping students' learning experiences Applying the notion of affordance adds to our understanding of how a given physical space can influence the learning and development of students when located in that environment.

- Affordances are not only provided by the structure of the space, but also by the design and configurations of furniture placed in a learning space. Whatever classroom you find yourself in as a teacher, try to see how you can creatively use and even enhance the affordances it provides.

REVIEW QUESTIONS

Guided responses

1. What is meant by 'analogy' and what do you think one brain scientist would have meant by saying that a characteristic of good teachers is their ability to create appropriate analogies for explanation and clarification (Geake, 2009, p. 95)?
2. Can you name and explain four characteristics of environmental ecosystems that can be applied as similarity relations in understanding classrooms as ecosystems?
3. What is meant by embodied cognition and how is that related to conceiving the brain as a feedback control system, employing perception/action feedback loops?
4. What, precisely, is meant by Gibson's notion of affordance and how is it applicable in understanding classroom learning environments?
5. What factors have contributed to a resurgence of research into open-plan classroom design and what has contributed to resolving the problem of noisy open-plan classrooms?

FOOD FOR THOUGHT

1. In what ways could picturing classrooms as highly dynamic and variable ecosystems help to explain why teaching can be very unpredictable, such that even the best planned lessons could fail to engage students? How might this help beginning teachers to reassess what can sometimes turn out to be chaotic lessons?
2. We use analogies and metaphors in all our thinking and also in science, but it is very important to note that some can prove to be very misleading if taken too literally. Would it be better if teachers did not use metaphors in their classroom teaching, or is that impossible – is language inherently analogical and metaphorical?

RESEARCH ACTIVITY: DESIGNING INNOVATIVE LEARNING SPACES

In many parts of the world, there is growing recognition that a physical learning environment is one of the critical parts of the school ecosystem. For example, the New Zealand Ministry of Education stresses that learning environment design should aim to align 'social', 'pedagogical' and 'physical' elements, and it should be 'capable of evolving and adapting as educational practices evolve and change – thus remaining future focused' (New Zealand Ministry of Education, 2018, para. 1).

Below is the list of government webpages that emphasise the importance of quality learning spaces. These webpages showcase photos of different schools that the government recognises as being equipped with well-designed learning spaces.

- NSW Department of Education: 'Learning Space Gallery'
- NSW Department of Education: 'Learning Space Videos'

Research activity links

- New Zealand Ministry of Education: 'Designing learning environments: case studies, research and other resources'
- New Zealand Ministry of Education: 'Innovative Learning Environments'

After checking each website, please think about following questions.

Questions

1. What kind of affordances are provided by the design of the physical elements of each school?
2. In what ways do the physical elements of a school support teaching and learning?
3. Are there any elements that potentially provide competing affordances that might prohibit teachers and students from meeting their pedagogical needs?

SCIENCE OF LEARNING, ASSESSMENT AND FEEDBACK

LEARNING OUTCOMES

By the end of this chapter, you will:

- Appreciate what the science of learning can bring to the debate about assessment in education, know that preparing for assessment employs Hebbian repetition and know the putative distinction between competence and performance
- Understand the need to reassess failure in education, know the importance of learning by trial and error in science and education, and understand what is implied by, and involved in, creating 'classrooms as safe places to be wrong'
- Understand how the notion of the predictive brain is challenging the classical view of perception and how feedback, in response to error, operates in the predictive brain
- Understand how one recent school-based EEG research project has started to shed new light on the students' 'emotional burden of error (EBE)' when undergoing a national summative assessment test

Learning and assessment in educational contexts

Teaching, learning and assessment are closely interrelated at all levels of education. Within schools, assessment is very much part of children's learning, and hence their cognitive and motor development. Good teachers (most teachers) fully appreciate that, in addition to delivering lesson content efficiently, they also need to monitor their students' progress. They also know the necessity of ensuring that *all* their students are keeping pace with the lesson content, and *testing* student progress is a core part of that ongoing teaching and monitoring process.

Moreover, there is an abundance of material available, nationally and internationally, on the many purposes of assessment, the many different kinds of assessment, how to use assessment, and the strengths and limitations of different assessment tools. National, state and local government authorities and other education organisations, such as Cambridge Assessment International Education, have produced informative websites to assist teachers on a range of matters related to assessment.

Assessment in education has also been driven by governments and policy-makers concerned with national *performative standards*, to the extent that they may eclipse other important educational priorities (Duncan & Sankey, 2019). In Australia and elsewhere, national and state government concerns have also been spurred by the Organisation for Economic Co-operation and Development (OECD) and its **Programme for International Student Assessment (PISA)** programme of national comparison tables, which aim to measure and internationally compare 15-year-old students' performance in mathematics, science and reading (OECD, 2019a).

Governments are understandably concerned that the massive investment they make in education is preparing students to contribute to the future economy (Grant, 2017) and maintaining social cohesion. But step inside a classroom and listen to good teachers talking about student assessment, and you will likely find that their main focus and priority are *how it can advance the learning and development* of their students. Some may also express genuine concerns about national testing regimes – for example, how a national performative assessment can distort educational priorities and the curriculum offered by the school (National Research Council, 2001) by focusing too much time and resources on preparing for the test. Some will also note how national summative assessments may adversely impact student wellbeing and confidence (Segool et al., 2013).

Programme for International Student Assessment (PISA): the OECD's international academic assessment program, which intends to evaluate the educational systems of different nations, including both members and non-members, by measuring 15-year-old students' academic performance on mathematics, science and reading

---

REFLECTION

What do you think should be the priorities of education and, given the important relationship between teaching, learning and assessment, do you agree with the critical points raised by some teachers, mentioned in the previous paragraph?

Taking a science of learning perspective

Testing in school aids learning at all age levels, especially when preparing for classroom tests and in the run-up to formal examinations. In the early years of schooling, in supporting young children's literacy and numeracy learning, teachers often use tests that the children are told about in advance, so that they can then prepare and practise. In the process of preparation for the test, although the children don't realise it, they are actually employing Hebbian repetition, building and strengthening synaptic connections between individual neurons, and forming and strengthening neuronal pathways (discussed in Chapter 2). The children are using brain plasticity to learn.

The same Hebbian principles apply in revising for an examination or when learning to play a musical instrument (Tavor et al., 2019), as well as when rehearsing for performance, such as in music, dance and drama, or practising to improve skills in sport. In each case, the learner is learning through repetition, which is strengthening synaptic connections and synaptic pathways in the brain. Although it may not feel that way when practising to kick a ball accurately, in addition to strengthening the related body parts, muscles, tendons and so on, the coordination and development of movement are actually occurring in the brain.

Providing a science of learning perspective on assessment will begin to address a major gap in the current literature and ongoing debate about teaching, learning and assessment in education. It is surely significant that much of the debate pays little or no regard to what the science of learning and development might have to offer. Textbooks purportedly designed for courses on educational psychology often provide lengthy and detailed chapters on every aspect of assessment, though it is far from clear what much of it has to do with psychology, let alone the brain. Moreover, and perhaps more disturbingly, there is also little or no recognition of the kinds of brain and developmental research that need to be conducted, preferably in schools, which could inform educational debate about assessment, learning and development, including the relationship between assessment and student wellbeing.

Basic forms of assessment in education

Assessment in education may take different forms, depending on a range of factors including the subject content being studied, but an important and recurring distinction is the difference between **summative assessment of learning** and **formative assessment for learning**. The distinction between formative and summative evaluation goes back to the 1960's, to the British-born Australian philosopher Michael Scriven (1967). At the end of the twentieth century, in the United Kingdom, Paul Black and Dylan William (1998a, 1998b) provided strong evidence that formative assessment, if used effectively, will increase student achievement. This century, in Australia, John Hattie has provided compelling evidence that feedback from formative assessment can have a significantly positive effect on learner achievement (Hattie & Timperley, 2007; Hattie, 2008).

Summative assessment of learning

Summative assessment typically comes at the end (summation) of the learning process and is designed to provide a summative (summing up) measure of achievement at the moment the test is given. In other words, assessment of learning is summative in terms of *timing* (when it occurs) and its *purpose*. Summative assessment can be used by teachers to simply check on what students seem to have learnt and, more importantly, what they have *failed* to learn. In this case, it is the purpose rather than the timing that makes it summative. This

summative assessment of learning: a type of assessment that is implemented at the end of a course or grade, and aims to gauge a student's achievement level against learning outcomes

formative assessment for learning: a type of assessment that is implemented within the process of learning; its purpose is to monitor student progress and assist in improving students' attainment

purpose becomes formative if it is used to reset the learning goals. Summative tests can also be used for pre-testing what students know before undertaking a curriculum initiative or research process, which can then be compared with results from a post-test as a way to measure improvement resulting from the curriculum initiative or research intervention.

Summative assessment of learning is particularly favoured by governments and policy-makers, who are keen to ensure that students are making the required progress. Thus, in Australia, there is a National Assessment Program-Literacy and Numeracy (NAPLAN) test taken by all students in Years, 3, 5, 7 and 9 to test their abilities in reading, writing (spelling, grammar and punctuation), language conversation and numeracy. Summative assessments also occur at the end (summation) of the educational process, in one form or another – for example, the New South Wales Higher School Certificate (HSC). Feedback to summative assessment normally takes the form of a quantitative final score and/or grade.

Needless to say, many students find that facing a summative assessment of learning is accompanied by varying degrees of stress and *resilience* to stress, particularly when doing a 'high stakes' summative assessment such as the HSC, because it has a bearing on future prospects. Stress may be experienced in many ways, but what a person experiences as stress starts in their brain with the release of a hormone called arginine-vasopressin (AVP) and another called corticotropin-releasing hormone (CRH), which are synthesised in the hypo-thalamus and released into the bloodstream from axon terminals in the pituitary gland. This triggers the release of cortisol, which is known as the body's main stress hormone, although it does much more than that. Vasopressin and cortisol act together to stimulate the body's stress response, which includes increasing blood sugar (along with increased heart rate and blood pressure), associated with stress (Beurel & Nemeroff, 2014; De Winter et al., 2003).

Stress, including a degree of high-stake assessment stress, is not necessarily bad; it is essential for triggering our response to threats, making sure the body is provided with the energy needed (Barrett, 2017). When taking high-stakes assessments, a small amount of stress can actually aid performance (Arent & Landers, 2004) – actors often speak of the importance of mild stage fright in honing performance – but high levels of stress can be damaging, leading to anxiety and depression (Beilock, 2011). Given the association between summative assessment and the kind of stress it can elicit, much more scientific research is needed on how to provide a summative measure of achievement while avoiding what some students experience as threats to their emotional wellbeing.

For example, would it help to replace formal examinations, taken on specific days with extended assessment of the whole of Year 12, or does that make it worse – prolonging the agony? What is actually happening in students' brains as they sit formal assessment exam-inations such as NAPLAN? How could that kind of assessment be more accurately tailored to what is happening in students' brains? We will return to this in the final section of this chapter.

Formative assessment for learning

Assessment is formative when it occurs within the learning process, and when its purpose is to monitor student progress and assist in improving students' attainment. Formative assessment is 'formative' in having a *shaping* or *moulding* influence on the processes of edu-cation (Figure 10.1). To that end, feedback given to students is usually qualitative, outlining

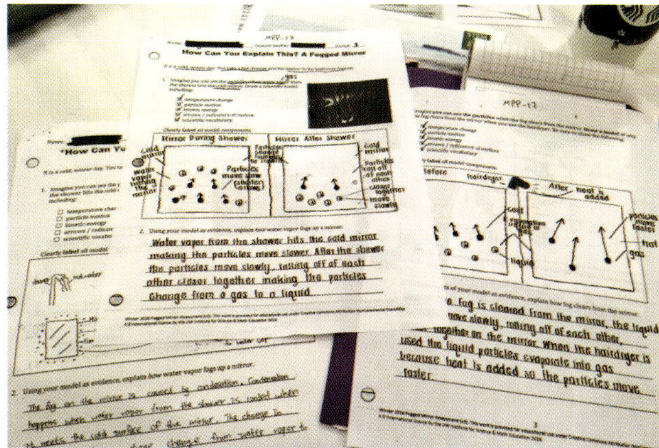

Figure 10.1 Examples of formative assessment: spelling test in a primary school and sample student response for the fogged mirror MS classroom assessment in a science classroom in the United States

students' strengths and weaknesses and what needs to be targeted in order to bring about improvement. When teachers employ formative assessment and provide carefully tailored feedback, they can have a significant impact on student success. However, this should not be a one-way process; it is essential that the learner provides feedback to the teacher and that the teacher listens attentively to the learner when explaining their difficulties and successes, so the teacher can adjust their teaching if necessary (Hattie, 2012).

Studies consistent with the principles of Hebbian learning have shown that **spaced/distributed practice** of formative assessment in the learning process can make a difference to learning outcomes (Cepeda et al., 2006; Cepeda et al., 2008). The importance of spacing was first recognised by the pioneering experimental psychologist Hermann Ebbinghaus (1850–1809) in the late nineteenth century. Ebbinghaus showed that we forget less when practising a skill or remembering what is learnt is spread over time, rather than simply repeated short term as in **massed practice**. Moreover, how the current learning being assessed relates to relevant past leaning, in the spacing of assessment, can make a difference.

Teachers often formatively test their students when completing a section of learning (perhaps a chapter in a textbook), and this may well be of value in checking whether students have understood that section in the short term. However, it may not be the best strategy for meeting the long-term goal of enabling students to access a wide range of knowledge and skills learnt over the whole course unit. This is supported by a study by Robert Lindsey et al. (2014), which compared three different spacing types of formative assessment in an eighth-grade Spanish foreign-language class in the United States (Figure 10.2). They found that when the formative assessment included both the current chapter studied and all previous chapters (*personalised spaced*), rather than just the chapter most recently studied (*massed*) or even that chapter and the previous chapter (*generic spaced*), students had better memory of what had been learnt across the different parts of the course over the long term.

Adopting a *formative mindset* is a core teaching skill that should be part of every teacher's professional toolkit. In practice, this means being constantly on the watch for formative learning and developmental opportunities as they emerge in each daily round of classroom practice – for example, noticing when a student or perhaps many students seem stuck, and

spaced/distributed practice: the idea that learning and retention of skills or knowledge is more efficient when practising (including formative testing) over an extended period of time

massed practice: the opposite of spaced/distributed practice; involves intensive and uninterrupted practice of skill or knowledge over a longer period of time

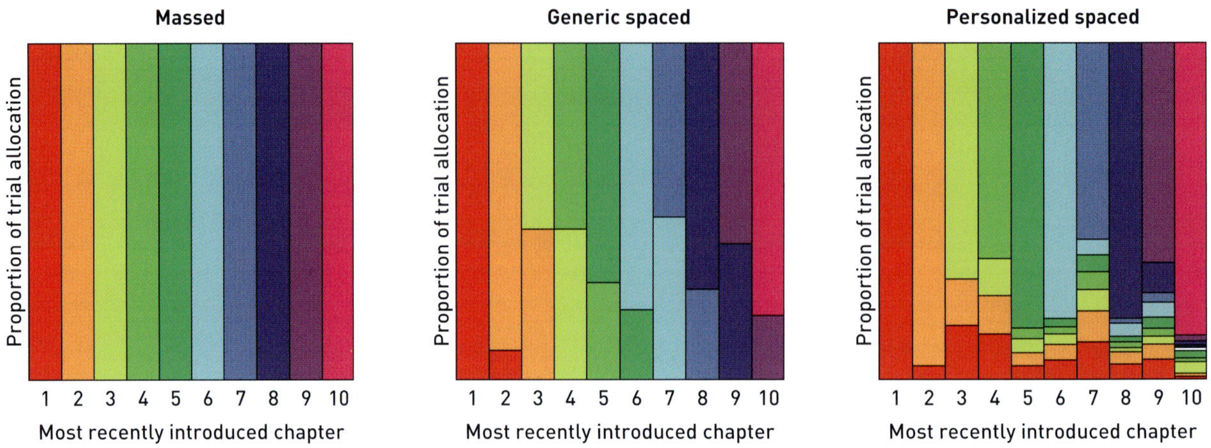

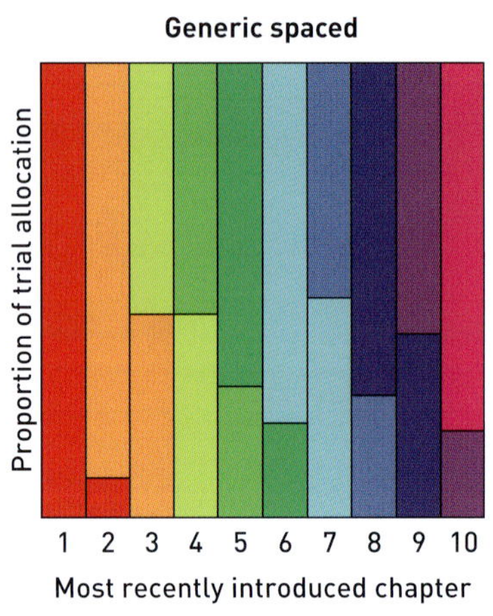

Figure 10.2 Different types of spacing in the schedule of formative assessments
Source: Lindsey et al. (2014, p. 640).

grasping the difficulty they are experiencing as a valuable learning opportunity. A formative mindset involves keeping in view where students are heading and making sure they are on track. It also means being critically aware of what precisely is being tested in any given assessment and ensuring it is fit for purpose.

Retrieval practice produces more learning than elaborative studying with concept mapping

Karpicke, J. & Blunt, J. R. (2011). Retrieval practice produces more learning than elaborative studying with concept mapping. *Science*, 331(6018), 772–5.

When we measure the length, capacity or weight of an object, the act of measuring does not bring any change to those features of chosen object. These features are invariable, so we are able to measure them precisely. Does this also apply to learning and memory? People often make the same assumption: that the act of assessment and testing simply measures students' learning and memory but does not change them. The learning and assessment procedures practised by many schools are premised on this simple assumption. Students learn new concepts in a lesson, and elaborate the learned contents via various strategies (e.g. drawing concept maps), then finally the precise amount of learning is measured via assessments.

However, recent research in memory and learning challenges this assumption. Retrieval exercises, such as the kinds of short-answer quizzes that often take place in classrooms, change memory and learning, and are a more powerful learning activity than what is often referred to as 'active, elaborative learning activities'.

Karpicke and Blunt (2011) examined the effectiveness of retrieval practice relative to one popular elaborative learning strategy: concept mapping. Concept mapping is 'considered an active learning task, and it serves as an elaborative study activity' in many classrooms worldwide (Karpicke & Blunt, 2011, p. 722). It is assumed that concept mapping encourages students to 'enrich the material they are studying and encode meaningful relationships among concepts within an organized knowledge structure' (p. 772). However, does it really serve as well as it is assumed? Could it be possible that short retrieval questions help students better?

In order to test these questions, Karpicke and Blunt (2011) randomly assigned 80 undergraduate students to four different groups, then asked them to study science materials in four different conditions. In the *study-once* condition, students studied the science materials in a single study period. In the *repeated study* condition, students studied the science materials across four successive study periods. In the *elaborative concept mapping* condition, students studied the materials in an initial study period and then were required to create a map of the scientific concepts they learned. The students in this condition received instruction on concept mapping, with an opportunity to watch an example concept map, then designed their own concept maps while reading the same science materials. Finally, in the *retrieval practice* condition, after studying the science materials in an initial study period, students participated in a free recall test. After completing the test once, these students studied the same materials again and repeated the recall. The total amount of learning time in the *retrieval practice* condition was exactly same as the amount as in the *concept mapping* condition.

In one week, students participated in a test. In order to assess meaningful learning, the test included both 'verbatim question' and 'inference questions'. The verbatim questions simply assessed factual understandings (e.g. what does sea otter fur consist of?) while the inference questions required students to identify relationships across multiple concepts (e.g. what would be the consequences of removing sea otters from their environment?).

The results (Figure 10.3) show that the students who participated in a recall test during the study period (retrieval practice) performed best among the four groups, 'better than elaborative studying with concept mapping' (Karpicke & Blunt, 2011, p. 773). Moreover, elaborative learning via concept mapping was 'not significantly better than spending additional time reading' (Karpicke & Blunt, 2011, p. 773); however, 'students tended to believe that elaborative concept mapping would produce the same or even greater learning than retrieval practice, even though the opposite was true' (Karpicke & Blunt, 2011, p. 773).

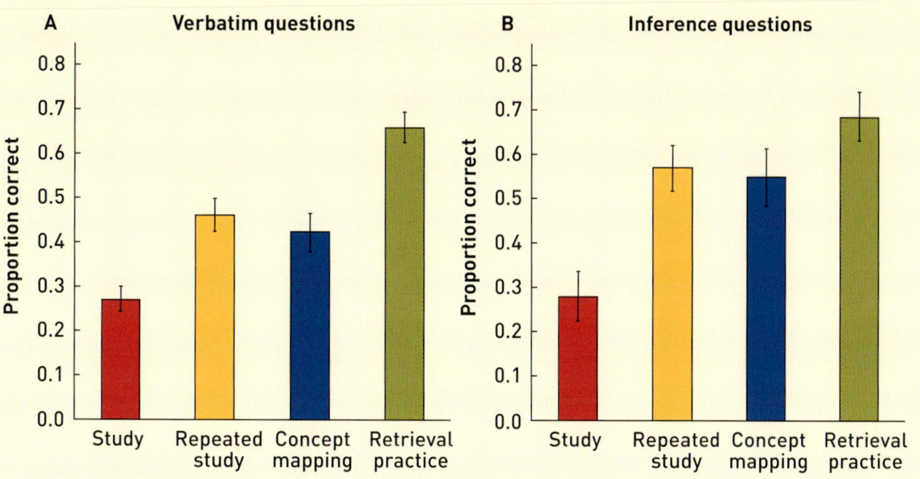

Figure 10.3 Assessment performances of four different groups
Source: Karpicke & Blunt (2011).

Competence versus performance in educational contexts

When we assess children and adolescents, either summatively or formatively, do the results provide data on their competence or their performance? At first sight, this might sound like a 'hair-splitting' issue, rather than one with any practical implications. However, within education the interface between competence and performance is relevant to a number of important issues, including the assessment of culturally and linguistically diverse (CALD) students in schools. It is also relevant to the assessment of student teachers when doing practical teaching in schools, especially if it is using what is called a 'competency-based' approach (Field, 1979).

With regard to CALD students, it seems there is an important distinction to be made: both 'cultural' and 'linguistic' factors might adversely impact *performance* in assessment, leading to an inaccurate measure of *competence*. How competent might a young Einstein have seemed if he had suddenly immigrated from Germany to live in Indonesia, for example? Culturally and linguistically, he would have been totally out of place when doing a physics assessment. The way the physics questions were conceived (perhaps using an Indonesian context) and the language in which they were written might both have substantially impacted his *performance* in a subject where he was highly *competent*. Clearly this is of central importance for teachers who work with immigrant children or adolescents. It is a very practical example of where teachers need to be critically aware of 'what precisely is being tested and ensuring it is fit for purpose', for all students involved in the assessment (Crowley, 2004).

The problem of using performance to assess competence raises issues regarding assessing teacher competence. Over the past half-century, the idea of **competency-based training (CBT)** has progressively taken hold across the whole range of vocational training (MCEETYA, 2001). In Australia, for example, it underpins the National Training Framework (NTF), the nationally recognised, competency-based training system. Within teacher education, listing the competencies of good teaching as a basis of assessing school practice became popular, as exemplified in the **competency-based teacher education (CBTE)** movement, which began in the early 1970s (Field, 1979). Sometimes the competencies listed in such approaches to student-teacher assessment are broadly based and generic, but sometimes they are designed as very detailed lists of 'can do' boxes to be ticked by the person (supervisor or mentor) observing the lesson. The core assumption in such approaches has been that an observer can accurately *assess teaching competence by observing performance*. But can they?

One problem with that assumption is that teacher performance – particularly a novice teacher's performance – may be very variable, depending on multiple factors such as the day of the week, the time of day, the particular class group being taught, what that group had previously experienced before the observed lesson and so forth. Indeed, it is entirely possible that what might be observed as an exemplary lesson with one particular group on one particular day, teaching a particular lesson content, might be followed by a disastrous lesson with the same group on a subsequent day when teaching different lesson content. In that case, which performance is the one that indicates the student teacher's actual competence? In short, context matters when assessing teacher performance.

competency-based
training (CBT):
a vocational/professional
training approach
that focuses on a set
of competencies a
person should be able
to demonstrate at the
completion of the training

competency-based
teacher education
(CBTE): a teacher
education or training
approach that is
underpinned by the
competency-based training
(CBT) model

Another problem, already alluded to in Chapter 9, is that many of the skills of teaching are tacit – they are embodied and learnt through doing, so are not easily articulated in formal competency statements. Scientist and philosopher Michael Polanyi strongly believed that we 'know more than we can tell' (Polanyi, 1983, p. 4), and this goes to the core of what he meant by the concept of *tacit knowledge*, which you encountered in Chapter 9. There is much that we know which cannot be adequately expressed in words, and sometimes we may not even be aware that we know what we know. But if you think about it, all that has been said in this sub-section on competence and performance raises a tricky philosophical problem. In educational contexts, are we ever assessing competence, much less innate ability – aren't we always, inevitably, assessing performance: on a given day, on limited content, extracted from the variabilities of the moment?

REFLECTION

What pedagogical and assessment strategies might teachers use when welcoming students who bring cultural and linguistic diversity to the classroom, and in ensuring that their performance is not mistakenly taken to be an indicator of competence?

Assessment, failure in education and learning through mistakes

Brains learn as a result of synaptic plasticity and Hebbian repetition (Chapter 2); we learn by association, analogy and metaphor and through interaction with the environment in our doings (Chapter 9); and we learn thorough our mistakes – through trial and error. Or, to put it another way, we learn through failure, by making mistakes and getting things wrong. In that sense, failure may be a key to success in learning, in life generally and in school.

However, often that is not how it feels. For the most part, failure (making mistakes, getting things wrong) is *not* something we enjoy and celebrate; it elicits very negative emotions and may even diminish our sense of self and our feelings of wellbeing. Yet it also seems to be very much part of the process of schooling. In school, we are always being tested and often fall short of what is benchmarked as success (by ourselves and others). Across different cultures and even within cultures, parents have different responses to perceived failure in schools. In the 'West', parents typically encourage children if they 'fail' a test: 'Never mind, well done – you gave it your best shot', whereas in many other cultures, anything short of a top score is simply not good enough.

REFLECTION

Given what has been said above, what was your experience of making mistakes and getting things wrong when you were at school? Did you experience a sense of failure and how was it viewed and responded to in your home environment? Did it matter to you?

The role of trial and error in science: Karl Popper

Figure 10.4 Sir Karl Popper (1902–1994)

Given the generally negative connotation of failure, especially as it often arises in education, it might come as a surprise to hear that the process of *trial and error* (trying and failing) plays a key role in scientific discovery and the evolutionary growth of science. According to Sir Karl Popper (1902–1994) (Figure 10.4), who is widely recognised as one of the leading philosophers of science of the twentieth century, science progresses through trial and error, or more technically through *conjectures and refutations* (Popper, 1963, 1978).

Popper was seeking a way to distinguish real science from false science (pseudo-science). The standard answer was that real science uses the *scientific method*, through inductive logic, observation and experiment, but that did not satisfy Popper at all. There are plenty of examples of theories that could claim they followed the scientific method, but that he felt were not really scientific, including Marx's theory of history and, in psychology, the theories of Sigmund Freud (1856–1939) and Alfred Adler (1870–1937). He had experienced this dilemma first-hand. As a university student, he had been attracted to Marxism and had discussed it often with other students. He had also worked with Adler, supporting his social work with working-class children and adolescents in Vienna, using Adler's theory of individual personality. Popper came to believe that these 'theories, though posing as sciences, had in fact more in common with primitive myths than with science; that they resemble astrology rather than astronomy' (Popper, 1978, p. 34).

It also seemed to him that these and other pseudo-scientific theories had a number of things in common. For example, those who believed in them seemed to view them as omnipotent, all powerful in explaining everything within their field. Also, for believers, the theory was like a revelation: 'Once your eyes were thus opened you saw confirming instances everywhere: the world was full of verifications of the theory' (Popper, 1978, p. 35). For those initiated into the theory, it seemed as if nothing could show that it might be wrong – the theory could always be adjusted to counter any attempted falsification. One wonders what he would have make of some current theories in education, which are deemed to be beyond dispute.

Popper believed that these characteristics made then unscientific, particularly their lack of openness to searching criticism and *falsification*. A truly scientific theory, he believed, should be open to the rigors of trial and error – it should be highly falsifiable, yet able to withstand whatever attempts are made to falsify it. **Scientific conjectures** (hypothesis and speculations) should be bold, they should take risks and scientists should be prepared to be wrong. In being wrong, you actually learn something; in simply confirming theories that already exist, you learn very little or nothing. Popper said, 'Confirming evidence should not count, except when it is the result of a genuine test of the theory; and this means it can be presented as a serious but unsuccessful attempt to falsify the theory' (Popper, 1978, p. 36).

scientific conjectures: a proposition, such as hypothesis, that is unproven

According to Popper, trial-and-error learning is an absolutely fundamental method of problem-solving. Indeed, in his view, 'All life is problem solving' (Popper, 1994) and *learning from our mistakes* (trial and error) is behind the growth of knowledge in every field of human discovery. Moreover, it is universal, not only across all human societies but also found in other animal species: all life is problem-solving and the method used in trial and error.

There is much in Popper's account of science that aligns with how practising scientists view science and how science *should* be conducted. However, in the philosophy and history of science, it seems not to match *how* it is actually conducted. In the 1960s and 1970s, Popper's *evolutionary* account of scientific growth was countered, at least in part, by Kuhn's (1962, 1970) *revolutionary* account, mentioned in Chapter 9. Popper was keen to defend the *rationality of science;* Kuhn argued that during periods of revolutionary change there is much irrationality. Between periods of revolutionary change, Kuhn argued there were periods of what he called 'normal science', when scientists explored the limits of the currently accepted paradigm. Popper believed that all science should be in revolutionary mode, constantly testing existing theories, and that what Kuhn called *normal science is actually bad science.*

It is beyond the scope of this book to follow this debate in any more detail. Those who are interested will enjoy the very readable second edition of Alan Chalmers' (1982) book *What is This Thing Called Science?* (The 1999 third edition is not so readable.)

Are classrooms safe places to be wrong?

We learn through our mistakes, through trial and error, but are classrooms in schools usually safe places to be wrong (Sankey, 1999) – not only for students, but also for teachers? Presumably, they ought to be, if students and teachers are to be allowed to learn from their mistakes. Do most classrooms in schools *celebrate error* alongside success? And what about the assessment regimes established by governments – are they tolerant of failure? In short, within classrooms and education generally, is there *freedom to fail* (Miller, 2015)? Is failure welcomed as a springboard to success, as Popper believed it is essential in real science (though avoided by false science), or is it usually treated negatively?

REFLECTION

In your own schooling, did you feel classrooms were safe places to be wrong, that there was freedom to fail and that you were encouraged to learn through trial and error, making mistakes? Or did the possibility of being wrong make you feel worried and anxious?

At the start of this chapter, we considered what is meant by summative assessment *of* learning, and formative assessment *for* learning. We might also view assessment *as* learning in education; however, arguably the science of learning should also be concerned about *assessment for failure*, the ways in which assessment in school might engender failure and associated feelings of anxiety, such that students may actually *learn to fail*? Perhaps this is something for policy-makers to consider when advocating assessment? The point is that all good teachers use assessment to drive learning: it is good Hebbian practice, repetition and reinforcement as the student prepares for the assessment. There is a strong case for it being a frequent and regular part of the learning process, but it comes with a health warning. Assessment can also be damaging if not planned and implemented with the greatest of care.

Why do so many children start their educational journey with such excitement and hope, in pre-school and kindergarten, and finish it feeling defeated and disenchanted? Why do students fail to learn and why do many, in the end, learn to fail? Could it be that in the context of education, and in studying the science of learning, the most important issue is not *how people learn* (NASEM, 2018), but *how people fail*? Moreover, what are the consequence of failure, not only for their lives – which is a very important educational concern – but also for their contribution to the good of the nation and the world, which is of social and political concern?

Making classrooms safe places to be wrong

In the context of education, there is a crying need for much more hands-on scientific research into how and why children fail to learn, and the relationship between failure and assessment. We need to know what is happening in the brains of children and adolescents when taking assessment tasks, including the role played by emotion in learning and assessment processes. We will consider one recent example in the final section of this chapter, which focused on the emotions involved when receiving feedback during assessment, but for the moment there remains a very practical question to be addressed: How can teachers and schools make classrooms safe places to be wrong?

First, some general points for consideration and then a practical example for discussion. One of the most important moves requires a change of mindset, but this is difficult. When Popper was trying to identify real science from pseudoscience, he was picking up on competing mindsets – those who were willing to embrace the possibility that their theory might be wrong and should be tested rigorously, and those who did all they could to protect their theory from rigorous testing. The problem for Popper was that even good scientists restrict what they are willing to falsify and what they will protect (Lakatos, 1970). The fact is that humans have a natural aversion to failure (Tversky & Kahneman, 1992), even while acknowledging that we learn through trial and error. Trying to create a classroom culture where failure is accepted, let alone encouraged, is going against the grain of what comes naturally. Still, it is one important way in which we learn; moreover, exposing children and adolescent to failure is preparing them for the disappointment they will experience in life. We need to change mindsets about failure in education.

As a first step, 'Establishing a culture of inquiry is critical if students are to embrace failure as a learning opportunity' (Miller, 2015, p. 6). Coupled with that is a learning culture of *metacognition* (Chapter 4). This applies equally to teachers and students. A classroom that is a safe place to be wrong must incorporate teacher error as well as student error and, for both, metacognitive appraisal and reflection on error (Chapter 4). Teachers make mistakes (sometimes!) and these should not be moments of embarrassment when brought to notice, but rather opportunities to model learning through failure. That involves talking to students about the positives of failure.

Students should learn that trial and error plays a major role in learning, so a productive learning environment in schools should embrace error as a valuable learning tool – for the teacher and students (Figure 10.5). Most importantly, classrooms that celebrate error as much as success aren't oriented to failure and low academic standards, but should be driven by high expectations and oriented toward high-quality achievement. Popper maintained

Figure 10.5 A teacher providing error feedback to a student

that trial and error is a crucial method of learning and discovery in science. It is equally crucial in every other human field of learning encountered in education.

Ultimately, within schools this means creating classrooms based on *trust*, the importance of which you previously considered in Chapter 8, particularly when considering the 'chemistry of care and trust' and the role of the neuropeptide oxytocin and related brain circuitry. Trust is thoroughly social, but it is also thoroughly biological. Classrooms can only be safe places to be wrong when students and teachers trust one another, as respectful members of learning communities that learn from their mistakes. As noted in Chapter 8, trust is not simply one moral value among others that schools might uphold; rather, 'trust *constitutes the precondition for a social life*. Without trust, a caring, sharing, considerate, compassionate, empathetic social life would be impossible' (Sankey, 2016, p. 184). In education, trust is a precondition for creating classrooms where students and teachers learn from their mistakes and there is genuine 'freedom to fail'.

REFLECTION

Is the idea of classrooms as safe places to be wrong really possible, or is it too idealistic? When planning lessons, marking students' work and using classroom space (using wall displays, arranging furniture), how might you incorporate respect for error?

When thinking about above Reflection point, you might like to critically consider an approach that was used in teaching maths (Years 7–12) in a school where curriculum time was reduced by the school's specialist focus on the performing arts, especially dance. Nevertheless, the school aimed for high academic success in maths, science and literacy, along with other academic subjects. Time pressures in maths were eased by adopting a *trial and error* approach to assessment.

Having introduced the maths topic (for example, Pythagoras's theorem) in a quite normal, imaginative way to catch the interest of the students, the students would then do the usual exercises (strengthening Hebbian synaptic learning through repetition). However, the teacher did not mark the students' work and did not ask to see their results. The students were *trusted* to do their own marking – they were *free to make errors* beyond the gaze of the teacher.

Once they had identified the examples that they had got wrong, their task was to work out why, carefully identifying the reasons for their mistakes. They could do this in discussion with classmates in time set aside for corrections. Each student had a workbook in which they briefly identified what they had done wrong and how it had now been corrected. What the teacher saw and marked was that notebook. The teacher was able to keep a close check on each student's learning and the students were using their errors as a springboard for successful learning in a way that reduced students' stress.

The predictive brain and feedback in response to error

In all that you have been reading so far about the science of learning and development, we have gradually been edging towards a radically new understanding of the brain, generally known as *predictive processing* or *predictive coding*. It is not only new; it is also very challenging – even for those working at the frontline exploring the fine detail, which is far from agreed. Nevertheless, there is a pervasive sense within brain science that over the last 15 years or so we have been standing on the verge of a paradigm change in brain science that may be as transformative as evolution by natural selection was in biology and quantum theory was in physics.

The implications for how, in education, we understand children as human beings, and how they learn and develop could be profound. The ground is beginning to shift beneath our feet, especially in regard to perception, cognition and action, and how they are related. Lisa Feldman Barrett states:

> We are, I believe, in the midst of revolution in our understanding of emotion, the mind, and the brain – a revolution that may compel us to radically rethink such central tenets of our society as our treatments for mental and physical illness, our understanding of personal relationships, our approaches to raising children, and ultimately our view of ourselves. (Barrett, 2017, p. xv)

The busy, embodied, predictive brain

In moving towards this new understanding of the brain, we have seen how the last quarter of a century or so, has already radically been challenging and changing established ideas. This has included new thinking about *synaptic plasticity* and how brain development is not fixed and finished by the onset of adulthood, but *always in flux* and ongoing. We have considered how the adoption of dynamics systems theory into our understanding of development, including brain development, has emphasised the *emergent, non-linear and self-organising properties* of dynamic system. We have seen how the view of human rationality as logical and emotionally detached, inherited from the Enlightenment, has been challenged and replaced by a view of human thought as necessarily *laced with emotion*. Furthermore, in Chapter 9 we saw how the notion of *embodied cognition* is challenging the classical cognitivist account of learning, and in this chapter we have seen the important role of *error* in learning. But is there a way of drawing these strands together to enrich our understanding of how the brain functions as a whole?

For example, it is possible to move beyond the classical three-step view of perception, in which (1) the brain receives information from the environment through perception, then (2) processes it in the higher cognitive regions of the brain, which then (3) directs action? This suggests a reactive, stimulus–response view of the brain, where neurons are generally inactive, waiting to receive a stimulus from the outside world that will trigger appropriate neurons into action.

Actually, that is how the nerve cells in your muscles (in your arms, legs, etc.) work, waiting to be stimulated before firing, but it is not compatible with what we now know about highly active dynamic systems, and the brain is the most dynamic system known to humankind. All that you have been learning about the science of learning and development suggests we need an image of the brain that is highly active and dynamic – a busy brain.

This is precisely what the notion of the **predictive brain**, with billions of *prediction loops* and the ever-present possibility of **prediction error**, is providing (Figure 10.6). It is a view of the brain, as a whole, which takes us back to something you were told in Chapter 1, that

predictive brain:
a theory of brain function in which the brain constantly makes refined predictions about the environment

prediction error:
a discrepancy or mismatch between the predictions produced by a brain and reality

Figure 10.6 The athlete's busy, predictive brain and the cost of prediction error

'the primary role of the brain is to constantly monitor all that is happening in the body keeping it all in metabolic balance – a process called homeostasis'. This is accomplished by a sensory system known as *interoception*, which communicates the *internal state* of the body to the brain. However, the role of interoception in the brain is increasingly being applied to the way the brain is processing *external stimuli*, providing a highly embodied and integrated model of brain functioning. **Predictive processing** is a 'new integrative framework for understanding perception, action, embodiment, and the nature of human experience' (Clark, 2015, p. 21).

This is the image of a busy brain, never off-duty day or night, always problem-solving, its neurons never waiting to be triggered by an internal or external stimulus. It is an image of the brain in which neurons are connected together in massive networks, constantly stimu-lating each other in a process known as *intrinsic brain activity* that continues from birth until death. Without any need for an external stimulus to trigger a reaction, the elements in the brain's vast array of neurons and neuronal networks are constantly signalling to each other in busy conversation. 'These neural conversations try to anticipate every fragment of sight, sound, smell, taste, and touch that you will experience and every action you will take' (Barrett, 2017, p. 59). Intrinsic brain activity is also 'the origin of dreams, daydreams, imagination [and] mind wandering' (Barrett, 2017, p. 58) – in fact, all the constant brain activity that makes us sentient, emotional and conscious human beings.

The remembered present and remembered future

Introducing the notion of interoception and prediction into our model of brain functioning is providing a profound challenge to the classical view of perception. In place of the linear 'three-step theory' mentioned above, the new theory suggests that neuronal networks are 'constantly active, trying to predict the steams of sensory stimulation before they arrive' (Clark, 2015, p. 2). If the brain were simply reactive, as the classical stimulus–response theory suggests, it would be far too slow to respond to immediate danger. The predictive, interoceptive, proactive brain is always one step ahead of your conscious deliberation, initiating your bodily response to events before you decide. Some psychologists refer to this as the 'illusion of free-will', but this assumes that the brain is somehow operating before the mind is free to act autonomously; it is a dualist conception of mind and brain. However, you *are* your brain – how you act is caused by your brain's predictions, based on past experience.

According to Gerald Edelman, who you met in Chapter 3, we always *live in the remem-bered present* (Edelman, 1989). The past creates the meanings and concepts that are held in memory and direct our present actions. In addition, the brain is also constantly looking forward, creating expectations of what will be and should be done based on past experience. In that sense, we also dwell in a remembered future. *All that you have experienced throughout your life so far provides a resource for prediction*; what you have remembered of your past ex-periences garnered through your senses, the emotional valence of those experiences, your previous thoughts and imagining that are remembered and your prior learnings are used by the predictive brain as a backdrop to and resource for working out your present experiences, including what you are currently trying to learn and your future thoughts and actions.

Prediction and prediction error

Andy Clark (2015, p. 2) believes that predictive processing 'represents the last and most radical step in the retreat from the passive, input dominated view of the flow of neural processing'. Prior to the arrival of an input signal in perception, proactive predictive systems in the brain are already 'predicting its most probable shape and implications ... And all they need to process are any sensed deviations from predicted states (known as *prediction errors*) that thus bear much of the information-processing burden, informing us of what is salient and newsworthy within the dense sensory barrage' (Clark, 2015, p. 2).

Feedback loops are therefore said to be top-down, influencing what you perceive and correcting (through feedback) for any mismatch between prediction and perception (Garrido et al., 2007). For the most part, thankfully, the brain's predictions sufficiently match the sensory input received and nothing more needs to be done. You have a new experience to add to your store. Where there is a mismatch, the brain has to get busy resolving the prediction error.

An important point to note is that prediction errors are normal and productive, just as the role of error and error feedback are a normal and productive part of learning. When prediction errors occur, the brain may change the prediction in preparation for a subsequent and somewhat similar event, or it might change the perception to come into line with the prediction, which is based on past experience. Ultimately, what the brain seeks is consistency and coherent meaning regarding how the world is understood (predicted) on the basis of past experience and what is perceived.

It is often said that 'seeing is believing' – in other words, visual evidence underpins what people normally believe. But in a very real sense it is the other way around: 'believing is seeing', so what we believe, based on past experiences, strongly influences what we see. In the context of education, children and adolescents come to school with predictions of what they will see and experience, based on their past experience. It is hardly surprising that if their past experience has been very negative, they will have negative expectations on entering your classroom – unless your classroom has been experienced as an oasis of positive experience. The same applies with regard to doing assessments in school. Past experience and present expectations (predictions in the brain), especially as these are valenced by emotion, will variously contribute to performance in assessment situations. We return to the predictive brain in Chapter 11.

Educational brain research on emotion and assessment

Assessment can sometimes be emotionally very stressful, causing anxiety and depression. In fact, not only assessment but also classroom learning can be very stressful for children and youngsters when experiencing a continued sense of failure to understand lesson content and keep pace with their peers. As previously noted, a 'certain amount' of stress may aid performance in assessments such as the summative assessment that occurs at the end of the schooling process, in Year 12 (Figure 10.7). However, it is not always clear how a 'certain amount' can be measured. What is the difference between a 'certain amount' and too much? The trouble with such statements is that they are often well intentioned, but also scientifically very vague.

Figure 10.7 Students taking a computer-based maths test in a class

What is currently 'known' about the emotional stresses experienced by students when learning and during assessment may come from research in which students use self-reporting and questionnaire research methods, as these are relatively easy to administer. However, as previously noted, a main problem with such methods is that our ability as human beings to be introspective and accurately self-assess feelings of stress is limited. We may feel the stress consciously, we may be able to describe what we are feeling, we may attribute those feeling to events past and present, but the neurophysiological causes are operating below the level of conscious awareness and are therefore not available to introspective reflection. To gain access to the subconscious emotions that emerge in the process of learning and assessment and when receiving feedback, researchers have to use brain imaging. This includes collecting and analysing EEG/ERP (event-related potential) data, to which you were introduced in Chapter 4.

Giving and receiving feedback in educational contexts

As noted earlier, it is well recognised that feedback is critical in the process of learning. Some empirical studies show that even the mere anticipation of rapid feedback improves performance (Kettle & Häubl, 2010), so a minor change such as altering the timing when the learners expect to receive feedback can cause significant differences in learning outcomes. Moreover, some authors provide helpful models and framework that one can consider when designing feedback for their classroom. One very helpful model of feedback is provided by John Hattie and Helen Timperley (2007). They believe effective feedback should answer three major questions, as illustrated in Figure 10.8:

- Where am I going? (What are the goals?)
- How am I going? (What progress is being made toward the goal?)

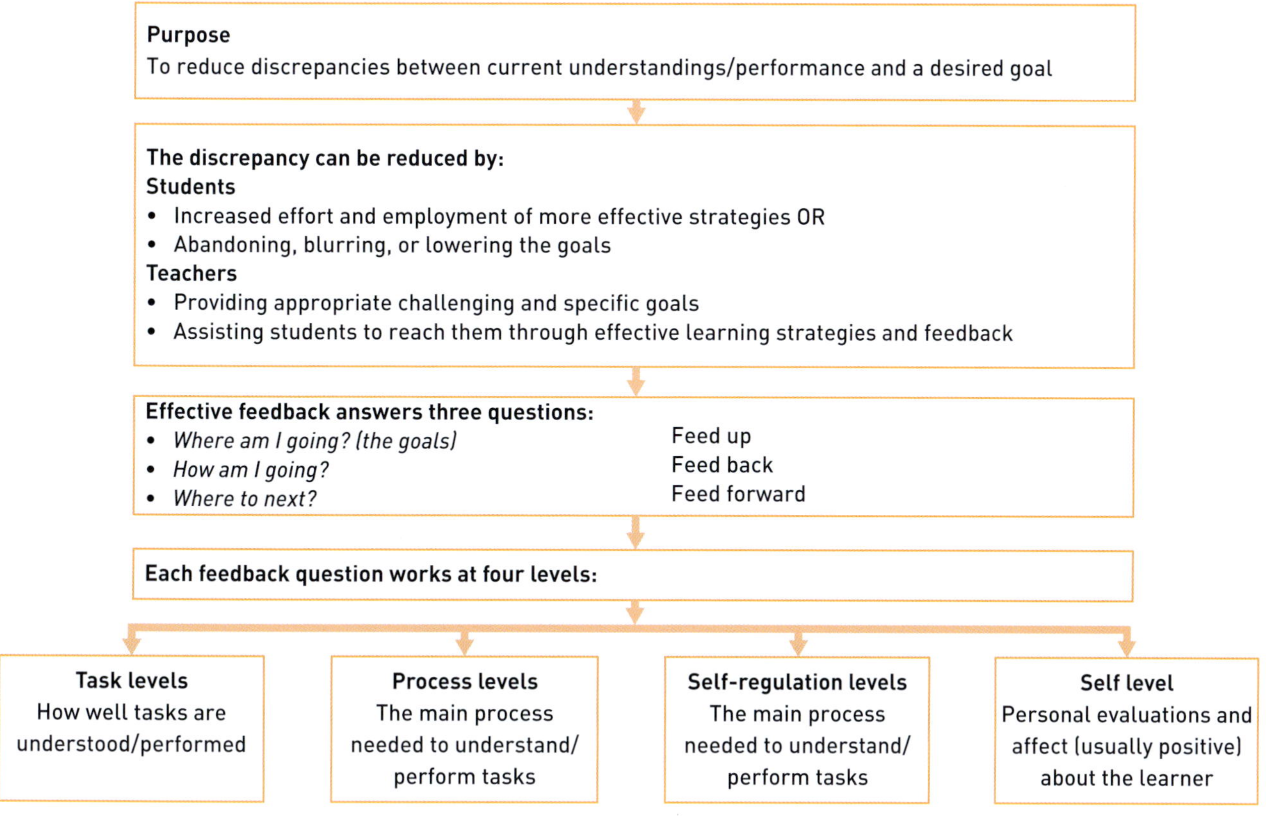

Figure 10.8 A model of feedback to enhance learning
Source: Hattie & Timperley (2007, p. 87).

- Where to next? (What activities need to be undertaken to make better progress?) (Hattie and Timperley, 2007, p. 86)

One can see at a glance that these questions are evoking meta-cognitive reflection (see Chapter 4) on the learner's current learning experiences. Along with this model, helpful strategies and guidance on feedback design are provided by different governmental and teacher accreditation websites. The Research Activity at the end of this chapter provides some helpful materials for your study, together with three questions for you to reflect upon.

Although a large body of educational literature on feedback exists, regarding how to design it and when to provide it, not much is known about how *students perceive* the feedback and how new learning occurs from the feedback they have received. In other words, educational literature on feedback tends to be 'teacher focused', concerned with what teachers should do and how, while little is known regarding how the feedback given is working. How *do students feel* when they find the given answer is wrong? Which type of feedback works better? Are there certain types of students who are more sensitive and vulnerable to negative feedback? These kinds of questions are oriented towards how feedback is *received by learners*. Arguably, in regard to research into feedback in learning, a shift of perspective, from teacher-centred to learner-centred, would start to provide new data.

In making such a shift, new insights drawn from the notion of the 'predictive brain' and the 'brain as a feedback control system' could start to provide unique perspectives in interpreting research data and in understanding how feedback is operating within the learner. From the perspective of the brain conceived as a *feedback control system*', feedback is 'an outcome of an action that is captured by the senses' (Di Bernardi Luft, 2014, p. 357); this includes both feedback that is arriving from the external world and internal feedback. Any one action often results in more than one sensory outcome. Imagine a child who has just started to learn to play the piano. When hitting a wrong key, the child receives diverse feedback, ranging from the direct auditory input of unfamiliar dissonance (internal) to more elaborated feedback such as the piano tutor's verbal feedback (e.g. 'this is F') and demonstration (external).

Although feedback plays a critical role in correcting, monitoring and improving performance, it cannot be assumed that everyone learns from feedback they receive. Actually, one's ability to *learn from feedback* is an important skill, not only for academic achievement in school settings but also in adapting to the ever-changing, everyday environment. As highlighted in Research link 10.2, how people learn from feedback is often investigated in controlled laboratory settings (Peters et al., 2017, p. 149); however, a strong case can be made for educational EEG research to be conducted in the 'natural setting' of participants' schools, as occurred in the case in the school-based EEG project reported below.

Feedback initiates a variety of distinct neural processes in the brain. Each process has distinct time-courses: some are fast while others occur slowly. Many studies have investigated neural correlates of distinctive feedback processing using EEG/ERPs techniques. The following two ERP components are particularly relevant educationally.

Feedback-related negativity (FRN): Early response to feedback

Feedback-related negativity (FRN), previously mentioned in Chapter 4, is the main ERP component investigated in EEG studies on feedback and learning. FRN is a negative deflection within the ERPs, which is observed between 200 and 300 ms following feedback, and is recognised as an 'index of the performance monitoring system activity in response to feedback' (Di Bernardi Luft, 2014, p. 357). FRN is usually observed in the mid-frontal part of the scalp and is known to be generated in the anterior cingulate cortex (ACC) (Di Bernardi Luft 2014; Holroyd et al., 2004), located in Broadman area 32 (Figure 10.9). The FRN component is triggered when the brain detects a 'prediction error, which is the difference between the expected and the obtained reward' (Di Bernardi Luft, 2014, p. 360). 'The ACC has, amongst its variety of functions, a role in integrating feedback related information and translating it into action. This brain structure has subgroups of neurons which differentially respond to positive and negative feedback' (2014, p. 365).

According to di Bernardi Luft (2014, p. 360), 'the ACC seems to be a key structure for understanding how we use feedback for learning. In animal studies (Kennerley et al., 2006) it was found that lesions in the ACC led to an impairment in integrating past feedback information, causing the animal to use only the immediate previous feedback as a guide for the next response.' In addition, the ACC has also been found to be 'more active when feedback indicated the need for revising previously learned information or when it represented new

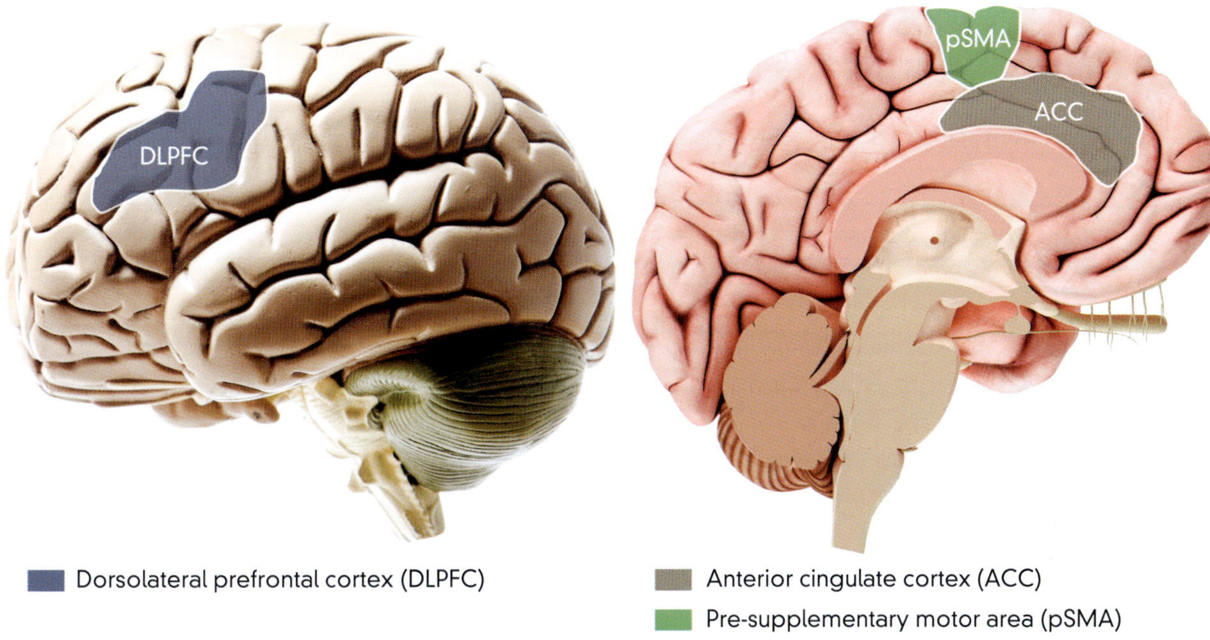

Figure 10.9 Areas of the brain where activity during the feedback learning predicts reading and maths performances in a two-year period

information' (di Bernardi Luft, 2014, p. 360). Indeed, the ACC seems to have 'a multitude of roles in using and *integrating* feedback information into the reinforcement history' (di Bernardi Luft, 2014, p. 360). In order to '*integrate* information, the ACC has to [synchronise] with other brain regions, as it was suggested in studies which found an increase in [synchronisation] centred in the midline electrodes' (di Bernardi Luft, 2014, p. 361). The ACC sends out signals to other parts of the brain via connections called *cingulum bundle* which reaches out to the variety of regions of the brain.

Late-positive potentials (LPP): Sustained contemplation

Sometimes sustained processing continues in the brain, especially if there is major prediction error or when the feedback is emotionally significant (e.g. negative feedback laced with the teacher's frustration and anger). This sustained processing is often captured via the late-positive potential (LPP). The LPP is a positive-going ERP component that shows sustained activation beginning from approximately 500 to 600 ms from the moment one received a feedback. The LPP is usually observed in the centro-parietal area (Hajcak et al., 2009; Glazer et al., 2018). For example, in one study (Mangels et al., 2006), undergraduate female students were given a computer-based maths test and EEG was recorded. EEG/ERP waveforms were compared between high-performing students and low-performing students. Although both groups show the FRN following the error feedback, it was only the low-performing student group that showed an LPP. In other words, in this study low-performing students tended to dwell on and give sustained attention to the negative feedback, resulting in an added burden of emotional regulation, or what might usefully be called the emotional burden of error (EBE).

RESEARCH LINK 10.2

Predicting reading and mathematics from neural activity for feedback learning

Peters, S., der Meulen, V., Zanolie, K. & Crone, E. A. (2017). Predicting reading and mathematics from neural activity for feedback learning. *Developmental Psychology*, 53(1), 149–59.

As illustrated above, feedback is crucial in one's learning process. One can assume learning should be improved if a child is sensitive to feedback and successfully uses that feedback either to correct their misunderstanding or to improve their skills. One question that should be asked is whether this assumption is supported by empirical evidence. Sabine Peters and her colleagues examined whether 'the degree to which participants could successfully use feedback was predictive of school learning measures' (Peters et al., 2017, p. 149). A total of 299 Dutch participants (aged between eight and 25 years) participated in data collection at two time points, which were about two years apart.

For the first data collection, participants performed a feedback learning task in the fMRI scanner. Participants had to sort each item into one of three squares. If the participant chose a correct square, positive feedback ('+') was shown. If the participant chose a wrong square, negative feedback ('–') was shown. Each participant did an average of 138 sorting trials. In order to gauge the extent to which each participant successfully used feedback, the research team calculated the percentage of trials where feedback was successfully used on the recurring trial.

The successful use of feedback was calculated as a percentage following two rules. If a participant received positive feedback following their response to a stimulus (image), and when that stimulus was repeated they also gave the same response, they were deemed to have applied positive feedback successfully. On the other hand, if a participant did not make the same response when given an identical stimulus (because they previously got the answer wrong) that incurred negative feedback, they were deemed to have applied negative feedback successfully. The research team calculated the feedback learning measure by dividing the number of successful feedback applications by the total number of trials during the learning phase. Subsequently, after two years, the same participants were tested for reading fluency and mathematical ability.

The results show that the extent to which a participant successfully applied the 'feedback learning predicted both reading and mathematics performance' in two years (Peters et al., 2017, p. 149). Also, 'activity during feedback learning in the left superior dorsolateral prefrontal cortex (DLPFC) predicted reading performance, whereas activity in presupplementary motor area/anterior cingulate cortex (pre SMA/ACC) predicted mathematical performance' (Peters et al., 2017, p. 149) (see Figure 10.9 above).

A school-based EEG project examining the brain's response to feedback in assessment

Conducting school-based research of any kind always comes with many challenges, and these become even greater when using EEG. However, conducting research with students in their own school setting, in which they can feel a sense of belonging, brings many advantages. For one thing, students participating in the study do not need to travel to be involved, but more importantly, they are more likely to be relaxed when engaging with the research.

Experience over a number of years has shown there is plenty of enthusiasm among students to participate in EEG studies, even though they are told in advance that they will wear a cap and that a salt-based gel will be applied to their scalp to ensure a very good contact. The researchers come armed with a supply of towels and hair shampoo for students to use after their involvement and the school needs to ensure that the students have ready access to washing facilities. All such research has to pass through the scrutiny of a university ethics committee, and schools and parents have to be willing to participate. Setting up the EEG on arrival at the school takes about half an hour and then it is time to start receiving students. In this project, each student's involvement lasted about one hour, which included being fitted with the cap and taking the test.

As previously noted, in Australia every year, a National Assessment Program-Literacy and Numeracy (NAPLAN) assessment is undertaken by all students in Years 3, 5, 7 and 9. NAPLAN was introduced nationwide in 2008, but prior to that literacy and numeracy testing had been carried out by individual states. As one might imagine, NAPLAN is criticised by many teachers and parents because it is summative, and it focuses entirely on literacy and numeracy, which provides a narrow view of any given student's abilities. Another major criticism is the results are received three or four months after students take the assessment.

In the processes of learning and assessment, timely feedback matters (Kettle & Häubl, 2010). If there is a too big gap between doing an assessment task and receiving the result, as happens with NAPLAN, it loses any formative learning value that it might possess. It simply becomes a summative statement of what happened on a particular day, at a given time, when the assessment task was attempted by a student who was experiencing a range of thoughts, feelings and concerns at that time. Would it be possible to provide immediate, supportive and meaningful feedback during a national assessment such as NAPLAN?

One school-based EEG project focused on testing the feasibility of providing feedback during a NAPLAN type of test. It was also especially interested in how different types of feedback affected students' emotional experiences and test performance (Duncan et al., in preparation). The preliminary findings from this study, currently being written up for publication, seem to confirm that it is very feasible to provide immediate feedback in a computer-based academic test, such as NAPLAN; moreover, different types of feedback lead to distinctive patterns of neural responses, yielding different emotional experiences.

In this school-based EEG study, Australian Year 9 (15-year-old) students were given a computer-based maths test, using questions from previous national NAPLAN tests, while their neural activities were recorded using EEG. They received an immediate feedback response on the screen after answering each maths question. One group of students received future-forward feedback with encouraging messages when a given answer was wrong. The other control group received feedback that simply said INCORRECT when the given answer was wrong.

In both groups, all students across many abilities managed to complete all 50 questions in the given time, which indicates that it is workable to provide immediate feedback in a computer-based academic achievement test. More importantly, it was found that the emotional burden of making an error (the experience of failing) can be reduced significantly by providing carefully phrased feedback. The group who received specially designed feedback showed a significantly smaller late-positive potential (LPP). As illustrated above, LPP indexes sustained appraisal and regulation of evoked arousal. Therefore, when carefully designed, the feedback given to a wrong answer does not necessarily impede the engagement with the following question.

These results are clearly interesting, but more needs to be done. The next step is to check whether different types of feedback actually affect students' performance over the course of a test engagement. Currently, what is called *structural equation modelling* (SEM) analysis is being undertaken to assess the relationship across three different measures: neural activities captured via EEG; self-reported attitudes on learning of mathematics; and performance change over the test period. There is a lot of research needed in applying the science of learning to our understanding of the emotional factors and burdens experienced by students in schools in the course of assessment, especially in regard to assessment and student wellbeing, which is the focus of Chapter 11.

STANDARDS — APST and ACECQA curriculum specifications

APST Standard 1.2 Understand how students learn

Demonstrate knowledge and understanding of research into how students learn and the implications for teaching.

APST Standard 2.3 Curriculum, assessment and reporting

Use curriculum, assessment and reporting knowledge to design learning sequences and lesson plans.

APST Standard 5.1 Assess student learning

Demonstrate understanding of assessment strategies, including informal and formal, diagnostic, formative and summative approaches to assess student learning.

APST Standard 5.2 Provide feedback to students on their learning

Demonstrate an understanding of the purpose of providing timely and appropriate feedback to students about their learning.

ACECQA curriculum specifications

3.1 Teaching methods and strategies

5.9 Curriculum planning, programming and evaluation

This chapter has elucidated what 'assessment' experiences offer and mean to learners by drawing on perspectives from the *science of learning and development*, including the principles of Hebbian repetition. When considering that trial and error (trying and failing) can play a key role in scientific discovery and the evolutionary, the experiences of error or failure are critical in the process of learning. In fact, empirical studies conducted in real classroom contexts show that formative assessment, especially when well spaced, enhances students learning more than other well-known learning-elaboration strategies. This understanding will help newly qualifying teachers design and plan assessments in their own classrooms, by providing an understanding of why assessments are necessary for learning.

This chapter has gone beyond what is generally available to newly qualifying teachers about feedback by providing a frontline overview of how feedback in response to error operates in the 'predictive brain'. It has provided readers with knowledge on neural mechanisms behind the learning afforded by feedback process. It has also added knowledge on how different types of feedback lead to different emotional experiences, which in turn may impact the learning process. Understanding gained by newly qualifying teachers can be applied in the plan of assessment strategies, and the design of the curriculum and reporting process.

SUMMARY

- Teaching, learning and assessment are closely interrelated at all levels of education. Within schools and education more generally, assessment is very much part of children's learning, and hence their cognitive and motor development. Testing in school aids learning at all age levels, especially when preparing for classroom tests and in the run-up to formal examinations.
- Assessment in education may take different forms, depending on a range of factors including the subject content being studied, but an important and recurring distinction is the difference between summative assessment *of* learning and formative assessment *for* learning.
- Many students find that facing a summative assessment of learning is accompanied by varying degrees of stress and varying degrees of *resilience* to stress. Stress starts in the brain with the release of a hormone called arginine-vasopressin (AVP) and another called corticotropin-releasing hormone (CRH), which are synthesised in the hypothalamus and released into the bloodstream from axon terminals in the pituitary gland.
- Stress is not necessarily bad; it is actually essential in triggering our response to threats, making sure that the body is provided with the energy needed. A small amount of stress can actually aid performance, but high levels of stress can be damaging, leading to anxiety and depression.
- Education is inherently invasive, physically changing the brains of students, for better or worse. This implies that adopting a science of learning perspective should not be an optional extra – something to bear in mind if you are so inclined. It should be front and centre of how each teacher understands what they are doing to the children in their care.
- We learn through our mistakes, through trial and error, which means we learn through failure, by making mistakes and getting things wrong. In that sense, failure may be a key to success in learning, in life generally and in school. A classroom that is a safe place to be wrong must incorporate teacher error as well as student error and, for both, metacognitive appraisal and reflection on error.
- According to Karl Popper, trial-and-error learning is an absolutely fundamental method of problem-solving. In science, it means opening science up to failure, making bold

conjectures and testing them to assess whether and where they are wrong. Popper claimed that all life is problem-solving.

- There is now a very new understanding of the brain, which is generally known as *predictive processing* or *predictive coding*. It is providing a new integrative framework that is challenging classical cognitive theory and providing new insights into the nature of human understanding, perception, action, embodiment and experience.
- This new view emphasises that the primary role of the brain is to constantly monitor all that is happening in the body, keeping it all in metabolic balance – a process called homeostasis – and this is accomplished by a sensory system known as *interoception*, which communicates the *internal state* of the body to the brain.
- Prior to the arrival of the input signal in perception, proactive predictive systems in the brain are already predicting its most probable shape and implications, based on past experience. If what is perceived matches what is predicted, all is well and experience is confirmed. It is the deviations from predicted states, known as *prediction errors*, that the brain has to resolve.
- New insights drawn from the new science of the interoceptive, predictive brain and the brain as a feedback control system could start to provide unique perspectives for interpreting research data and understanding how feedback in educational contexts is operating within learners.

KEY POINTS FOR TEACHERS

- In addition to delivering lesson content efficiently, research shows that teachers also need to monitor their students' progress and ensure that *all* their students are keeping pace with the lesson content.
- Currently, there is a major gap in the research literature and in educational debate on the interplay between teaching, learning and assessment in education, much of which pays little or no regard to what the science of learning and development research might have to offer.
- Although feedback plays a critical role in correcting, monitoring and improving performance, teachers should be aware that not everyone learns from the feedback they receive. In fact, one's ability to *learn from feedback* is an important skill for teachers to nurture in their students.
- Adopting a *formative mindset* is a core teaching skill that should be part of every teacher's toolkit. In practice, this means being constantly on the watch for formative learning and developmental opportunities as they emerge in each daily round of classroom practice. A formative mindset involves keeping in view where students are heading and making sure they are on track.
- Good teachers uses assessment to drive learning. It is good Hebbian practice, repetition and reinforcement as the student prepares for the assessment. There are therefore strong scientific reasons to make testing a frequent and regular part of the learning process. However, there is also abundant evidence that assessment can be damaging if not planned and implemented with the greatest of care.

- Establishing a culture of inquiry is critical if students are to embrace failure as a learning opportunity, as is a learning culture of metacognition. This applies equally to teachers and students. A classroom that is a safe place to be wrong must incorporate teacher error as well as student error and, for both, metacognitive appraisal and reflection on error.
- Research is increasingly showing that the conventional stimulus–response view of the brain, where neurons are generally inactive, waiting to receive a stimulus from the outside world that will trigger appropriate neurons into action, is wrong. It is being replaced by an image of a busy brain, never off-duty day or night, always problem-solving, its neurons never waiting to be triggered by an internal or external stimulus. Children in school are busy brains.
- There is a large body of educational literature on feedback, regarding how to design it and when to provide it. Teachers should be aware that not much is known about how students perceive the feedback and how new learning occurs from the feedback they have received. We need to know more about *how students feel* when they find the given answer is wrong and what type of feedback works best. These are important questions to consider in classrooms.

REVIEW QUESTIONS

Guided responses

1. How are Hebbian learning and plasticity related to what is occurring in the brain when students are preparing for classroom tests and more formal summative and formative assessments in education?
2. What processes are likely to be occurring in students' brains when experiencing stress, when facing a demanding summative assessment of learning?
3. What did Karl Popper mean by conjectures and refutations in science, and how is that related to our everyday experience of learning by trial and error? Why do some commentators believe that Popper's notion of conjectures and refutations was itself refuted?
4. What is meant by the classical *three-step* view of perception, and how is that being challenged by the notion of a dynamic, interoceptive and predictive brain?
5. What is meant by feedback-related negativity (FRN) and late-positive potentials (LPP) in the context of EEG research using event-related potentials, and how are these able to enhance our understanding of students' brain processes when undertaking assessments?

FOOD FOR THOUGHT

1. Despite the clear relationship between assessment and learning, do you think the massive growth of interest in assessment by policy-makers, governments and international bodies such as the OECD over the past two decades or so is entirely helpful in the context of classroom educational practice? What actually matters in regard to assessment?
2. Are there ways to bridge the distinction between summative assessment *of* learning and formative assessment *for* learning, or are these distinctions as absolute in educational contexts as they are often taken to be? In light of the school-based EEG research project reported in this chapter, do you think NAPLAN could be made more formative, by giving feedback?

3. Do you have any conscious sense of the 'remembered present' – the idea that your past creates the meanings and concepts that you hold in memory and that direct your present actions, as the notion of a predictive brain suggests? Does it make any sense to you that you also dwell in a 'remembered future'?

RESEARCH ACTIVITY: PROVIDING FEEDBACK IN ONLINE LEARNING

As online and remote learning is becoming more prevalent, educators need to invent strategies to provide timely and meaningful feedback through the technology. Now educators should understand what good feedback looks like, and then think about how such feedback can be conveyed to students who are located remotely, but still connected via a technology.

Below is the list of webpages that emphasise the importance of quality feedback in classrooms. These webpages recommend specific principles of feedback design with relevant examples:

- AITSL: 'Effective feedback animation'
- Victorian Department of Education and Training: 'Feedback and reporting'
- NSW Education Standards Authority: 'Effective feedback'
- DER @ Monash University: 'Technology mediated assessment feedback'.

Research activity links

After checking each website, please think about the following questions.

Questions

1. What kind of principles are recommended by these resources? Is there any commonality across difference resources?
2. Reflect on given principles and strategies, drawing on what you learned from this chapter, especially the importance of errors and the predictive process in the brain. How much do you think each principle and strategy is consistent with what we know about how brains learn?
3. Can you think of a technology-based learning experience in your subject area? Choose one principle of effective feedback design. How can you implement the chosen principle in a technology-based learning context?

STUDENT WELLBEING AND THE SCIENCE OF LEARNING AND DEVELOPMENT

LEARNING OUTCOMES

By the end of this chapter, you will:

- Understand how learning and development are intimately related to student wellbeing in school and appreciate the need for teachers to be constantly watchful in regard to their students' wellbeing
- Understand what is meant by learning and development being value-embedded and the importance of creating value-based learning environments in schools
- Know that wellbeing occurs in the brain and appreciate that the recent model of the brain as interoceptive and predictive is providing new insights into student wellbeing
- Appreciate that a teacher's ability to assist student wellbeing depends, in part, on how the teacher values students as sentient, emotional and relational human beings

Student wellbeing and schooling

It is a very salutary experience to observe parents and children on the first day of compulsory education, which in Australia means kindergarten. There is a buzz of excitement and most children seem to be taking it in their stride, but it is not uncommon to find one or two who seem deeply distressed and are crying. Parents do their best to reassure their children and give them confidence, but it comes with a degree of hope that their child (or children) will be okay, especially if the child is displaying distress. It is easy to say that children soon get over it and adjust, but do they always? Aren't some primary and high school students *afraid* of school and the people they will meet at school (other students and teachers), as suggested in Chapter 10? Is it not the case that some students at all age levels are bullied and/or marginalised at school – even in the 'best' schools?

Are schools safe places to be?

Beyond popular wisdom and anecdote, is there any evidence that some children do not feel safe at school? In 2020, the Australian Institute of Health and Welfare (AIHW) published a major report, *Australia's Children*, which found that some 14 per cent of students, aged from six to 12 years said 'they did not feel safe at school … a lot of the time' (AIHW, 2020, p. 15). Stop and think about this for a moment. That amounts to one in every seven children surveyed, or four children in every average class of 28 students. This is quite startling because children in this age group (six to 12 years) are largely pre-puberty, before the onset of the emotional rollercoaster years of adolescence. It is this time at school that is usually marked out as the 'best years', yet one in every seven children in Australia does not feel safe at school *a lot of the time*.

Needless to say, that does not mean that in every classroom there will be four students who do not feel safe at school; it is an average number, and there may be fewer or there may be more. However, this statistic also suggests that while 14 per cent do not feel safe at school, the other 86 per cent surveyed *do* feel safe a lot of the time. Teachers can certainly take comfort in that positive figure, but the fact that some children and adolescents do not feel safe should alert all teachers to be constantly watchful in regard to their students' wellbeing. Feeling safe, physically and mentally, at home, in school and in the neighbourhood, is one important marker of student wellbeing.

One of the main threats to student wellbeing in school is the prevalence of bullying. The Australian Institute for Teaching and School Leadership (AITSL) states that bullying 'occurs in many forms, including overt physical aggression, covert teasing, harassment online or in the classroom' (AITSL, 2021, p. 3). It has lasting effects, physically impacting the brains of victims, their bullies, peer groups, families and school communities). AITSL points out that in a large-scale survey conducted in 2009, it was 'found that one in four Australian year four to nine school children had been bullied' (AITSL, 2021, p. 3).

What happens to you when you are socially excluded?

RESEARCH LINK 11.1

Social exclusion is a form of bullying that is prevalent among children and young people. Some children experience being socially isolated, either by being physically excluded from activities (e.g. sports activities or games) or being ignored or excluded emotionally during social relationships and communications. Social exclusion can happen in different environments where children and young people interact, ranging from the school classroom to the online environment. According to the data provided by the Office of the eSafety Commissioner (2018) in Australia, approximately 16 per cent of children aged from eight to 12 years who participated in the survey reported that they had experienced social exclusion in an online environment.

social exclusion: a process that impedes or prohibits a person's access to various rights, relationships, opportunities and activities

Those who experience social exclusion report they have 'painful' or 'hurtful' feelings (Figure 11.1). Actually, neuroimaging studies have shown that similar areas in the brain are involved when we experience physical and social pain (Eisenberger & Lieberman, 2004; Eisenberger et al., 2003). In order to study which areas of the brain react when being excluded, neuroscientists often use the Cyberball paradigm (Williams et al., 2000). Imagine you are in a group of players waiting to play with a ball. The ball game starts and you expect other players will pass you a ball to *include* you as a member of the team. However, you notice that the other players constantly *exclude* you by not tossing the ball to you.

Figure 11.1 Social exclusion in an online environment is a form of painful and hurtful bullying

Eisenberger et al. (2003) invited participants to play Cyberball inside a fMRI scanner and identified which region of the brain becomes active when the person experiences exclusion. Participants showed 'increased dorsal anterior cingulate cortex (dACC) activity during the exclusion [when being] compared with the inclusion episode' (Eisenberger & Lieberman, 2004, p. 296). Moreover, participants' self-reported distress correlated highly with dACC activity during the exclusion episode. Another area that was found to be active when experiencing the exclusion was the right ventral prefrontal cortex (rvPFC). The rvPFC is the area well known to the scientists who study placebo effects on physical pain.

For example, one study monitored whether participants felt improvement of pain symptoms while being given placebo therapy over three months. Increases in right rvPFC activity over the course of placebo therapy predicts self-reported symptom improvement (Lieberman et al., 2004). This suggests that rvPFC might serve a self-regulatory function when experiencing physical pain, and this regulatory function might be happening when a person experiences social pain stemming from exclusion (Eisenberger & Lieberman, 2004). In other words, physical and social pain overlap in their underlying neural networks and processing. Pain coming from bullying and social exclusion really 'hurts' children in their brains.

Student wellbeing and the Australian context

The core notion of student wellbeing has come to the fore in educational debate and policy concerns over the past two decades or so. However, it is difficult to find an agreed definition. Gwyther Rees and colleagues have pointed out that 'it is generally used within the research literature as an over-arching concept regarding the quality of people's lives' (Rees et al., 2010, p. 8). Citing Michaelson et al. (2009), they also say that wellbeing 'is best thought of as a dynamic process, emerging from the way in which people interact with the world around them' (Rees et al, 2010, p. 8). You will recognise that in referring to *dynamic emergence*, this definition is drawing on a dynamic systems framework.

Rees et al. (2010) also refer to a distinction in the literature between **hedonic wellbeing**, which include notions of happiness, life satisfaction, and positive and negative affect or feelings, and **eudaimonic wellbeing,** which include having a sense of purpose or meaning in life, and personal growth. The terms 'hedonic' and 'eudaimonic' are derived from ancient Greek. *Hēdonē* (ἡδονή) literally means pleasure, so *hedonism* is the pursuit of pleasure. *Eudaimonia* (εὐδαιμονίᾱ) means something like human flourishing, though it is often translated into English as 'happiness'. In 'positive psychology', both hedonic wellbeing and eudaimonic wellbeing may be linked to the English notion of *happiness*, but that does not really capture the meaning of the Greek words. The word 'happiness' usually refers to a quite transitory state of mind; we drift in and out of happiness depending on ever-shifting circumstances, but neither of the Greek terms carries that kind of transitory connotation.

Moreover, eudaimonia is much broader but also more nuanced than the English word 'happiness'. In fact, while recognising the risk of providing a circular definition, eudaimonia

hedonic wellbeing: the quality of life based on happiness, life satisfaction, and positive and negative affect or feelings

eudaimonic wellbeing: the quality of life that is enabled by having a sense of purpose or meaning in life, and personal growth

is more accurately translated as wellbeing – the sense that 'all is well and all manner of things will be well', to paraphrase the English mediaeval mystic, Julian of Norwich, cited by the Nobel prize-winning poet T. S. Eliot (1888–1965) in his poem 'Little Gidding', one of the *Four Quartets*. Eudaimonia is the opposite of feeling unpleasant, troubled, anxious, fearful, unloved, unwanted, isolated, bullied and so forth; however, it is more than the absence of these negative factors – it is a holistic concept that includes positive feelings. Perhaps, in an Australian context, it amounts to the feeling 'I'm good', when asked 'How are you?'

The AIHW's (2020) *Australia's Children* report adopted what it called a people-centred data model and an *ecological approach to child development* (Figure 11.2). This same ecological approach can be used in identifying the various factors that bear on a child's sense of wellbeing. It is not just school that matters, although school is a very immediate part of each school student's experience. Children are certainly impacted by parental employment, household income and family support, but they experience housing health and education much more directly.

Figure 11.2 AIHW people-centred data model and an ecological approach to child development
Source: AIHW (2020, p. 7).

In the survey of 43,700 children and adolescents aged from six to 12 years reported in *Australia's Children,* some 63 per cent (close to two out of three) reported 'feeling happy lots of the time', although 76 per cent (three out of four) said they 'felt scared or worried at least some of the time' (AIHW, 2020, p. 15). They also said that the 'things most likely to cause worry, include the future (73%), issues with friends (68%), issues with family (69%) and their health (69%)' (AIHW, 2020, p. 15).

However, it should also be noted that many people, including and especially traditional Aboriginal and Torres Strait Islander peoples, 'define their health and wellbeing beyond physical health and social and emotional aspects, to include spiritual and/or cultural aspects and the need for a connection to environment, community and family' (AIHW, 2020, p. 15). In 2014–15, a National Aboriginal and Torres Strait Islander Social Survey focused on how approximately 173,000 Aboriginal and Torres Strait Islander children aged between four and 14 years perceived their cultural identity and wellbeing. It found that 50 per cent identified with a clan, tribal or language group, 75 per cent had been involved in cultural events, ceremonies or organisations over the past 12 months and 44 per cent spent time with a leader or elder each week. These figures increased to 71 per cent, 80 per cent and 66 per cent respectively for children living remotely (ABS, 2016).

Student wellbeing globally: The sense of belonging and the quest for meaning

All these factors relate to what is often referred to globally as a *sense of belonging* or being *at home* in one's physical or social and cultural environment. They also relate to the basic and universal human quest for meaning or meaningfulness in life, which the distinguished Viennese psychiatrist Victor Frankl (1905–1997) found to be the difference between surviving or not surviving the terrifying ordeal of being a prisoner in a Nazi concentration camp, which he personally endured (Frankl, 1985).

Within the context of education, whether students feel they 'belong' at school and what 'meaning' school holds for them were included as one item in a very large international comparison report conducted by the Organisation for Economic Cooperation and Development (OECD, 2019b). The survey, conducted in 2018 across all 38 OECD member nations and 43 other educations systems in non-OECD partner nations, sampled 15-year-old students, which is the OECD's Program for International Student Achievement (PISA) target population. The report, published in 2019 as the *PISA 2018 Results,* provides very extensive and detailed statistical data, presented in six volumes. Volume III, *What School Life Means for Students' Lives*, which contains 14 chapters, is concerned with student wellbeing and what it refers to as 'school climate'.

What meaning do children and youngsters attach to their experience of school and, as an important component of that, their sense of belonging at school? The report (OECD, 2019b) found that:

- Across OECD countries, the majority of students reported that they felt socially connected at school. For instance, three out of four students agreed or strongly agreed that they can make friends easily at school (2019b, p. 130).

- Students in socioeconomically disadvantaged, rural and public schools were more likely to report a weaker sense of belonging at school than students in advantaged, city and private schools (2019b, p. 130).
- On average across OECD countries, students who reported a greater sense of belonging scored higher in the reading assessment, after accounting for socioeconomic status (2019b, p. 130).
- Students reported a greater sense of belonging when they also reported higher levels of cooperation among their peers, whereas students' perception of competition was not associated with their sense of belonging at school (2019b, p. 130).
- Students who reported a greater sense of belonging were also more likely to expect to complete a university degree, even after accounting for socioeconomic status, gender, immigrant background and overall reading performance (2019b, p. 130).
- On average across OECD countries, 67 per cent of students reported being satisfied with their lives (students who reported scores of between 7 and 10 on the 10-point life-satisfaction scale). Between 2015 and 2018, the share of satisfied students shrank by five percentage points (2019b, p. 188).
- More than 80 per cent of students reported sometimes or always feeling happy, cheerful, joyful or lively, and about 6 per cent of students reported always feeling sad, on average across OECD countries (2019b, p. 188).

REFLECTION

Pause for a moment to study these findings. Notice, for example, the links found between a sense of belonging and scoring higher in reading assessment, and expecting to complete a university degree. Did you have a strong sense of belonging to your school? To what extent did your school experience relate to your sense of meaning and purpose in life?

The words 'meaning and purpose in life' can easily trip off the tongue, with little reflection on how, for many people – especially traditional and Indigenous peoples – they cut to the core of what it means to be human, including the centrality of each individual person's sense of wellbeing, or lack of it. Thus, the loss of meaning and purpose is closely associated with a loss of 'self' resulting in *giving up*. As teachers are very well aware, a child or adolescent who ceases to find meaning and purpose in school will likely give up.

As noted above, teachers have a very special duty to monitor the wellbeing of children in their care, and to do that effectively they need to employ empathy (Chapter 8). They need to try to *feel what the student might be feeling*, especially if they sense a child in their care is nearing the point of giving up. Needless to say, if loss of meaning and purpose extends to life itself, it can lead to giving up on life. The AIHW's *Australia's Children* report states that the suicide rate for children age 0–14 remained stable from 2010 to 2017 at '0.4 per 100,000 with little difference between boys and girls' (AIHW, 2020, p. 111), so the numbers are not high. However, every child lost to suicide is one too many, and teachers need to be constantly watchful.

Although we cannot conduct scientific experiments to understand the depth of distress that can lead to giving up, there have been a number of cases over the past century where human beings have been subjected to the most terrible cruelty, including the Nazi Holocaust. Victor Frankl was a scientist and doctor who became a prisoner and endured, with his fellow prisoners, the full force of Nazi dehumanisation. Frankl survived the experience and, after it was over, applied his expertise as a scientist and doctor to provide a summary account and analysis of his experience. The short book he wrote, called *Man's Search for Meaning* (Frankl, 1985) became a classic.

Frankl's book is highly readable and of interest to teachers because it provides clues into the loss of wellbeing, meaning and purpose that students and colleagues might suffer – admittedly not under those particular extreme conditions, although their own circumstances will often feel extreme. Experience of working with so-called *disaffected students* over many years suggests that a child or adolescent who ceases to find meaning and purpose in school, which so often is associated with a sense of failure, will also be anxious about their future. In the concentration camps, Frankl recalled that the loss of wellbeing and meaning was often associated with the loss of faith in the future. He wrote:

> The prisoner who had lost faith in the future – his future – was doomed. With his loss of belief in the future, he also lost his spiritual hold; he let himself decline and became subject to mental and physical decay. Usually this happened quite suddenly, in the form of a crisis, the symptoms of which were familiar to the experienced camp inmate. We all feared this moment – not for ourselves, which would have been pointless, but for our friends. Usually it began with the prisoner refusing one morning to get dressed and wash or to go out on the parade grounds. No entreaties, no blows, no threats had any effect. He just lay there, hardly moving. If this crisis was brought about by an illness, he refused to be taken to the sick-bay or to do anything to help himself. He simply gave up. There he remained, lying in his own excreta, and nothing bothered him anymore. (Frankl, 1985, p. 95)

Students' fear of failure and concern for their future

One of the statements students were asked to respond to in the 2018 OECD survey, discussed above was, 'When I am failing, this makes me doubt my plans for the future' (OECD, 2019b, p. 189). The results make very interesting reading. Of all nations surveyed, both OECD countries and non-OECD (total 78), 19 countries scored more than 60 per cent of their 15-year-old students indicating concern for their future when they felt they were failing (see OECD, 2019b, p. 189):

- For OECD member nations, the United Kingdom was highest (70 per cent), then Australia, Canada and New Zealand (68 per cent), then Ireland and the United States (65 per cent), Turkey (64 per cent, France (62 per cent) and Japan (61 per cent).
- For non-OECD nations, Singapore was highest (78 per cent), then Chinese Taipei (77 per cent), Brunei (73 per cent), Hong Kong (72 per cent), Malaysia (67 per cent), Macao, China (66 per cent), Thailand and the United Arab Emirates (64 per cent), the Philippines (63 per cent) and Kosovo (61 per cent).

In Chapter 10, we critically examined the problem of failure in school, in the context of assessment, and argued for a reassessment of failure in education – the idea that within classrooms and education generally, there should be *freedom to fail*. The 2018 OECD report devotes a whole chapter (Chapter 13) to student failure. Three findings in particular stand out:

- Students in many Asian countries and economies expressed the greatest fear of failure, while students in many European countries expressed the least fear.
- In almost every education system, girls expressed greater fear of failure than boys, and this gender gap was considerably wider among top-performing students.
- In a majority of school systems, students who expressed a greater fear of failure scored higher in reading, but reported less satisfaction with life, than students expressing less concern about failing, after accounting for the socioeconomic profile of students and schools. (OECD, 2019b, p. 188)

These survey findings provide statistical comparisons between nations, but they do not explain the reasons for these differences. When viewing these results, what thoughts do you have that might go some way towards explaining possible causes? For example, why do you think fear of failure is greatest in many Asian countries and least in many European countries? Bearing in mind that the survey sampled 15-year-olds, does it surprise you that girls expressed greater fear of failure than boys, and that this gap was wider among high-performing students? Why, in a majority of school systems, do you think students who expressed a greater fear of failure scored higher in reading? Does this make any sense to you? If possible, discuss these findings with classmates and share your ideas and perspectives.

Wellbeing and the affordance of value-based learning environments

In Chapter 8, a strong link was drawn between education and values. It was pointed out, for example, that education, seeks what is *good* and *intrinsically worthwhile* in attending to the development of children's minds (brains) and their *mental and physical wellbeing*. There is also 'a strong consensus in the educational literature that teaching is an inherently moral endeavour, that the moral work of teachers is of central importance to education', and has important implications for 'the practice of teacher education' (Sanger, 2008, p. 169). What Sanger refers to as 'the moral work of teachers' is very much concerned with students' mental and physical wellbeing. That point is emphasised by Ron Toomey et al. (2010) in their detailed and thought-provoking book.

Values pedagogy and student wellbeing

Teacher Education and Values Pedagogy: A Student Wellbeing Approach (Toomey et al. (2010) was published in an Australian context that was very much concerned with the development of a National Framework for Values education in Australian Schools (DEST, 2005), as a key product of the government-funded Australian Values Education Program (AVEP). In the opening chapter of the book, Terry Lovat and Ron Toomey provide 'a novel conception of values education as the basis of a new student wellbeing pedagogy' (Lovat & Toomey, 2010, p. 1). The core idea is that values education is not restricted to being a curriculum subject, such as moral education, or character education, but rather 'primarily a conception of values education as pedagogy, with effective teaching and learning being enhanced by the positive human relationships and explicit values-oriented transactions that are forged within quality values education programs. As its heart is student personal, academic, social and emotional wellbeing' (Lovat & Toomey, 2010, p. 2). So values education pedagogy is linked directly to student wellbeing as 'student wellbeing pedagogy' (Clement, 2010, p. 15).

Moreover, as Neville Clement (2010, p. 15) notes, the notion of student wellbeing pedagogy 'evolved with the realisation that learning is an interaction between individual endowments and the learning environment'. Taken together, it seems that what is being presented is a holistic and dynamic values-centred notion of student wellbeing (incorporating personal, social, academic and emotional wellbeing), where student learning in school is contingent on a values-based pedagogy and learning environment. Clement also provides an account of how much of the thinking involved in this endeavour was informed by neuroscience research literature, though the AVEP project itself did not conduct the kind of school-based EEG brain research described in the last section of Chapter 10.

However, that EEG research project was partly conceived in response to the notion of values in education and student wellbeing provided by Toomey et al. (2010). Quite clearly, this highly funded AVEP government initiative was on to something very important. Surprisingly, though, AVEP is hardly remembered and seldom referred to in teacher education nowadays (Duncan & Sankey, 2019). It was eclipsed by the so-called Education Revolution, introduced by the incoming Rudd Labor government in 2007, which introduced the National Assessment Program – Literacy and Numeracy (NAPLAN) and My School website initiatives.

There is a treasure house of high-quality materials that were produced by the program and are continuing to be produced that are available for teacher education and research posted on the Australia Government website. You might like to consider, for example, the 'Ten Principles for Good Practice' that emerged from the project. These are listed on Page 5 of the Final Report of the values in action schools project, called *Giving Voice to the Impacts of Values Education* (Hamston et al., 2010) (Figure 11.3). You will find the summary image of the 'Impacts of Values Education' on p. 11 of the same document.

Figure 11.3 *Giving Voice to the Impacts of Values Education*, October 2010
Source: Hamston et al. (2010).

Value-embedded learning: Value, emotion and salience in the learning process

The school-based EEG research project described in Chapter 10 interpreted AVEP as an attempt to advance what it called an *embedded values* vision of education and contrasted that with the *performative vision* of the Education Revolution (Duncan & Sankey, 2019, p. 1). The notion of 'embedded values' was subsequently developed by the researchers into a broader and more encompassing notion of **value-embedded learning (VEL)**, (Duncan et al., 2021). This went beyond a concern with values education, as undertaken within AVEP. It encompassed a reappraisal of the role that *value* (not just values) plays in learning and development, especially 'the close interrelationship between value, emotion and salience in the learning process' (Duncan, et al. 2021, p. 8).

The concept of VEL was conceived within the overall framework of dynamic systems and brain plasticity. Technically, VEL seeks to provide 'a philosophically non-reductive and biologically plausible account of the dynamic relationship between value and emotional salience in the learning process, and how value might best be afforded in school and classroom learning environments' (Duncan et al., 2021, p. 1). Something that is salient is important, noticeable, prominent and thus meaningful. The key idea is that what we learn and remember best is what we find salient or meaningful – whether positively or negatively. At school, what we learn and remember best are the things we find salient and that matter to us. What matters is what we value and that comes with emotional valence: we enjoy it, it gives pleasure – 'We feel, therefore we learn' (Immordino-Yang & Damasio, 2016).

You may remember that in Chapter 5 it was said that those who excel in maths are first and foremost those who enjoy it – it is valued, and is therefore salient or meaningful. In Chapter 3, we saw Gerald Edelman (2006, p. 11) argue that *perceived value and salience drives the selective process in the brain*, that value is 'imposed in the brain by the brain' as a core component of brain plasticity. In this research project, VEL was not just applied to provide a scientific, neurobiological framework; it was incorporated into the methodology of the study, which used EEG brain measurement to probe what is occurring in the students' brains as they meet the emotional challenges of summative assessment. Without such probing technology, all researchers would have at their disposal would be self-reporting methods – and these can only tell us what students *believe and say* they are thinking and feeling. As we have noted before, that is limited by the fact that much of our thinking is occurring below the level of conscious awareness, and so unable to be probed by conscious introspection.

value-embedded learning (VEL): a philosophically non-reductive and biologically plausible account of the dynamic relationship between value and emotional salience in the learning process

self-efficacy: in the PISA study, self-efficacy is defined as 'the extent to which individuals believe in their own ability to engage in certain tasks including academic tasks' (OECD, 2019b, p. 188)

Students' fear of failure and self-efficacy

Organisation for Economic Co-operation and Development (OECD) (2020). *'Students' self-efficacy and fear of failure'* in PISA 2018 Results (Volume III): What school life means for students' lives. Paris: OECD Publishing.

The PISA 2018 survey collected indices of students' wellbeing, focusing on students' belonging at school, life satisfaction and meaning in life, feelings, **self-efficacy** and fear of failure.

RESEARCH LINK 11.2

The PISA survey's self-efficacy items are posited on Bandura's (1977) conceptualisation of self-efficacy – in other words, 'the extent to which individuals believe in their own ability to engage in certain activities and perform specific tasks, especially when facing adverse circumstances' (OECD, 2019b, p. 188).

In addition to students' self-efficacy, the PISA survey asked questions on participants' perceptions of risks or consequences of failure. Figure 11.4 provides each nation's average percentage of students who either agreed or strongly agreed with the questions related to their self-efficacy and fear or failure.

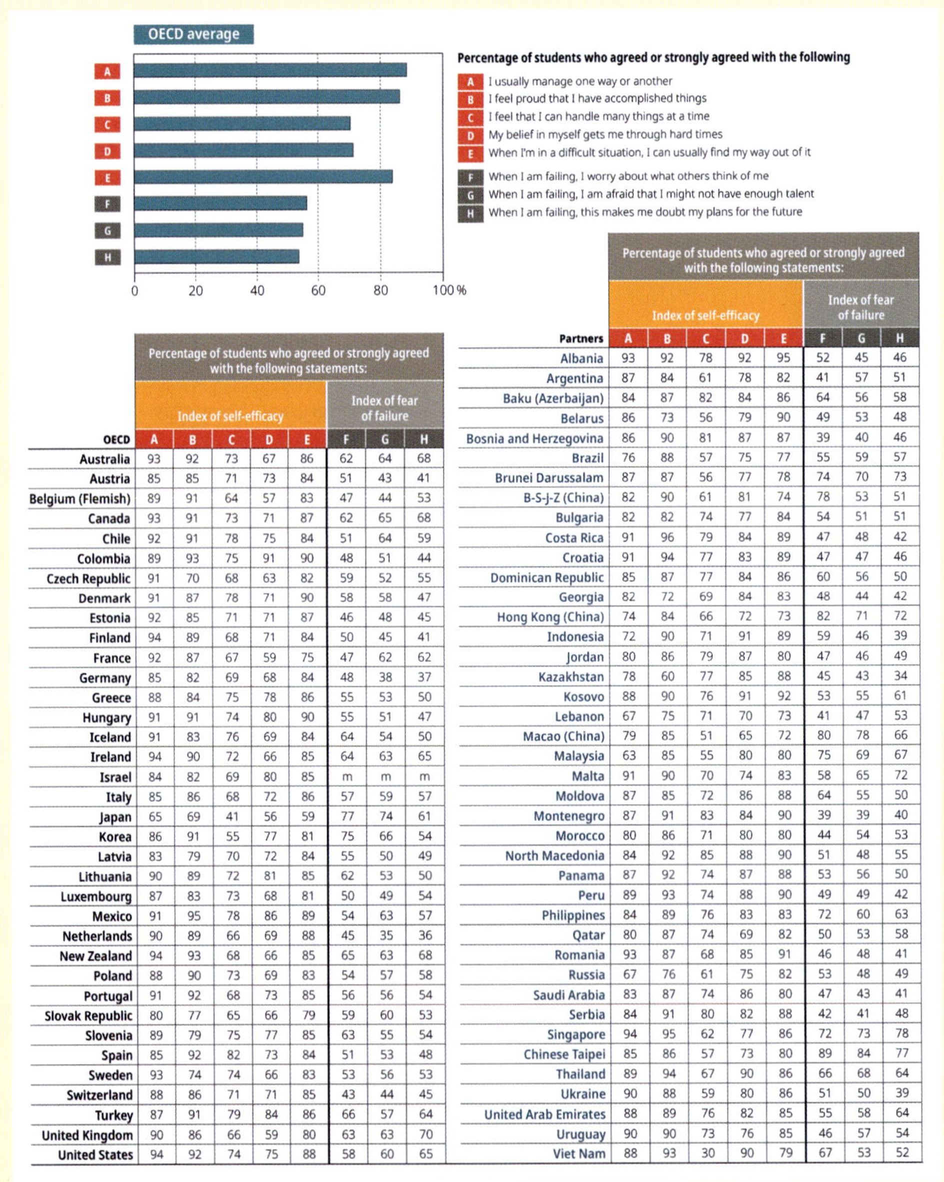

OECD	Index of self-efficacy					Index of fear of failure		
	A	B	C	D	E	F	G	H
Australia	93	92	73	67	86	62	64	68
Austria	85	85	71	73	84	51	43	41
Belgium (Flemish)	89	91	64	57	83	47	44	53
Canada	93	91	73	71	87	62	65	68
Chile	92	91	78	75	84	51	64	59
Colombia	89	93	75	91	90	48	51	44
Czech Republic	91	70	68	63	82	59	52	55
Denmark	91	87	78	71	90	58	58	47
Estonia	92	85	71	71	87	46	48	45
Finland	94	89	68	71	84	50	45	41
France	92	87	67	59	75	47	62	62
Germany	85	82	69	68	84	48	38	37
Greece	88	84	75	78	86	55	53	50
Hungary	91	91	74	80	90	55	51	47
Iceland	91	83	76	69	84	64	54	50
Ireland	94	90	72	66	85	64	63	65
Israel	84	82	69	80	85	m	m	m
Italy	85	86	68	72	86	57	59	57
Japan	65	69	41	56	59	77	74	61
Korea	86	91	55	77	81	75	66	54
Latvia	83	79	70	72	84	55	50	49
Lithuania	90	89	72	81	85	62	53	50
Luxembourg	87	83	73	68	81	50	49	54
Mexico	91	95	78	86	89	54	63	57
Netherlands	90	89	66	69	88	45	35	36
New Zealand	94	93	68	66	85	65	63	68
Poland	88	90	73	69	83	54	57	58
Portugal	91	92	68	73	85	56	56	54
Slovak Republic	80	77	65	66	79	59	60	53
Slovenia	89	79	75	77	85	63	55	54
Spain	85	92	82	73	84	51	53	48
Sweden	93	74	74	66	83	53	56	53
Switzerland	88	86	71	71	85	43	44	45
Turkey	87	91	79	84	86	66	57	64
United Kingdom	90	86	66	59	80	63	63	70
United States	94	92	74	75	88	58	60	65

Partners	Index of self-efficacy					Index of fear of failure		
	A	B	C	D	E	F	G	H
Albania	93	92	78	92	95	52	45	46
Argentina	87	84	61	78	82	41	57	51
Baku (Azerbaijan)	84	87	82	84	86	64	56	58
Belarus	86	73	56	79	90	49	53	48
Bosnia and Herzegovina	86	90	81	87	87	39	40	46
Brazil	76	88	57	75	77	55	59	57
Brunei Darussalam	87	87	56	77	78	74	70	73
B-S-J-Z (China)	82	90	61	81	74	78	53	51
Bulgaria	82	82	74	77	84	54	51	51
Costa Rica	91	96	79	84	89	47	48	42
Croatia	91	94	77	83	89	47	47	46
Dominican Republic	85	87	77	84	86	60	56	50
Georgia	82	72	69	84	83	48	44	42
Hong Kong (China)	74	84	66	72	73	82	71	72
Indonesia	72	90	71	91	89	59	46	39
Jordan	80	86	79	87	80	47	46	49
Kazakhstan	78	60	77	85	88	45	43	34
Kosovo	88	90	76	91	92	53	55	61
Lebanon	67	75	71	70	73	41	47	53
Macao (China)	79	85	51	65	72	80	78	66
Malaysia	63	85	55	80	80	75	69	67
Malta	91	90	70	74	83	58	65	72
Moldova	87	85	72	86	88	64	55	50
Montenegro	87	91	83	84	90	39	39	40
Morocco	80	86	71	80	80	44	54	53
North Macedonia	84	92	85	88	90	51	48	55
Panama	87	92	74	87	88	53	56	50
Peru	89	93	74	88	90	49	49	42
Philippines	84	89	76	83	83	72	60	63
Qatar	80	87	74	69	82	50	53	58
Romania	93	87	68	85	91	46	48	41
Russia	67	76	61	75	82	53	48	49
Saudi Arabia	83	87	74	86	80	47	43	41
Serbia	84	91	80	82	88	42	41	48
Singapore	94	95	62	77	86	72	73	78
Chinese Taipei	85	86	57	73	80	89	84	77
Thailand	89	94	67	90	86	66	68	64
Ukraine	90	88	59	80	86	51	50	39
United Arab Emirates	88	89	76	82	85	55	58	64
Uruguay	90	90	73	76	85	46	57	54
Viet Nam	88	93	30	90	79	67	53	52

Figure 11.4 Each nation's average percentage of students who either agreed or strongly agreed with the questions related to their self-efficacy and fear or failure
Source: OECD (2020, p. 189).

The PISA report claims that 'fear of failure and self-efficacy go hand-in-hand: students who believe they are not capable of performing adequately in certain situations are more likely to be fearful of such situations' (OECD, 2020, p. 188). But is this correct? Do the figures for self-efficacy actually correlate with those given for fear of failure, across a significant number of participating nations, and what do you notice about those nations that do correlate?

If the OECD claim is valid, there should be a correlation between the responses to statements A–E (indexing self-efficacy) and F–H (indexing fear of failure). For those who find statistical tables a bit daunting, one way to approach it is to choose a benchmark number for fear of failure – say, 60 – and check out how many nations score more than 60 across at least one of the responses F–H. Start with Australia. Answer 'Yes'. Do they also score above 60 for responses to A–G? 'Yes'. Tick them off. How many and which other nations returned scores over 60 for F–H and also A–G? How many correlations do you find?

You may find reason to doubt that self-efficacy and fear of failure go 'hand in hand'. If they don't correlate, you may also wonder whether self-efficacy is even a helpful research item in regard to student's fear of failure?

Research link 11.2

Value-embedded learning and motivation theory in education

The notion of VEL offers an alternative way to conceptualise ideas that have been key planks of *motivation theory* in education for the past half-century: *self-efficacy theory*, *expectancy-value theory* and *self-determination theory* (Duncan et al., 2021). For example, the 2018 OECD survey discussed above aligned its survey of students' *fear of failure* with Bandura's (1977) notion of *self-efficacy*. The OECD's central claim is that 'fear of failure and self-efficacy go hand in hand' (OECD, 2019b, p. 188), but do they? In Research link 11.2, you are given the opportunity to critically assess the statistics produced for this item.

While conceding that the term 'motivation' (with its accompanying notions of self-efficacy and value expectancy) may very well be appropriate within mainstream psychology, Duncan et al. (2021) question whether 'motivation' is an appropriate term (concept) to use in educational contexts. One major problem is that, 'Whether intended or not, the word motivation implies deficit, students are lacking something – they have no motivation' (Duncan et al., 2021, p. 7), and deficit models in educational contexts are generally viewed as problematic. In this case, the problem is that

> diagnosing children as lacking motivation tends to blame the student when not wanting to engage with what the teacher wants them to learn, there's little or no recognition, for example, that the troublesome student may still be learning and developing, though not learning what the school requires, or that the affordances of the learning environment might be partly or wholly to blame. (Duncan et al., 2021, p. 7)

If one looks more closely at **expectancy-value theory (EVT)**, the core idea is that a student's expectancy for success in studying a subject or topic and whether it is interesting, worthwhile and of value are key 'determinants of their motivation and subsequent academic outcomes (i.e. grades, test scores, choices of course taking and career paths, etc.)' (Rosenzweig et al.,

expectancy-value theory (EVT): a theory of motivation; the core idea is that a student's expectancy for *success* is a key determinant of their motivation and subsequent academic outcomes

2019, p. 617). The basic assumption of *expectancy-value* is that students rationally and deliberatively choose to engage in academic tasks and activities that have high subjective values and in which they expect a high probability of success. But is this actually correct? Are these choices made rationally, consciously and deliberatively, as EVT assumes, or are they often intuitive and perhaps even operating below the level of conscious awareness? The other main concern is that motivation in general and expectancy value in particular belong very much within *classical cognitive theory* and, as we have seen, that is now facing increasing criticism.

It is not denied that children who are reluctant to learn what they are required to learn at school may well need encouraging, and that helping them see its benefits (expectancy-value) may well play a part of that encouragement – every teacher is aware of that. However, the notion of value-embedded learning takes a very different starting point, arguing that, 'Each and every student is equipped with a brain that is plastic and value-orientated towards learning – and will learn when nested in suitably affordant learning environments' (Duncan et al., 2021, p. 8). The processes involved in learning are many and the challenges of schooling are complex, 'but the recognition that students have brains that are predisposed by evolution to learn should be the starting point for all teachers – not the notion that students won't learn unless they are motivated' (2021, p. 8).

There are a number of fundamental differences between motivation theory and the concept of value-embedded learning that are laid out in Table 11.1, adapted from Duncan et al. (2021). Check it through and then consider the Reflection that follows.

Table 11.1 From motivation and expectancy value to dynamic systems and value-embedded learning

	Motivation/expectancy value	Dynamic systems/value-embedded learning
Theoretical foundation	Rational cognitive theory incorporating dualisms of organism/environment, inner/outer, intrinsic/extrinsic, autonomous/controlled.	Dynamic systems (complexity) theory that eschews dualisms and dichotomies. Stability and change arise from the interweaving of many parts and causes.
Core analogy	Machine – mechanistic	Ecosystem – organismic
Process	Linear, sequential over time	Non-linear, feedback loops
Mind and brain	Informational/computational semantic processing. Mind, conceived as modular and inner, providing a mirror representation of the outer (external) world.	Dynamic neutral network in brain responding to salience and creating meaning. Brain mediates emotion with cognitive demands and actions.
Perception	Perception, cognition and action are separate linear entities. Passive reception of stimuli subsequently processed to inform action.	Dynamic continuity between perception, thought and action. To perceive is necessarily to perceive-as (categorisation).
Value	Notions of 'expectancy' and 'task value' are central explanatory concepts. Value derived from learner's estimate of anticipated outcome/benefit	Value is neurobiologically embedded and perceived directly by learner. Dynamic relationship between value and emotional salience in learning.

Table 11.1 (*cont.*)

	Motivation/expectancy value	Dynamic systems/value-embedded learning
Learner choices	Learner rationally and deliberatively chooses to engage in learning by anticipating success and calculating value derived from task completion.	Implicit/tactic awareness of what is valuable in learning. Choices and actions are often intuitive and may precede conscious reasoning.
The view of the learner	Learner viewed as detached, complex mind with simple body, processing stimuli derived from the learning environment. Learner requires a motivational push to engage with learning.	Learner-in-learning environment is the core unit of analysis. Learner viewed as embodied, with a plastic brain that is value-oriented towards learning, and will learn when nested in a sufficiently affordant learning environment
The learning environment	Separate stable outer entity to be replicated in the inner mind of the learner.	Learning environments as dynamic, shifting ecosystems that ideally invite and afford learning.

Source: Adapted from Duncan et al. (2021).

Look at Table 11.1. There is much you will recognise from studying the previous chapters of this textbook, but do you see any issues with the comparisons it is drawing? For example, regarding learner choices, do you carefully calculate the subject and topic choices you make as a learner, or are they more intuitive, based on what you sense will be best for you?

Student wellbeing and the interoceptive, predictive brain

The human brain was *not designed* to allow us to go to school, to learn to read and write and study all the subjects on the school curriculum – these abilities are *exaptations* (Dehaene, 2020), co-opting areas of the brain not originally naturally selected or adapted for that purpose (Chapter 5). Nor was the brain designed to allow some adults to choose to become teachers. *The mammalian brain was not designed at all*; it evolved from earlier precursors (fish, birds, reptiles), and the main purpose it serves is the *wellbeing* of the organism and its survival. Having said that, it should be noted, contrary to what some authors insist at this point in the discussion, that denying the notion of design and insisting that brains evolved says nothing about whether life on Earth, or the cosmos as a whole, is divinely ordained; that is a separate and very complex issue, well beyond the scope of this text.

The key point for us is that wellbeing and survival are not optional extras for the brain, something it does once it has taken care of all its other competing demands; rather, wellbeing and survival *are* the main demands. As noted in Chapters 1 and 10, this process is called *homeostasis*, and is concerned with maintaining sufficiently stable conditions within the living organism necessary for survival. The word 'homeostasis' is derived from the Greek words *homeo* (ὅμοιος), meaning similar, and *stasis* (στάσις), meaning standing still or stable.

Figure 11.5 We are all embodied brains that are concerned primarily with maintaining homeostasis, wellbeing and flourishing

However, as Antonio Damasio (2018) points out, this implies a very static notion, whereas life is highly dynamic, future oriented and concerned with *flourishing*. Therefore, he insists, homeostasis should not only connote that 'life is regulated within a range that is not just compatible with survival but also conducive to flourishing, to a projection of life into the future of an organism or a species' (Damasio, 2018, p. 26). Within the context of education, this extension to the meaning of homeostasis implies that the embodied brains of students and their teachers are *concerned primarily with wellbeing and flourishing* (Figure 11.5).

Children and their teachers are functioning brains. Strip away all that we see as the different physical features of each individual child or teacher, and what we take to be their personality or character – all those indicators we normally use to identify ourselves and others – and, at base, we find an embodied brain that is *concerned primarily with wellbeing and flourishing*. It is desperately hard to value and show love for children and youngsters who cause teachers (and often their parents) so much trouble, and the same can apply to some colleagues with whom one has to engage in the workplace. Yet all children and colleagues are ultimately embodied brains that are concerned primarily with maintaining homeostasis, wellbeing and flourishing. Surely this is something that needs to go into every teacher's professional toolkit (Chapter 1), along with the many other insights it is hoped you have gained from this book. Use it, empathetically, whenever you are frustrated with a child in your care.

Can you think of three or four ways in which stripping away all the usual indicators we use in viewing children and adolescents (their physical appearance and what we take to be their personality or character), and instead conceptualising them as embodied brains seeking wellbeing and flourishing, might change how you value them as behaving persons?

As you learnt in Chapter 10, maintaining homeostasis is achieved through the sensory system known as *interoception*, which constantly communicates the *internal state* of the body to the brain. More precisely, 'Interoception, as the sense of the physiological condition of the body, supports homeostatic control and allostatic adaptation, ensuring the stability of the organism and by driving behaviour through feelings such as hunger, thirst and dyspnoea' (Tsakiris & Critchley, 2016, p. 1), which means shortness of breath. These processes are all concerned with internal bodily, metabolic regulation – it is constant, day and night, day in, day out, and lifelong.

However, as you also learnt in Chapter 10, the concept of interoception is currently being employed in brain science, well beyond the maintenance of internal metabolic balance, to include the way the brain processes *external stimuli*. Although the details are far from complete, 'interoceptive signals are increasingly recognised to have a pervasive … impact on cognition, influencing attention and perception, guiding decision-making and memory and emotion processing' (Tsakiris & Critchley, 2016, p. 1), all of which are highly relevant to education and teaching. Moreover, the notion of the interoceptive, predictive brain is starting to provide a highly embodied and unified model of brain functioning (Clark, 2015). If this new understanding of the brain takes hold, it will rewrite the textbooks used in teacher education. It is very possible that we are standing on the brink a of a major paradigm shift in our understanding of brain dynamics, with significant implications for education.

It is surely entirely fitting, if not ironic, that after the strictures of behaviourism and the subsequent cognitive revolution in the twentieth century, with its many problems in accounting for perception, cognition and action, brain scientists are now returning to this core idea of *maintaining wellbeing* to provide a radically new model of how the brain works as a whole. It is a new and alternative model, moreover, that is turning the behaviourist and cognitivist models, including and especially 'the traditional picture of perception on its head' (Clark, 2015, p. 21). In this new and alternative interoceptive, predictive model, the same overall processes that are keeping the organism alive are also those that account for our thoughts, feelings, emotions, cognition and behaviour, along with all that we call mental (brain) health and wellbeing (Tsakiris & Critchley, 2016).

The interceptive, predictive brain: Feelings, emotions and wellbeing

Mental health, or brain health (wellbeing and flourishing), is closely associated with feelings and emotions, as evident in childhood but particularly in the often-turbulent years of puberty and adolescence. Much of the psychological stress experienced by some children and adolescents is directly related to the associated feelings and emotions, and in your role as a teacher (or parent), you will spend a lot of time talking to students about their feelings and emotions – particularly if they seem especially anxious or even depressed.

However, it should be noted that as a teacher you may occasionally come across students who seem unable to describe their feelings and emotions that they are clearly experiencing. The word psychiatrists use to describe that condition is *alexithymia,* which is another of those terms derived from Greek – in this case, putting three Greek words together – α (*a*), meaning *not*, λέξις (*lexis*), meaning *words* and θῡμός (*thýmós*), meaning *emotions* (or heart). Together they become 'alexithymia', which literally means having no word for one's own emotions.

It is increasingly being realised that feelings (affect) and emotions are the product of the *interoceptive, predictive brain* – a brain that is constantly making predictions and responding to perceived prediction errors (Barrett, 2017). If correct, this has massive implications for how you view yourself. Thoughts, feeling and emotions are not things that happen to you. They are not things at all – they are processes occurring in your thoroughly embodied, interoceptive, predictive brain. There is no independent 'you' that exists apart from your thoughts, feeling and emotions, that keeps having thoughts and experiencing feelings and emotions in response to triggering events that occur in your life. Your brain is not composed of neurons waiting to be triggered by events that make you feel up or down, pleasant or unpleasant (affect), or more complex emotions such as those used in Ekman's theory of emotional facial expressions that you met in Chapter 6 – anger, fear, disgust, surprise, sadness and happiness.

On the contrary, thoughts, feelings and emotions are created in the embodied brain as an integral part of the ceaseless predicting and interoceptively sensing processes that are keeping you alive – checking your internal bodily states and sensing the external environment, comparing it with its predictions of what should be, based on all that you have experienced in the past. And that doesn't just apply to what might be called the flow of everyday feelings and emotions, which we all experience on our trajectory through life's winding pathways, but also the more fearful feelings associated with mental illness, such as anxiety and the onset of depression, and hopefully its treatment. It is also associated with alexithymia, the inability to describe one's own feelings and emotions.

The core idea is that 'chronic prediction errors of a certain magnitude and frequency could constitute a transdisorder vulnerability to illness' (Barrett et al., 2016, p. 7). Indeed, it is speculated that 'both alexithymia and depression are linked to diminished interoceptive awareness, which itself has recently been characterized as an inability to calibrate precision estimates for interoceptive prediction errors' (2016, p. 8). In other words, it is hoped that this new understanding of the brain will start to shed fresh light on what is happening in the brain when patients (including children and adolescents) suffer from anxiety, depression and related problems. If our thinking, feelings, emotions and actions depend on comparisons between the brain's predictions (based on past experiences) and actual ongoing experiences, then chronic prediction errors are likely to be highly disruptive of 'normal' behaviour. Moreover, dysfunction 'of interoception is increasingly recognised as an important component of different mental health conditions including anxiety disorders, mood disorders, eating disorders, addictive disorders and somatic symptom disorders' (Khalsa et al., 2016, p. 501).

The dynamics of perceptual categorisation, thinking, feelings and emotion

In Chapter 6, you were introduced to key ideas in a dynamic systems account of emotion; that they are 'emergent, self-organised processes' and, just like cognitions, they are 'fluid, context-sensitive, non-linear, and contingent' (Thelen & Smith, 1994, p. 320) and are constructed by a complex and dynamic process of interaction among many components, in response to changes in the social and physical context (Fogel et al., 1992). You also encountered Antonio Damasio, who drew a strong connection between thinking and feeling, logic and emotion and, moreover, the idea that these 'systems are interwoven with *those that regulate the body*' (Damasio, 1994, p. 245, emphasis added). He has continued to explore the embodied nature of thought and action and the relationship between feelings, emotions and homeostasis in his subsequent publications, including *The Strange*

Order of Things (Damasio, 2018), and he has also related this to education in conjunction with Mary Helen Immordino-Yang (2016).

The other scientist you met in Chapter 6 was Lisa Feldman Barrett, who we saw was highly critical of the 'classical' theory of emotion and especially critical of Ekman's theory of emotional facial expressions. Her own theory of how emotions are made (Barrett, 2017) relies very heavily on notion of the interoceptive, predictive brain. The human brain, she says, 'is anatomically structured so that no decision or action can be free of interoception and affect, no matter what fiction people tell themselves about how rational they are' (Barrett, 2017, p. 82) and interoception, as a fundamental part of the predictive process, is a key ingredient of emotion (Barrett, 2017, p. 83).

Originally, Barrett called her own approach the 'conceptual act model of emotion', but has since changed it to 'the theory of constructed emotion'. The emphasis she now places on 'construction' is intended to signal that emotions are not genetically endowed, but rather are constructed by multiple interwoven brain networks working together in tandem. They are therefore allied to affect, feelings related to wellbeing that we experience continuously, throughout each and every day, although emotions are more complex. Emotions such as anger or sadness are far more complex than simple feelings of unpleasantness or arousal. Also, in the original version, she did not use the notion of interoception, instead referring to what she called 'core affect'. In currently stressing the role of interoception, she is not saying that interoception on its own provides a full account of emotion; the missing ingredient (and this is where her ideas become more controversial) is the role played by categorisation and concepts in her theory.

The important role perceptual categorisation plays in learning, and indeed human consciousness, was emphasised by Gerald Edelman, who you met in Chapter 3. Using concepts, your 'brain learns to categorise, construct phonemes in tens of milliseconds within all this variable, noisy information, ultimately permitting you to communicate with others' (Barrett, 2017, p. 85). The brain categorises what it perceives via the senses and uses concepts to make the sensory perceptions meaningful. Without concepts, 'you'd experience a world of ever-fluctuating noise. Everything you'd ever encountered would be unlike anything else. You'd be experientially blind ... You'd be incapable of learning' (Barrett, 2017, p. 85). That would seem to be entirely correct, but critics such as Ralph Adolphs are keen to point out that emotions aren't concepts – at least not in the way that traditional cognitivist theory has understood concepts, as 'mental representations by which we think about something (i.e. semantic knowledge)', but he says, 'I don't understand Lisa's concept of a concept' (Adolphs, 2017, p. 33).

This debate takes us back to the problem of representations and meanings in the head, which we encountered in the 'Affordance, Embodied Cognition and Perception–Action Loops' section of Chapter 9. To be consistent, and to clarify what precisely she means by aligning emotions with concepts, Barrett might do better to align herself with Walter Freeman's position that representations exist only in the world and have no meaning, whereas, 'meanings exist only in brains without being represented there' (Freeman, 2000, p. 16). In stressing that concepts make the sensory perceptions meaningful, she seems to come close to Freeman's position, but she doesn't seem to be aware of his account of representations and meanings.

Valuing students as sentient, emotional and relational human beings

Almost a century ago, a Canadian teacher of English and innovative school principal wondered what was going on in the heads of his student when they learned or failed to learn. As you read in Chapter 2, Donald Hebb left teaching, dispirited by the very restricted curriculum he had to teach, and went on to be one of the most influential brain scientists of the twentieth century. Most of the learning that occurs in schools is Hebbian learning, or Hebbian plasticity. Over the past century, there have been tremendous advances in our understanding of the dynamics of brain functioning but, as the debate mentioned in the previous paragraph illustrates, there is still much we do not know and much that is contested about how that thing in our heads makes us who we are.

The interoceptive, predictive Bayesian brain

Figure 11.6 Karl Friston (1959–)

So how exactly does the brain work? How does it do all the wonderful things it does that allow you to learn, develop and flourish? To be entirely honest, no one really knows for sure. However, there is a growing number of brain scientist and especially those who are interested in creating artificial brains who believe the concept of the predictive brain will provide the breakthrough they are looking for. One of those is Karl Friston, at University College, London (Figure 11.6), who advocates the concept of a *predictive, Bayesian brain*, the idea that the brain makes Bayesian probabilistic inferences – a kind of 'best guess' – about the world it is perceiving.

Briefly, Thomas Bayes (1701–61), a contemporary of Isaac Newton, was an English philosopher, statistician and a minister in the Presbyterian Church. In his lifetime, he did not actually publish the theorem that is named after him; it remained as a set of notes that were written up and published after his death by another philosopher, mathematician and non-conformist minister, Richard Price. The theorem is therefore often referred to as the Bayes-Price theorem. It is concerned with calculating the *probability of a given event*, based on a prior knowledge of factors or conditions that might be related to the occurrence of that event, and 'how probabilities are to be changed in the light of new evidence' (Chalmers, 1999, p. 175). So, for example, a punter at the races will place a bet on a horse, based on the evidence of the past performances of the horses in the race, but the punter's bet might change if they notice in the paddock that their fancied horse is sweating badly and looking unwell.

Friston says that 'to understand the Bayesian brain one needs to understand connectivity and the distributed processing that it supports' (Friston, 2012, p. 1231). Now that begins to sound very much like Hebbian plasticity (Chapter 2) and even more like Gerald Edelman's emphasis on connectivity and neuronal group selection (Chapter 3). Indeed, throughout this book considerable emphasis has been placed on the massive interconnectivity of neurons – neurons to neurons and across neuronal maps. Friston asks the simple question, 'Why does the brain have (axonal and synaptic) connections? Many other functionally specialised organs like the liver or blood do not have a delicate connectivity, so why does the brain?' (Friston, 2012, p. 1231). In Chapter 2, we went into some considerable detail about neuronal connectivity, but we didn't stop to ask Friston's simple question 'Why'?'

Friston says that from the point of view of the Bayesian brain, the answer is also quite straightforward. Connectivity is supporting distributed processing, which is making Bayesian inferences about the causes of its sensations. To do that, it 'must have a model of the causal relationships (connections) among (hidden) states of the world that cause sensory input. It follows that neuronal connections encode (model) causal connections that conspire to produce sensory information' (Friston, 2012, p. 1231). As we saw Andy Clark explaining in Chapter 10, contrary to the input-dominated view of neural processing, 'naturally intelligent systems (human and other animals) do not passively await sensory stimulation. Instead they are constantly active, trying to predict streams of sensory stimulations before they arrive' (Clark, 2015, p. 2). One might say the brain is always one step ahead: it is not waiting for input to arrive in order to begin processing, rather it is actually processing deviations from prior predictions – it is 'looking for' prediction errors.

We have just mentioned Gerald Edelman and his very influential theory of neuronal group selection, which you met in Chapter 3. Friston explains the very demanding but also exciting time he spent working with Edelman in the early 1990s. He recalled a major point of conflict between himself and Edelman. Friston had been a student of probability theory and quantum physics at Cambridge University in the United Kingdom, believing that ultimately all formal theories should be stated mathematically. On the other hand, Edelman – this great and original neuroscientist and Nobel Laureate in medicine – thought of this as a kind of undesirable disease he called 'mathematosis'.

When they eventually collaborated in writing a joint paper on value-dependent selection in the brain (Friston et al., 1994), Edelman insisted that all the maths should be relegated to an appendix. (This is surely a very interesting story for those who find science fascinating, but don't like mathematics very much!) Friston concedes that in a sense Edelman was right: the deep questions are 'embedded in selectionist thinking, population dynamics and self-organisation', but this leaves others (like himself) to 'meet the challenge of relating value to more formal treatments in information and probability theory. An approach one could now understand as the Bayesian brain' (Friston, 2012, p. 1232).

Humanising winds of change

But where is the human heart in all this? The notion of the Bayesian brain is not only being used to explain the working of the human brain; it is also being applied enthusiastically in the field of artificial intelligence by Friston and many others worldwide. Imagine that one day it will be possible to produce a machine that is able to replicate human thought and action, to the point where it is impossible to tell whether you are speaking to a human being or a robot (called passing 'The Imitation Game' or the 'Turing Test', after Alan Turing) – will that finally undermine the widely shared human belief that we are fundamentally special beings? And what would that mean for education?

At the close of Chapter 1, we raised the question of the human 'heart'. We recounted the story of a presentation that had focused on brain anatomy and learning but had also included machine learning. Following the presentation, one member of the audience asked, 'Where's the human heart in all this?' Using the 'heart' as a metaphor, she was pointing out that in the lecture the human, thinking, feeling dimensions of learning were missing. In this textbook, you have covered a lot of content related to the science of learning and development, but we also hope that in reading this book, you have been very aware that we have been keeping the 'human heart' fully in view. The reason is that this study of the science of learning and development, like the presentation just referred to, is *situated in education*.

First and foremost, that demands valuing students as sentient, emotional and relational human beings. And, at least in part, this entails viewing education as a values-based endeavour and viewing learning as value embedded. In education, we should never lose sight of the human dimension. In that regard, education shares something important with medicine: it should always be oriented towards human growth and flourishing, even though firmly based on science. Jean Piaget said, 'Pedagogy is like medicine: an art, but one which is based – or should be based – on precise scientific knowledge' (Piaget, 1949). The problem, as Piaget implies, is that often the study of education has not been founded on 'precise scientific knowledge', and evidence of that has been provided in the pages of this book. All too often, the study of education has emphasised the 'art' of teaching and the social context of education, and sidestepped relevant scientific, biological knowledge. The hope is that this book might help to restore a balance, but not swing the pendulum to the opposite extreme, which is equally undesirable.

Arguably, the opportunity to strike a balance is now being provided by science itself. Within science over the past 30 or so years, humanising winds of change have been blowing. They started in the mid-twentieth century, with the work of Popper, Polanyi and Kuhn, all of whom you have encountered in this book. Each, in their own way, steered science away from the often dehumanising view of science and the scientist that emerged in the eighteenth and nineteenth centuries, culminating in so-called logical positivism in the early twentieth century, which provided the philosophical foundation for behaviourism in psychology, but which Popper, Polanyi and Kuhn strongly opposed.

The humanising winds of change in science continued and gained pace with the seminal work of Antonio Damasio in reuniting human reason, feelings and emotion. This continued with Edelman, and the strong emphasis he placed of the role of value in selectionist learning, and the seminal work of Esther Thelen and Linda Smith, introducing dynamic systems theory to human development. In the last decade or so, it has progressed with the introduction of the interoceptive, predictive brain – at least as it is interpreted by commentators such as Andy Clark, Karl Friston and Lisa Feldman Barrett. This is not to deny that science should act as a necessary check on human sentimentality and challenge the grand views we have constructed of humanity as somehow standing over and against our biological origins – for example, the idea that, while we start off as biological beings, through socialisation, we nevertheless grow out of biology and become social beings – a view criticised by Tim Ingold (2008), as we saw in Chapter 7. On the other hand, it entails adopting what might be called a *humanitarian view* of both education and the neurobiological self – understood as a belief in the intrinsic value and dignity of human life and a concern with promoting human wellbeing.

So, after all our explorations in the science of learning and development, within the pages of this text there is a very real sense that we are returning to where we started, with Donald Hebb's questions about what is happening in the heads of children as they learn. But we have learnt a lot along the way, and in light of our explorations, we now see that question anew, as though for the first time, through the notion of the interoceptive, predictive brain. In education, as teachers, we are concerned with the children and adolescents who *we* teach, and with their voices – which may often be only partly heard; yet their learning and development are intimately interwoven with their wellbeing. Schools can be scary places for some and the fear of failure is ever-present. This does not call for sentimentality; instead, it requires a humanised science, as modelled so eloquently by Antonio Damasio in his emphasis on homeostasis and *human flourishing*, bringing what might be viewed as the 'fire' of science and the 'rose' of humanity together as one. We teach in the earnest hope that, for *our* students, all shall be well and all manner of things shall be well, to cite Julian of Norwich (Figure 11.7) again. You will see these closing reflections captured in a poem (quoted earlier) by Nobel Laureate T. S. Eliot, inspired by a small village in England, near Cambridge, called Little Gidding. Check it out using keywords 'Eliot' and 'Little Gidding'.

'Little Gidding'

Figure 11.7 Statue of Julian of Norwich (1342–c. 1416), an inspirational mystic who is thought to be the first woman to write a work in English

APST Standard 1.1 Physical, social and intellectual development and characteristics of students

Demonstrate knowledge and understanding of physical, social and intellectual development and characteristics of students and how these may affect learning

APST Standard 4.3 Manage challenging behaviour

Demonstrate knowledge of practical approaches to manage challenging behaviour

APST Standard 4.4 Maintain student safety

Describe strategies that support students' wellbeing and safety working within school and/ or system, curriculum and legislative requirements

ACECQA curriculum specifications

1.3 Social and emotional development

1.4 Child health, wellbeing and safety

▶ ▶ This chapter has provided a frontline overview of recent research into student wellbeing that is highly relevant in advancing newly qualifying teachers' quality engagement with their students. Given the focus on the science of students' wellbeing, this chapter has gone beyond what is generally available (e.g. behaviour management, motivation, etc.) in many other textbooks. In doing so, it adds considerably to their understanding of how wellbeing emerges in the brain and helps newly qualifying teachers appreciate that the recent model of the brain as interoceptive and predictive is providing new insights into student wellbeing. Moreover, this chapter has offered new insights on what is meant by learning and development being value-embedded and the importance of creating value-based learning environments in schools

The chapter has provided sufficient grounding for newly qualifying teachers to address some of the challenges students might experience, including anxiety, depression and fear of failure, by providing informed perspectives about why such states emerge, drawing on insights from the notion of the predictive brain. This will help newly qualifying teachers to create engaging and safe classrooms, while at the same time supporting students' wellbeing.

SUMMARY

- For many parents or primary carers, the wellbeing of their children is closely interwoven with their learning and development at school. In 2020, the Australian Institute of Health and Welfare published a major report, *Australia's Children* (AIHW, 2020), which found that some 14 per cent of students aged from six to 12 years said they did not feel safe at school a lot of the time.
- One of the main threats to student wellbeing in school is the prevalence of bullying. Bullying occurs in various forms (e.g. overt vs covert) and in different contexts (e.g. schools vs online spaces outside actual classrooms). It has lasting effects, impacting the brains of victims, peer groups, families and school communities, including the bullies themselves.
- Wellbeing has come to the fore in educational debate and policy concerns over the past two decades, although it is difficult to find an agreed definition. A distinction is often drawn between *hedonic wellbeing*, concerned with the pursuit of pleasure, and *eudaimonic wellbeing*, which means something like human flourishing.
- Definitions of health and wellbeing sometimes include more than physical health and social and emotional safety, involving spiritual and/or cultural aspects and the need for a connection to land and environment, community and family (AIHW, 2020).
- All these factors relate to what is often referred to globally as a *sense of belonging* or being *at home* in one's physical or social and cultural environment. They also relate to the basic and universal human quest for meaning or meaningfulness in life. Across all OECD countries, 67 per cent of students reported being satisfied with their lives, although that figure shrank by five percentage points between 2015 and 2018.
- According to a major Australian government initiative that developed a National Framework for Values Education in Australian Schools, there is a strong link between

student wellbeing and what is called *values pedagogy*. This places a strong emphasis on what may be called embedded values, where human relationships and explicit values-oriented transactions are forged within quality values education programs.

- Subsequently, researchers developed the notion of *value-embedded learning,* (VEL), conceived as a philosophically non-reductive and biologically plausible account of the dynamic relationship between value and emotional salience in the learning process, and how value and salience are afforded in school learning environments (Duncan et al., 2021).
- The human brain was *not designed* to allow us to go to school, to learn to read and write and study all the subjects on the school curriculum – these abilities are *exaptations*, co-opting areas of the brain not originally naturally selected or adapted for that purpose. In fact, the mammalian brain was not designed at all; it evolved from earlier precursors.
- Wellbeing and survival are not optional extras for the brain, something it does once it has taken care of all its other competing demands; wellbeing and survival *are* the main demands.
- The notion of the interoceptive, predictive brain is starting to provide a highly embodied and unified model of brain functioning that is turning previous models of perception and cognition upside down – the brain does not begin with 'lower-level' perceptions, but with 'higher-level' predictions that are checked against perceptions.

KEY POINTS FOR TEACHERS

- One helpful suggestion to be found in research into student wellbeing is that it is best thought of as a dynamic process, emerging from the ways in which people interact with the world around them. The implication of this for educational practice is that a student's wellbeing is not a fixed entity, but always subject to change.
- Teachers should be aware the 2020 AIHW report has emphasised that many 'Indigenous people define their personal health and wellbeing beyond physical health and social and emotional aspects, to include spiritual and/or cultural aspects and the need for a connection to environment, community and family' (AIHW, 2020, p. 15).
- In a survey conducted in 2018, the Organisation for Economic Cooperation and Development (OECD) found that 'students in socioeconomically disadvantaged, rural and public schools were more likely to report a weaker sense of belonging at school than students in advantaged, city and private schools' (OECD, 2019b, p. 130).
- The same OECD report also found that students in many Asian countries and economies expressed the greatest fear of failure, while students in many European countries expressed the least fear. Also, in almost every education system, girls expressed greater fear of failure than boys, and this gender gap was considerably wider among top-performing students.
- Teachers should be aware that there can be a strong correlation between student feelings of failure at school and their concern about the future. Of all nations surveyed by the OECD in 2018, both OECD countries and non-OECD countries (total 78), 19 countries scored more than 60 per cent of their 15-year-old students indicating concern for their future when feeling they were failing.
- Although it may be hard to value and show love for children and youngsters who cause teachers so much trouble, and the same can apply to some colleagues with whom one has

to engage in the workplace, the message of research is that all children and colleagues are ultimately embodied brains that are concerned primarily with maintaining homeostasis, *wellbeing and flourishing.*

- Recent research is showing a strong relationship between wellbeing and brain function. Thoughts, feelings and emotions are created in the embodied brain as an integral part of the ceaseless predicting and interoceptively sensing processes that are keeping us alive.
- The processes involved in learning are many, and the challenges of schooling are complex, but arguably the recognition that students have brains that are predisposed by evolution to learn should be the starting point for all teachers – not the notion that students won't learn unless they are motivated.
- In education, in policy and in curriculum and pedagogy research, we should never lose sight of the human dimension. In that regard, teachers should be aware that education shares something important with medicine. It should always be oriented towards human growth and flourishing, even though firmly based on *science.*
- Over the past century, there have been tremendous advances in our understanding of the dynamics of brain functioning, but teachers should be aware that there is still much we do not know and much that is contested about how that thing in our heads makes us who we are, and the children in our care in school who they are.

Guided responses

REVIEW QUESTIONS

1. What is meant by the notion of 'schools as safe places to be', and what is currently known about Australian primary students' feelings about their safety when at school?
2. How might student wellbeing be defined and what is the difference between hedonic wellbeing and eudaimonic wellbeing?
3. What is meant by the claim that thoughts, feelings and emotions are created in the embodied brain as an integral part of the ceaseless predicting and interoceptively sensing processes that are keeping you alive?
4. Why are wellbeing and survival not optional extras for the brain – something that the brain does once it has taken care of all its other competing demands?
5. What is meant by the saying that the brain is an interoceptive, predictive system and how does this begin to provide a very new account of the relationship between student learning and student wellbeing?

FOOD FOR THOUGHT

1. If it is the case, as evidence suggests, that one in every seven children in Australian schools does not feel safe at school a lot of the time, by drawing on all that you have studied in this textbook, suggest how teachers can begin to address this quite alarming statistic. How can we make schools safer places for students to be?
2. According to the OECD's PISA 2018 results, 'students reported a greater sense of belonging when they also reported higher levels of co-operation amongst their peers, whereas

students' perception of competition was not associated with their sense of belonging at school' (OECD, 2019b, p. 130). What messages does that finding have for teachers regarding cooperative versus competitive learning environments in schools and classrooms?

3. As a teacher or prospective teacher, what messages do you take from the idea that when we strip away all that we see as the different physical features of each individual child or teacher, and what we take to be their personality or character – all those indicators we normally use to identify ourselves and others – at base, we find an embodied brain that is *primarily concerned with wellbeing and flourishing*?

RESEARCH ACTIVITY: INTEROCEPTION, EMBODIMENT AND WELLBEING IN SCHOOLS

As we learned from this chapter, maintaining homeostasis – that is, maintaining a healthy state of being – is achieved through the sensory system known as *interoception,* which constantly communicates the *internal state* of the body to the brain. More precisely, 'Interoception, as the sense of the physiological condition of the body, supports homeostatic control and allostatic adaptation, ensuring the stability of the organism and by driving behaviour through feelings such as hunger, thirst and dyspnoea' (Tsakiris & Critchley, 2016, p. 1).

The South Australian Department of Education (2021) has recommended embracing the notion of 'interoception' to enhance students' self-regulation and wellbeing, noting that 'without interoception social skills are just the application of rules and not a meaningful way of interaction'.

Check the information and resources provided by different Australian states to support schools in the practical application of interoception, then think carefully about three questions below.

- South Australian Department of Education: 'Applying interoception skills'
- New South Wales Department of Education: 'Interoception as a proactive tool to decrease challenging behaviour'

Research activity links

Questions

1. Carefully check the definitions of 'interoception' provided on each website. Do you think the definitions provided on each website are sound and in line with the contemporary understandings of the predictive brain?
2. Please check the 'Wellbeing, Stress, Distress Questionnaire' provided on the South Australian Department of Education website titled 'Applying Interoception Skills'. Do you think statements included in this questionnaire capture the important indices of students' wellbeing? If you were to construct a questionnaire for your own classroom, which statement would you add or remove?
3. How do you think you could monitor students' wellbeing status efficiently? What kind of classroom-based approaches do you think you could take?

action potential: an electrical signal that travels from a neuron's cell body to the axon terminal, where it triggers the release of neurotransmitters (BrainU, 2021)

affect: a collective term referring to a broad range of feelings that people experience. Emotion builds on affect.

affordance: a term coined by James J. Gibson to draw attention to the necessary relationship between an animal and its environment and what that environment 'offers the animal, what it provides or furnishes either for good or ill' (Gibson, 1979, p. 127)

alexia: a partial or complete inability to read

amygdala: almond-shaped part of the limbic system that is involved in emotional reactions, particularly fear

aphasia: 'an impairment of language, affecting the production or comprehension of speech and the ability to read or write' (National Aphasia Association, 2021)

arousal: a property of affect; a basic feeling that we experience, which can range from calm to agitated

axon: a long nerve fibre that transmits action potentials away from the neuron's cell body towards the axon terminal

axon terminals: the 'very end part of an axon that makes a synaptic contact with another cell; the point where neurotransmitters are released' (BrainU, 2021)

behaviourists: those who accept and work within the doctrine of behaviourism that 'psychology is the science of behaviour and is not the science of the inner mind – as something other or different from behaviour' (Graham, 2019)

biological determinism: see *genetic, or biological, determinism*

brain plasticity: the ability of brains to be physically moulded and changed in response to experience

Broca's aphasia: see *expressive aphasia (or Broca's aphasia)*

Broca's area: a region in the frontal lobe of the dominant hemisphere, usually the left for right-handed people. It is conventionally linked to speech production.

Brodmann areas: functional areas of the cortex assigned by a German anatomist Korbinian Brodmann. Until recently, this was the most commonly used system of brain reference to facilitate communication among scientists regarding the regions implicated in different functions.

broken mirror hypothesis: a belief that malfunctioning mirror neuron systems could explain the social deficits experienced by people with autism spectrum condition

category error: mistaken application of things that belong to one distinct category to a different distinct category to which they do not apply

cell body: also called the soma (using the Greek word for body); it is the neuron's powerhouse, providing energy to drive the cell. The cell body also houses the cell nucleus, which in turn houses the DNA that encodes the cell's genes.

cell migration: movement of newly generated neurons along radial glial fibres to their specialist location in the brain

cerebellum: a subcortical part of the brain that is tucked under the temporal and occipital lobes

cerebral cortex: the largest part of the human brain that largely comprises up to six layers of cell bodies and is traditionally referred to as 'grey matter'

classical view of emotion: the contentious view that basic discrete emotions have unique, biological foundations that are universal because they are passed down from our evolutionary, biological past

cognitive subconscious: an alternative term for the cognitive unconscious

cognitive unconscious: much of our thinking is operating below the level of conscious awareness and is therefore not available to conscious introspection

cognitivism or representationalism: the belief that 'thinking can be understood in terms of the representational structures in the mind and computational procedures that operate on those structures' (Thagard, 2014, para. 11)

competency-based teacher education (CBTE): a teacher education or training approach that is underpinned by the competency-based training (CBT) model

competency-based training (CBT): a vocational/ professional training approach that focuses on a set of competencies a person should be able to demonstrate at the completion of the training

complex system: see *dynamic (or complex) system*

complexity science: the study of dynamic, complex systems and how they give rise to emergent, dynamic, self-organising behaviours

consolation: 'reassurance behaviour directed at a distressed party, such as a victim of aggression' (de Waal & Preston, 2017)

Copernican revolution: a major shift in astronomy from an Earth-centred view of the universe to a

Sun-centred view, as proposed by Polish astronomer Nicolaus Copernicus in the sixteenth century

critical realism: accepts that science is a human construct, but insists scientific claims are informed and constrained by the nature of the world

cross-sectional sampling: a data-collection approach that gathers data from individuals of various ages or developmental levels so as to study the differences among them

culture: highly complex systems of values, conventions and artefacts that constitute daily social realities. Culture includes symbolic resources, such as icons and scripts that are accumulated and transmitted across generations.

declarative memory: memory for facts, ideas, and events; it is also called explicit memory, conscious memory and memory with record

deliberate emotion regulation: slower, reflective (but not necessarily conscious) modulation of emotional states

dendrites: the main input component of the neuron that receives signals from other neurons close by

deontological choice: from deontology (from Greek: δέον, 'obligation, duty' + λόγος, 'study'), the belief that the morality of an action should focus on the rights of the individual rather than trying to satisfy the utilitarian concern for the greater good

determinism: a philosophical view that every event is determined completely by pre-existing causes. One example is the idea that genes determine who we are and our destiny (genetic determinism).

distributed practice: see *spaced/distributed practice*

dynamic (or complex) system: a system of many components whose interactions give rise to the collective emergent patterns of the entire system

dynamic systems theory (DST): a theoretical framework in developmental science that focuses on the origins and self-organisation processes of novel functions

dyslexia: an impairment with reading and spelling despite having the ability to learn. The 'central difficulty for a student with dyslexia is to convert letter symbols to their correct sound (decode) and convert sounds to their correct written symbol (spell)' (Australian Dyslexia Association, 2021).

electroencephalogram (EEG): a brain imaging technique that detects electrical activity in the brain using small metal discs (electrodes) attached to the scalp

embodied cognition: the belief that human thinking and reasoning are not an autonomous, disembodied mental process, but rather deeply dependent on features of the physical body

embrained culture: a process whereby recurrent, active and long-term engagement in cultural practices shapes and modifies brain pathways

emergence: the ability of individual components within a large system to work together to give rise to novel patterns of behaviour that are often not predicted by an understanding of the behaviour of each constituent part

emotion contagion: an automatic tendency to share another's emotional state

emotion regulation: the 'suite of executive functions directed at the modulation of emotional states and emotional impulses' (Woltering & Lewis, 2009, p. 160)

empathic concern: 'concern about another's state and attempts to ameliorate this state' (de Waal & Preston, 2017)

empathy: the ability to understand and share the feelings of another person

epigenetic: biological processes that do not depend directly upon gene expression. Development emerges in response to what the organism is currently experiencing and what has happened previously.

error-related negativity (ERN): a negative deflection in an EEG waveform that is observed between 50 and 150 ms after perceiving that an error is made. It reflects the brain's activity as a monitoring system.

eudaimonic wellbeing: the quality of life that is enabled by having a sense of purpose or meaning in life, and personal growth

eugenics: the 'science' of selective mating of those deemed to possess desirable genes, in order to advance humanity

event-related potentials (ERP): a time-locked electrophysiological response that arises as a direct response to a specific event

exaptation: a biological feature that has emerged later, by co-opting mechanisms that existed for other purposes, rather than being 'built into' the organism as a result of evolutionary adaptation

excitatory post-synaptic potential (EPSP): a shift in membrane potential when the change of the receiving cell is *more* likely to fire its own action potential after receiving the neurotransmitter

expectancy-value theory (EVT): a theory of motivation; the core idea is that a student's expectancy for *success* is a key determinant of their motivation and subsequent academic outcomes

experiential learning: 'the process whereby knowledge is created through the transformation of experience' (Kolb, 1984, p. 38)

expressive aphasia (or Broca's aphasia): an impairment in the production of language

false-belief test: a critical theory-of-mind task that assesses whether an observer knows what another knows, even if the reality is inconsistent with what the observed person knows

falsification: the act of disproving a proposition, hypothesis or theory

feedback control system: a system that is able to keep itself controlled within a given range via monitoring congruence and incongruence between the reference signal and perceived signal (feedback). The brain is proposed to be a feedback control system.

feedback-related negativity (FRN): an electrical brain signal measured with an electroencephalogram when a person engages with performance feedback (e.g. right vs wrong). It reflects the brain's activity as a monitoring system.

Flynn effect: the tendency of IQ scores to change over time

formative assessment for learning: a type of assessment that is implemented within the process of learning; its purpose is to monitor student progress and assist in improving students' attainment

functional magnetic resonance imagining (fMRI): a brain imaging technique that detects the changes in blood oxygenation and flow that occur in response to neural activity

genetic, or biological, determinism: the doctrine that what we inherit genetically plays a significant, if not total, role in determining human nature, personality and intelligence

glial cells: non-neuronal cells in the brain; radial glia cells in particular provide a fibre pathway that physically and chemically guides the migration process of new neurons

global neural workspace: 'a temporary conscious memory within which we can maintain, for a short period, practically any piece of information that seems relevant to us and relay it to any other module' (Dehaene, 2020, p. 160)

grapheme: a letter or a group of letters representing a sound (phoneme) in a word

gyrus (plural gyri): a bulging part of each fold on the cerebral surface of the brain

Hebb's postulate (or Hebbian learning): Donald Hebb's theory of learning by association, when adjacent cells are excited simultaneously and repetitively, thus forging a connection that is strengthened through further repetition

hedonic wellbeing: the quality of life based on happiness, life satisfaction, and positive and negative affect or feelings

hemispherectomy: a neurosurgical procedure in which a cerebral hemisphere is removed, disconnected from the rest of the brain

hemispheres: the two halves of the cerebral cortex that are joined together by a thick body of nerve fibres called the corpus callosum

hippocampus: 'the oldest part of the cerebral cortex responsible for spatial localisation, formation of declarative memory, and transfer of short-term to long-term memories' (BrainU, 2021)

homeostasis: 'the coordinated and largely automated physiological reactions required to maintain steady internal states in a living organism' (Damasio, 1999, p. 39)

incommensurability of scientific theory: the idea that there are no shared, objective standards for comparing or evaluating competing theories

inhibitory post-synaptic potential (IPSP): a shift in membrane potential when the change of the receiving cell is *less* likely to fire its own action potential after receiving the neurotransmitter

innovative learning space: a specially designed learning environment that aims to provide optimal conditions for teaching and learning

intelligence quotient (IQ): a total score derived from a set of standardised tests designed to assess intelligence

interoception: the 'brain's representation of all sensations from your internal organs and tissues, the hormones in your blood, and your immune system' (Barrett, 2017, p. 56)

Kohlbergian paradigm of moral development: cognitive developmental perspective of moral development that comprises six distinctive, invariant stages

larynx: the 'voice box', located in the top of the neck

lesion: damage to any part of brain

limbic structure: a set of subcortical brain structures including the hippocampus, amygdala, thalamus, hypothalamus, fornix and cingulate gyrus

linear: progressing from one stage to another in a single series of steps; or a straight line-like relationship between two variables

literacy: the ability to both comprehend texts through listening, reading and viewing and compose texts through speaking, writing and creating

lobes of the brain: the four large divisions of the cerebral cortex

localisation and specialisation: the idea that brain is highly localised, containing many specialised parts such that language and other main functions occur in highly specific regions

logographic (or pictorial): children's tendency to recognise words as 'wholes' because they haven't as yet developed a set of alphabetic or orthographic spelling rules

long-term memory: memories that are stored in a variety of places in the brain and are retained over long periods of time

long-term potentiation (LTP): a long-lasting synaptic strength that has been induced by high frequency electrical stimulus

massed practice: the opposite of spaced/distributed practice; involves intensive and uninterrupted practice of skill or knowledge over a longer period of time

mathematics anxiety: feelings of tension, nervousness or worrisome thoughts about one's ability that impede engagement with numbers and mathematical problems

metacognition: the mental process of monitoring and regulating one's cognitive activities

mind/body dualism: a philosophical view that the mind is non-physical, and the mind and body are distinct substances or properties

mind/brain identity theory: the philosophical view that processes of the mind are identical to states and processes of the brain

mirror neurons: neurons that fire when a person performs an action but also fire as a result of watching (mirroring) another person perform the same action. They are thought to play a role in imitation, speech and language, and are also associated with emotion and empathy.

monitoring: any activity aimed at evaluating or regulating one's own cognitions (Flavell, 1979), which includes checking errors, self-testing, assessing one's progress, etc.

moral reasoning: the process by which individuals try to determine the difference between what is right and what is wrong

moral relativism: the belief that moral beliefs are personal and there is no touchstone to ground morality – it's just personal opinion, your truth versus my truth

moral values: 'guiding principles, beliefs, sensitivities, held and displayed by individuals or groups in respect to how they relate to and deal with others and the world' (Sankey & Kim, 2016)

motherese: a term used in the study of child language acquisition to describe the way mothers often talk to their young children

naïve realism: the view that our senses are able to give us direct access to the world as it objectively exists; we perceive the world exactly as it is

nature/nurture dichotomy: a view of development that separates the biological and environmental influence and postulates how they are said to interact

neurodiversity: the range of variation across all human brains

neuromyth: mistaken beliefs about the human brain and its development (Kim & Sankey, 2018)

neuron: a brain cell that is specialised for the transmission of information and characterised 'by long fibrous projections called axons, and shorter, branch-like projections called dendrites' (BrainU, 2021)

neuronal recycling: 'the partial or total invasion of a cortical territory initially devoted to a different function' (Dehaene, 2009, p. 147); see also *exaptation*

neurotransmitters: chemicals that are released from the tip of the axon or axon terminal, including glutamate, serotonin, dopamine, norepinephrine, acetylcholine and gamma-aminobutyric acid (GABA)

non-declarative memory: a type of memory related to performance and the subconscious use of skills, such as riding a bicycle or playing the piano

non-linear: a term referring to a relationship where the change of the outcome is not proportional to the change of the input

non-symbolic numeracy: the ability to discriminate the numerical magnitudes or perform numerical operations without involving numeric symbols

number conservation: awareness that the number of objects remains the same, even if the objects are rearranged or hidden

number sense: a primitive ability innate to infants to appreciate quantities, numbers and their relationships

numeracy: the ability to perform and use mathematics with an appreciation of its social relevance and value

open-plan classrooms: classrooms that have removed or reconfigured walls within or between classrooms

organism-in-environment: an alternative approach to unite an organism and the environment in which it is situated, in opposition to pure 'organicism' or 'environmentalism' in developmental science

orthographic: recognising a set of conventions for writing a language, including norms of spelling, word breaks, emphasis and punctuation

oxytocin: a hormone produced by the hypothalamus and secreted by the pituitary gland

perception-action mechanism (PAM): spontaneous activation of an individual's own personal representations for the other person, their state and their situation when perceiving the other's state (de Waal & Preston, 2017)

perspective taking: perceiving or understanding a situation from an alternative perspective, including that of another person

phonemes: discrete sound constituents of *spoken* words

phonics: a language teaching method that explicitly demonstrates the relationship between phonemes and graphemes

phonological: phonological awareness enables children to identify different units (e.g. syllables, rimes) in the words

pictorial: see *logographic (or pictorial)*

post-synaptic potential: change in the membrane potential of the post-synaptic terminal of a synapse

prediction error: a discrepancy or mismatch between the predictions produced by a brain and reality

predictive brain: a theory of brain function in which the brain constantly makes refined predictions about the environment

predictive processing: a new integrative 'framework for understanding perception, action, embodiment, and the nature of human experience' (Clark, 2015, p. 1)

predilection to value: an in-born value bias within organisms, a fundamental ability to discriminate what is better and what is worse, what is beneficial and what is harmful

pre-synaptic action potential: depolarisation that occurs when the inside of the nerve cell fibre becomes positively charged

primary repertoire: large numbers of variant neuronal groups (or local circuits) within a given anatomical region that are formed via the epigenetic generation of diversity (Edelman, 1989, p. 44)

Programme for International Student Assessment (PISA): the OECD's international academic assessment program, which intends to evaluate the educational systems of different nations, by measuring 15-year-old students' academic performance on mathematics, science and reading

proto-letters: shapes that resemble symbols, or elementary Chinese characters. Neurons on the inferior temporal lobe are particularly responsive to these shapes.

pure alexia: impaired reading without impaired writing

pyramidal cells or pyramidal neurons: excitatory neurons in the cerebral cortex, hippocampus and amygdala

reactive emotion regulation: relatively automatic processes of modulating an emotional state over which we do not exert intentional or conscious control

reductionism: reducing a phenomenon to its basic constituent parts. Science is therefore methodologically reductionist when analysing a phenomenon in terms of its constituent molecules, atoms and sub-atomic 'particles', although that does not necessarily entail ontological reductionism – the claim that a phenomenon is *nothing but* its parts.

reification of intelligence: turning the abstract concept of intelligence into a concrete thing

resting membrane potential: the difference in electrical potential between the interior and exterior of the neuron

scientific conjecture: a proposition, such as hypothesis, that is unproven

secondary repertoire: neuronal groups (or local circuits) that have been selected among populations of synapses by strengthening some synapses and weakening others. Certain circuits and neuronal groups in such repertoires are more likely to be favoured over others in future encounters with signals or similar types (Edelman, 1989, p. 46).

selection: a process in which environmental or biological influences decide which type of constituent parts succeed better than others

self-efficacy: in the PISA study, self-efficacy is defined as 'the extent to which individuals believe in their own

ability to engage in certain tasks including academic tasks' (OECD, 2019b, p. 188)

self-organisation: 'describes characteristic relationships in emergent systems created by feedback mechanisms that either amplify an effect (positive feedback) or dampen an effect (negative feedback)' (University of Southampton, n.d.)

sensory receptors: endings of the sensory cells that are distributed across different parts of the body (e.g. touch and temperature receptors in the skin)

short-term memory: memory over a short period of time that cannot be retained for more than a day or so

social constructivist (or constructionist) theories: sociological theories of knowledge according to which knowledge is constructed through interaction with others. Social constructivists tend to ignore contributions from biology.

social determinism: a contentious view that social research should systematically exclude biological explanations from its account of human behaviour, and only employ social factors such as social interactions and constructs

social exclusion: a process that impedes or prohibits a person's access to various rights, relationships, opportunities and activities

socialisation (classic view of): the process whereby a social child acquires rules for categorising and positioning other people in the social environment and guidelines for appropriate action within the social world (Fogel, 2008)

socialised learning: learning in social contexts. In opposition to the view that learning is 'socialisation' or 'internalisation', the learner is an active generator of theory arising out of their practice.

somatic (body) markers: bodily feelings that act as emotional warning signals focusing attention on the possible negative outcomes of an intended action

spaced/distributed practice: the idea that learning and retention of skills or knowledge is more efficient when practising (including formative testing) over an extended period of time

sub-cortical structures: a set of structures located below the cerebral cortex; include limbic structures and the basal ganglia

sulcus: a groove that runs between two gyri

summative assessment of learning: a type of assessment that is implemented at the end of a course or grade, and aims to gauge a student's achievement level against learning outcomes

superior colliculus: 'multisensory midbrain structure that integrates visual, auditory, and somatosensory spatial information to initiate orienting movements of the eyes and head toward salient objects in space' (Myoga, 2020, p. 601)

synaptic plasticity: activity-dependent modification of the strength or efficacy of synaptic transmission at pre-existing synapses (Citri & Malenka, 2008)

synaptic potential: the potential difference across the postsynaptic membrane that is triggered by neurotransmitters at a synaptic gap or cleft

telencephalon: from the Greek words *telos* (end) and *enkephalos* (brain), telencephalons literally mean 'end-brain'; it includes all that we normally mean by the brain, the cerebral cortex and subcortical structures

thalamus: a part of the brain situated very close to the centre of the brain and at the top of the brain stem; acts as a kind of 'relay station' in the brain

theory of mind: 'the ability to attribute mental states to others, such as knowledge, intentions and beliefs' (de Waal & Preston, 2017); the mental process of monitoring and regulating one's cognitive activities

theory of neuronal group selection (TNGS): an account of brain function proposed by Gerald Edelman (1987) that emphasises the role of value and salience in the formation of, and subsequent modifications to, the strengthening of synaptic connections and synaptic maps (or pathways)

theory/practice dichotomy: a view in education that theory and practice are two distinct entities

triune brain: Paul MacLean's controversial idea that the human brain evolved in three distinct stages that are now represented in the current structure, comprising a primitive reptilian brain and then a limbic brain that appeared with the first mammals, both of which were wrapped around by a cortex that evolved with higher mammals

under-determination of theory by evidence: the observational evidence used to support any hypothesis or theory will often (some say always) be less than required and, moreover, it will always be compatible with more than just that theory

universality of emotional facial recognition: the contentious view that people from around the world exhibit and recognise facial expressions of emotion without any training whatsoever

utilitarian choice: a choice based on the belief that the consequences of any action are the only standard of right and wrong

valence: the extent to which feelings or emotions are experienced, from pleasant to unpleasant, and the value associated with any given feeling or emotion

value: within Edelman's TNGS, the word 'value' refers to evolutionarily derived constraints favouring behaviour that fulfil homeostatic requirements or increases fitness in an individual species (Edelman, 1989, pp. 287–8). These constraints are provided by a selectional system in the brain, comprising diffuse ascending systems such as the dopaminergic system (Edelman, 2004, p. 180).

value-embedded learning (VEL): a philosophically non-reductive and biologically plausible account of the dynamic relationship between value and emotional salience in the learning process

variability: differences or inconsistencies observable from both *within* and *across* individuals. During the processes of learning and development, both the amount and the quality of variability can change. Often, increased variability provides flexibilities in performance.

ventromedial prefrontal cortex (VMPFC): a broad area in the lower (ventral) central (medial) region of the prefrontal cortex; a key region supporting the decision-making process

Visual Word Form Area (VWFA): or brain's 'letter-box'; an area in the left occipito-temporal area (where the occipital and temporal lobes meet) within Brodmann area 37. It receives the incoming letters we read before sending them to other parts of the brain concerned with pronunciation and meaning.

vocal apparatus: the organs, including the pharynx, larynx, teeth, tongue and lips, that produce sounds and speech

vocalisation: the act or process of producing sounds with the voice; also a sound thus produced

Wernicke's aphasia: an impairment of language resulting in the loss of ability to understand spoken or written language

Wernicke's area: a region lying close to the auditory cortex in the temporal lobe of the dominant hemisphere. This area is traditionally considered to be a centre for the comprehension of language.

whole-language approach: a literacy teaching approach in English, based on the belief that language should *not* be broken down into letters and combinations of letters and 'decoded'. Instead, learners should focus on whole meanings.

zone of proximal development (ZPD): the putative 'distance between the actual developmental level as determined by independent problem solving and the level of potential development as determined through problem solving under adult guidance or in collaboration with more capable peers' (Vygotsky, 1978a, p. 86)

Adolphs, R. (2017). Reply to Barrett: Affective neuroscience needs objective criteria for emotions. *Social Cognitive and Affective Neuroscience*, 12(1), 32–3.

Ahmed, S., Bittencourt-Hewitt, A. & Sebastian, C. (2015). Neurocognitive bases of emotion regulation development in adolescence. *Developmental Cognitive Neuroscience*, 15(C), 11–25.

Allman, J. (2000). *Evolving brains*. New York: Scientific American Library.

Amalric, M. (2017). *Study of the brain mechanisms involved in the learning and processing of high-level mathematical concepts*. Neurons and Cognition [q-bio.NC]. Université Pierre et Marie Curie – Paris VI, 2017. English. NNT: 2017PA066143.

Amalric, M. & Dehaene, S. (2016). Origins of the brain networks for advanced mathematics in expert mathematicians. *Proceedings of the National Academy of Science Sciences of the United States of America*, 113(18), 4909–17.

American Psychological Association (APA) (2020). Genetic determinism. *Dictionary of Psychology*. Retrieved from: https://dictionary.apa.org/genetic-determinism

Anisfeld, M. (1996). Only tongue protrusion modeling is matched by neonates. *Developmental Review*, 16(2), 149–61.

Ansari, D. (2016). The neural roots of mathematical expertise. *Proceedings of the National Academy of Science Sciences of the United States of America*, 113(18), 4887–9.

Arent, S. & Landers, D. M. (2004). Arousal, anxiety, and performance: A reexamination of the inverted-U hypothesis. *Research Quarterly for Exercise and Sport*, 74(4), 436–44.

Arlin, P. K. (1981). Piagetian tasks as predictors of reading and math readiness in grades K–2. *Journal of Educational Psychology*, 73(5), 712–21.

Armstrong, T. (2012). *Neurodiversity in the classroom: Strength-based strategies to help students with special needs succeed in school and life*. Alexandria, VA: ASCD.

Ashcraft, M. H. (1992). Cognitive arithmetic: A review of data and theory. *Cognition*, 44, 75–106.

Australian Bureau of Statistics (ABS) (2016). *National Aboriginal and Torres Strait Islander Social Survey, 2014–15*. Cat. no. 4714.0. Canberra: ABS.

Australian Children's Education and Care Quality Authority (ACECQA) (2021). *National Quality Standard*. Retrieved from: https://www.acecqa.gov.au/nqf/national-quality-standard

Australian College of Educators (2018). *The great literacy debate*. Special edition of *Professional Educator*. Melbourne: Australian College of Educators.

Australian Curriculum, Assessment and Reporting Authority (ACARA) (2017). Numeracy. Retrieved from: www.australiancurriculum.edu.au/f-10-curriculum/general-capabilities/numeracy

Australian Dyslexia Association (2021). What is dyslexia? Retrieved from: https://dyslexiaassociation.org.au/dyslexia-in-australia

Australian Institute of Health and Welfare (AIHW) (2020). *Australia's children*. Cat. no. CWS 69. Canberra: AIHW.

Australian Institute for Teaching and School Leadership (AITSL) (2011). *Australian Professional Standards for Teachers* (revised 2018). Melbourne: AITSL. Retrieved from: www.aitsl.edu.au/docs/default-source/national-policy-framework/australian-professional-standards-for-teachers.pdf

Australian Institute for Teaching and School Leadership (AITSL) (2021). Spotlight – bullying in Australian schools. Retrieved from: www.aitsl.edu.au/docs/default-source/research-evidence/spotlight/spotlight_bullying.pdf?sfvrsn=613bf73c_6

Bandura, A. (1977). Self-efficacy: Toward a unifying theory of behavioral change. *Psychological Review*, 84(2), 191–215.

Baron-Cohen, S. (2017). Editorial perspective: Neurodiversity – a revolutionary concept for autism and psychiatry. *Journal of Child Psychology and Psychiatry*, 58(6), 744–7.

Barr, J. J. (2011). The relationship between teachers? Empathy and perceptions of school culture. *Educational Studies*, 37(3), 365–9.

Barrett, L. F. (2017). *How emotions are made: The secret life of the brain*. Boston, MA: Houghton Mifflin Harcourt.

Barrett, L. F., Adolphs, R., Marsella, S., Martinez, A. M. & Pollak, S. D. (2019). Emotional expressions reconsidered: Challenges to inferring emotion from human facial movements. *Psychological Science in the Public Interest*, 20, 1–68.

Barrett, L. F., Quigley, K. S. & Hamilton, P. (2016). An active inference theory of allostasis and interoception in depression. *Philosophical Transactions of Royal Society B*, 371. http://doi.org/10.1098/rstb.2016.0011

Bedford, O. & Yeh, K. H. (2019). The history and the future of the psychology of filial piety: Chinese norms to contextualized personality construct. *Frontiers in Psychology*, 10, 100.

Beilock, S. (2011). *Choke: What the secrets of the brain reveal about getting it right when you have to*. New York: The Free Press.

Berkovich, I. (2020). Conceptualisations of empathy in K–12 teaching: A review of empirical research. *Educational Review*, 72(5), 547–66.

Berkowitz, T., Schaeffer, M. W., Maloney, E. A., Peterson, L., Gregor, C., Levine, S. C. & Beilock, S. L. (2015). Math at home adds up to achievement in school. *Science*, 350, 196–8.

Beurel, E. & Nemeroff, C. B. (2014). Interaction of stress, corticotropin-releasing factor, arginine vasopressin and behaviour. *Current Topics in Behavioral Neurosciences*, 18, 67–80.

Bhaskar, R. (1978). *A realist theory of science*. London: Routledge.

Biggs, J. B. (1994). Asian learners through Western eyes: An astigmatic paradox. *Australian and New Zealand Journal of Vocational Educational Research*, 2(2), 40–63.

Black, P. & William, D. (1998a). *Inside the black box: Raising standards through classroom assessment*. London: School of Education, King's College.

Black, P. & William, D. (1998b). Assessment and classroom learning. *Assessment in Education, Policy & Practice*, 5(1), 7–74.

Blair, R. J. R. (2011). Should affective arousal be grounded in perception–action coupling? *Emotion Review*, 3, 109–10.

Bliss, T. V., Collingridge, G. & Morris, R. (2003). *LTP: Long-term potentiation*. Oxford: Oxford University Press.

Bliss, T. V. & Lømo, T. (1973). Long-lasting potentiation of synaptic transmission in the dentate gyrus of the anesthethized rabbit following stimulation of the perforant path. *Journal of Physiology*, 232, 331–56.

Boatman, D., Freeman, J., Vining, E., Pulsifer, M., Miglioretti, D., Minahan, R., Carson, B., Brandt, J. & McKhann, G. (1999). Language recovery after left hemispherectomy in children with late-onset seizures. *Annals of Neurology*, 46(4), 579–86.

Boseley, S. (2019). Work of renowned UK psychologist Hans Eysenck ruled 'unsafe'. *The Guardian*, 11 October. Retrieved from: www.theguardian.com/science/2019/oct/11/work-of-renowned-uk-psychologist-hans-eysenck-ruled-unsafe

Boyle, R. (1661). *The sceptical chemist*. London: F. Crooke.

Brainerd, C. J. (1978). *Piaget's theory of intelligence*. Englewood Cliffs, NJ: Prentice Hall.

BrainU (2021). Glossary of neuroscience terms. Retrieved from: http://brainu.org/glossary-neuroscience-terms

Brentano, F. (1973 [1874]). *Psychology from an empirical standpoint*. London: Routledge and Kegan Paul.

Bruer, J. T. (1997). Education and the brain: A bridge too far. *Educational Researcher*, 24, 4–16.

Bruer, J. T. (1999). *The myth of the first three years: A new understanding of early brain development and lifelong learning*. New York: The Free Press.

Burgess, P. W. & Wu, H.-C. (2013). Rostral prefrontal cortex (Brodmann area 10): Metacognition in the brain. In D. T. Stuss & R. T. Knight (eds), *Principles of frontal lobe function*. New York: Oxford University Press, pp. 524–34.

Burt, C. (1921). *Mental and scholastic tests*. London: London County Council.

Butterworth, B., Reeve, R. & Reynolds, F. (2011). Using mental representations of space when words are unavailable: Studies of enumeration and arithmetic in Indigenous Australia. *Journal of Cross-Cultural Psychology*, 42, 630–8.

Butterworth, B., Reeve, R., Reynolds, F. & Lloyd, D. (2008). Numerical thought with and without words: Evidence from Indigenous Australian children. *The Proceedings of National Academy of Sciences of the USA*, 105(35), 13179–84.

Cajal, S. R. (1888). Sobre las fibras nerviosas de la capa molecular del cerebelo. *Revista Trimestral de Histologia Normal y Patológica*.

Cajal, S. R. (1894). The Croonian lecture: La fine structure des centres nerveux. *Proceedings of the Royal Society of London*, 55, 444–68.

Cajal, S. R. (1975). History of the synapse as a morphological and functional structure. In M. Santini (ed.), *Golgi Centennial Symposium: Perspectives in Neurobiology*. New York: Raven Press, pp. 39–50.

Cantlon, J. F. (2012). Math, monkeys, and the developing brain. *Proceedings of the National Academy of Science Sciences of the United States of America*, 109, 10725–32.

Cantor, G. (1975). The reception of the wave theory of light in Britain. *Historical Studies in the Physical Sciences*, 6, 109–32.

Cantor, P., Osher, D., Berg, J., Steyer, L. & Rose, T. (2019). Malleability, plasticity, and individuality: How children learn and develop in context. *Applied Developmental Science*, 23(4), 307–37.

Casey, B. J., Duhoux, S. & Cohen, M. M. (2010). Adolescence: What do transmission, transition, and translation have to do with it? *Neuron*, 67(9), 749–60.

Casey, B. J., Tottenham, N., Liston, C. & Durston, S. (2005). Imaging the developing brain: What have we learned about cognitive development? *Trends in Cognitive Sciences*, 9, 104–10.

Castles, A., Rastle, K. & Nationa, K. (2018). Ending the reading wards: Reading acquisition from novice to expert. *Psychological Science in the Public Interest*, 19(1), 5–51.

Cepeda, N. J., Coburn, N., Rohrer, D., Wixted, J. T., Mozer, M. C. & Pashler, H. (2008). Optimizing distributed practice: Theoretical analysis and practical implications. *Experimental Psychology*, 56(4), 236–46.

Cepeda, N. J., Pashler, H., Vul, E., Wixted, J. T. & Rohrer, D. (2006). Distributed practice in verbal recall tasks: A review and quantitative synthesis. *Psychological Bulletin*, 132, 354–80.

Chalmers, A. (1982). *What is this thing called science?* (2nd ed.). Milton Keynes: Open University Press.

Chalmers, A. (1999). *What is this thing called science?* (3rd ed.). Milton Keynes: Open University Press.

Chomsky, N. (2006). *Language and mind*. Cambridge: Cambridge University Press.

Churchland, P. S. (2011). *Braintrust: What neuroscience tells us about morality*. Princeton, NJ: Princeton University Press.

Cisek, P. (2007). Cortical mechanisms of action selection: The affordance competition hypothesis. *Philosophical Transactions of The Royal Society B*, 362, 1585–99.

Cisek, P. & Pastor-Bernier, A. (2014). On the challenges and mechanisms of embodied decisions. *Philosophical Transactions of The Royal Society B*, 369.

Citri, A. & Malenka, R. C. (2008). Synaptic plasticity: Multiple forms, functions, and mechanisms. *Neuropsychopharmacology: Official publication of the American College of Neuropsychopharmacology*, 33(1), 18–41.

Clark, A. (2015). Embodied prediction. In T. Metzinger & J. M. Windt (eds), *Open MIND*, 7(T). Frankfurt am Main: MIND Group. Retrieved from: https://doi.org/10.15502/9783958570115

Clark, C. (2001). *The affordances of adolescents' environments*. Unpublished PhD thesis, University of Surrey.

Clement, N. (2010). The first pillar of the student wellbeing pedagogy: The neuroscience research. In R. Toomey, T. Lovat, N. Clement, & K. Dally (eds), *Teacher education and values pedagogy: A student wellbeing approach*. Terrigal: David Barlow Publishing, pp. 15–31.

Cloud, J. (2010). Why your DNA isn't your destiny: The new field of epigenetics is showing how your environment and your choices can influence your genetic code – and that of your kids. *Time*. Retrieved from: http://content.time.com/time/subscriber/article/0,33009,1952313,00.html

Cole, P. (1997). Constructivism and scientific realism? Which is the better framework for educational research. *Australian Journal of Teacher Education*, 22(1).

Collaborative for Academic, Social and Emotional Learning (CASEL) (2020). CASEL's SEL Framework: What are the core competence areas and where are they promoted? Retrieved from: https://casel.org/wp-content/uploads/2020/12/CASEL-SEL-Framework-11.2020.pdf

Conway, B. R. (2018). The organization and operation of inferior temporal cortex. *Annual Review of Vision Science*, 4, 381–402.

Copernicus, N. (1543) *De revolutionibus orbium coelestium* (On the Revolutions of the Celestial Spheres) (2nd ed.). Basel: Officina Henricpetrina.

Crick, F. (1994). *The astonishing hypothesis: The scientific search for the soul*. New York: Touchstone Books.

Crowley, C. J. (2004). The ethics of assessment with culturally and linguistically diverse populations. *American Speech-Language-Hearing Association Leader*, 9.

Damasio, A. R. (1989). The brain blinds entities and events by multiregional activation from convergence zones. *Neural Computation*, 1, 123–32.

Damasio, A. R. (1994). *Descartes' error: Emotion, reason and the human brain*. New York: Putnam.

Damasio, A. R. (1999). *The feeling of what happens: Body and emotions in the making of consciousness*. New York: Harcourt.

Damasio, A. R. (2018). *The strange order of things: Life, feeling, and the making of cultures*. New York: Pantheon.

D'Andrade, R. G. (1995). *The development of cognitive anthropology*. Cambridge: Cambridge University Press.

Dantzig, T. (1967). *Number: The language of science*. New York: The Free Press.

Darling-Hammond, L., Flook, L., Cook-Harvey, C., Barron, B. & Osher, D. (2020). Implications for educational practice of the science of learning and development. *Applied Developmental Science*, 24(2), 97–140.

Darwin, C. (1965 [1872]). *The expression of the emotions in man and animals*. Chicago: University of Chicago Press.

Davidov, M., Zahn-Waxler, C., Roth-Hanania, R. & Knafo, A. (2013). Concern for others in the first year of life: Theory, evidence, and avenues for research. *Child Development Perspectives*, 7(2), 126–31.

de Waal, F. B. M. (2009). *The age of empathy: Nature's lessons for a kinder society*. New York: Harmony Books.

de Waal, F. B. M. & Preston, S. D. (2017). Mammalian empathy: Behavioural manifestations and neural basis. *Nature Reviews Neuroscience*, 18, 498–509.

De Wied, M., Gispen-De Wied, C. & Van Boxtel, A. (2009). Empathy dysfunction in children and adolescents with disruptive behavior disorders. *European Journal of Pharmacology*, 626, 97–103.

De Winter, R. F. P., van Hemert, A. M., DeRijk, R. H., Zwinderman, K. H., Frankhuijzen-Sierevogel, A. C., Wiegant, V. M. & Goekoop, J. G. (2003). Anxious-retarded depression: Relation with plasma vasopressin and cortisol. *Neuropsychopharmacology*, 28, 140–7.

Decety, J. & Cowell, J. M. (2014). Friends or foes: Is empathy necessary for moral behavior? *Perspectives on Psychological Science*, 9(5), 525–37.

Decety, J. & Michalska, K. J. (2010). Neurodevelopmental changes in the circuits underlying empathy and sympathy from childhood to adulthood. *Developmental Science*, 13, 886–99.

Decety, J. & Michalska, K. J. (2012). How children develop empathy: The contribution of developmental affective neuroscience. In J. Decety (ed.), *Empathy: From bench to bedside*. Cambridge, MA: MIT Press, pp. 167–90.

Dehaene, S. (1992). Varieties of numerical abilities. *Cognition*, 44(1–2), 1–42.

Dehaene, S. (1997). *The number sense: How the mind creates mathematics*. New York: Oxford University Press.

Dehaene, S. (2009). *Reading in the brain: The new science of how we read*. New York: Penguin.

Dehaene, S. (2020). *How we learn: The new science of education and the brain*. London: Penguin Random House.

Dehaene, S. & Cohen, L. (1995). Toward an anatomical and functional model of number processing. *Mathematical Cognition*, 1(1), 83–120.

Dehaene, S., Molko, N., Cohen, L. & Wilson, A. J. (2004). Arithmetic and the brain. *Current Opinion in Neurobiology*, 14(2), 218–24.

Dehaene, S., Piazza, M., Pinel, P. & Cohen, L. (2003). Three parietal circuits for number processing. *Cognitive Neuropsychology*, 20(3), 487–506.

Dehaene-Lambertz, G., Monzalvo, K. & Dehaene, S. (2018). The emergence of the visual word form: Longitudinal evolution of category-specific ventral visual areas during reading acquisition. *PLoS Biology*, 16(3). https://doi.org/10.1371/journal.pbio.2004103

Delazer, M. & Benke, T. (1997). Arithmetic facts without meaning. *Cortex*, 33(4), 697–710.

Denham, S. A., Bassett, H. H. & Wyatt, T. (2015). The socialization of emotional competence. In J. Grusec & P. Hastings (eds), *Handbook of socialization* (2nd ed.). New York: Guilford Press, pp. 590–613.

Department of Education, Employment and Workplace Relations (DEEWR) (2009). *Belonging, being and becoming: The early years learning framework for Australia*. Canberra: Commonwealth of Australia.

Department of Education, Science and Training (DEST) (2005). *National framework for Values Education in Australian schools*. Canberra: Commonwealth of Australia.

Descartes, R. (1637). *Discourse on the method*. Paris.

Descartes, R. (1641). *Meditations on first philosophy*. Paris.

Dewey, J. (1938). *Experience and education*. New York: Macmillan.

Di Bernardi Luft, C. (2014). Learning from feedback: The neural mechanisms of feedback processing facilitating better performance. *Behavioural Brain Research*, 261, 356–68.

Djebbara, Z., Fich, L. B., Petrini, L. & Gramann, K. (2019). Sensorimotor brain dynamics reflect architectural affordances. *Proceedings of the National Academy of Sciences of the United States of America*, 116(29), 14769–78.

Doidge, N. (2007). *The brain that changes itself: Stories of personal triumph from the frontiers of brain science*. New York: Penguin.

Donahue, C. J., Glasser, M. F., Preuss, T. M., Riling, J. K. & Van Essen, D. C. (2018). Quantitative assessment of prefrontal cortex in humans relative to nonhuman primates. *Proceedings of National Academy of Science USA*, 115, E5183–E51992.

Dovey, K. & Fisher, K. (2014). Designing for adaptation: The school as socio-spatial assemblage. *The Journal of Architecture*, 19(1), 43–63.

Doyle, W. (1977). Learning the classroom environment: An ecological analysis. *Journal of Teacher Education*, 28(6), 51–5.

Dubinsky, J. M. (2010). Neuroscience education for prekindergarten–12 teachers. *Journal of Neuroscience*, 30(24), 8057–60.

Duncan, C., Kim, M., Baek, S., Wu, K. Y. Y. & Sankey, D. (2021). The limits of motivation theory in education and the dynamics of value-embedded learning (VEL). *Educational Philosophy and Theory*. https://doi.org/10.1080/00131857.2021.1897575

Duncan, C., Kim, M. & Sankey, D. (n.d.). Untitled manuscript, in preparation.

Duncan, C. & Sankey, D. (2019). Two conflicting visions of education and their consilience. *Educational Philosophy and Theory*, 51(14), 1454–64.

Edelman, G. M. (1987). *Neural Darwinism: The theory of neuronal group selection*. New York: Basic Books.

Edelman, G. M. (1989). *The remembered present*. New York: Basic Books.

Edelman, G. M. (2004). *Wider than the sky: The phenomenal gift of consciousness*. New Haven, CT: Yale University Press.

Edelman, G. M. (2006). *Second nature: Brain science and human knowledge*. New Haven, CT: Yale University Press.

Eisenberg, N., Fabes, R. A., Bernzweig, J., Karbon, M., Poilin, R. & Hanish, L. (1993). The relations of emotionality and regulation to preschoolers' social skills and sociometric status. *Child Development*, 64, 1418–38.

Eisenberg, N., Smith, C. L., Sadovsky, A. & Spinrad, T. L. (2004). Effortful control: Relations with emotion regulation, adjustment, and socialization in childhood. In R. F. Maumeister, & K. D. Vohs (eds), *Handbook of self-regulation: Research, theory, and applications*. New York: Guilford Press, pp. 259–82.

Eisenberger, N. I. & Lieberman, M. D. (2004). Why rejection hurts: A common neural alarm system for physical and social pain. *Trends in Cognitive Sciences*, 8, 294–300.

Eisenberger, N. I., Lieberman, M. D. & Williams, K. D. (2003). Does rejection hurt? An fMRI study of social exclusion. *Science*, 302, 290–2.

Ekman, P. (1972). Universals and cultural differences in facial expressions of emotions. In J. Cole (ed.), *Nebraska Symposium on Motivation*. Lincoln, NE: University of Nebraska Press, pp. 207–83.

Ekman, P. (1980). *The face of man*. New York: Garland Publishing.

Elliott, L. L. (1979). Performance of children aged 9 to 17 years on a test of speech intelligibility in noise using sentence material with controlled word predictability. *The Journal of the Acoustical Society of America*, 66(3), 651–3.

Ernest, M., Nelson, E. E., Jazbec, S., McClure, E. B., Monk, C. S., Leibenluft, E., Blair, J. & Pine, D. S. (2005). Amygdala and nucleus accumbens in responses to receipt and omission of gains in adults and adolescents. *Neuroimage*, 25, 1279–91.

European Association for Research on Learning and Instruction (EARLI) (2021). SIG 22 – neuroscience and education. Retrieved from: www.earli.org/node/45

Ferrari, P. F. & Rizzolatti, G. (2014). Mirror neuron research: The past and the future. *Philosophical Transactions of the Royal Society of London. Series B, Biological Sciences*, 369(1644). https://doi.org/10.1098/rstb.2013.0169

Ferrero, M., Garaizar, P. & Vadillo, M. A. (2016). Neuromyths in education: Prevalence among Spanish teachers and an exploration of cross-cultural variation. *Frontiers in Human Neuroscience*, 10, 496. https://doi.org/10.3389/fnhum.2016.00496

Field, H. (1979). Competency based teacher education (CBTE): A review of the literature. *Journal of In-Service Education*, 6(1), 39–42.

Field, T. M., Woodson, R., Greenberg, R. & Cohen, D. (1982). Discrimination and imitation of facial expression by neonates. *Science*, 219, 179–81.

Flavell, J. H. (1976). Metacognitive aspects of problem solving. In L. B. Resnick (ed.), *The nature of intelligence*. Hillsdale, NJ: Lawrence Erlbaum, pp. 231–6.

Flavell, J. H. (1979). Metacognition and cognitive monitoring: A new area of cognitive–developmental inquiry. *American Psychologist*, 34(10), 906–11.

Flynn, J. R. (1984). The mean IQ of Americans: Massive gains 1932–1978. *Psychological Bulletin*, 95(1), 29–51.

Flynn, J. R. (2012). *Are we getting smarter? Rising IQ in the twenty-first century*. Cambridge: Cambridge University Press.

Fogarty, R. (1994). *How to teach for metacognition*. Palatine, IL: IRI/Skylight Publishing.

Fogel, A. (2008). Relationships that support human development. In A. Fogel, B. J. King & S. G. Shanker (eds), *Human Development in the twenty-first century: Visionary ideas from systems scientists*. Cambridge: Cambridge University Press, pp. 57–64.

Fogel, A., Nwokah, E., Dedo, J., Messinger, D., Dickson, K., Matusov, E. & Holt, S. (1992). Social process theory of emotion: A dynamic systems approach. *Social Development*, 1(2), 122–42.

Frankl, V. (1985). *Man's search for meaning*. New York: Washington Square Press.

Freeman, W. J. (2000). *How brains make up their minds*. New York: Columbia University Press.

Friston, K. J. (2012). The history of the future of the Bayesian brain. *NeuroImage*, 62, 1230–3.

Friston, K. J., Tononi, G., Reeke Jr., G. N., Sporns, O. & Edelman, G. M. (1994). Value-dependent selection in the brain: Simulation in a synthetic neural model. *Neuroscience*, 59(2), 229–43.

Frith, C. D. (2012). The role of metacognition in human social interactions. *Philosophical Transactions of The Royal Society B*, 367, 2213–23.

Frith, U. (1985). Beneath the surface of developmental dyslexia. In K. E. Patterson, J. C. Marshall & M. Coltheart (eds), *Surface dyslexia: Neurological and cognitive studies of phonological reading*. London: Lawrence Erlbaum, pp. 301–30.

Frith, U. (1986). A developmental framework for developmental dyslexia. *Annals of Dyslexia*, 36, 69–81.

Gage, N. M. & Baars, B. J. (2018). *Fundamentals of cognitive neuroscience: A beginner's guide* (2nd ed.). Cambridge, MA: Elsevier.

Galvan, A., Hare, T. A., Parra, C. E., Penn, J., Voss, H., Glover, G. & Casey, B. J. (2006). Earlier development of the accumbens relative to orbitofrontal cortex might underlie risk-taking behavior in adolescents. *Journal of Neuroscience*, 26, 6885–92.

Gardner, H. (1983). *Frames of mind: The theory of multiple intelligences*. New York: Basic Books.

Garrido, M. I., Kilner, J. M., Kiebel, S. J. & Friston, K. J. (2007). Evoked brain responses are generated by feedback loops. *Proceedings of the National Academy of Sciences of the United States of America*, 104(52), 20961–6.

Gazzaniga, M. S. (2011). *Who's in charge? Free will and the science of the brain*. New York: HarperCollins.

Geake, J. (2009). *The brain at school: Educational neuroscience in the classroom*. Maidenhead: McGraw Hill/Open University Press.

Gehring, W. J., Goss, B., Coles, M. G., Meyer, D. E. & Donchin, E. (1993). A neural system for error detection and compensation. *Psychological Science*, 4, 385–90.

Gershkoff-Stowe, L. & Thelen, E. (2004). U-shaped changes in behavior: A dynamic systems perspective. *Journal of Cognition and Development*, 1(5), 11–36.

Gibson, J. J. (1979). *The ecological approach to visual perception*. Boston: Houghton-Mifflin.

Glazer, J. E., Kelley, N. J., Pornpattananangkul, N., Mittal, V. A. & Nusslock, R. (2018). Beyond the FRN: Broadening the time-course of EEG and ERP components implicated in reward processing. *International Journal of Psychophysiology*, 132(Pt B), 184–202.

Gogtay, N., Giedd, J. N., Lusk, L., Hyashi, K. M., Greenstein, D., Vaituzis, A. C., … Tompson, P. M. (2004). Dynamic mapping of human cortical development during childhood through early adulthood. *Proceedings of the National Academy of Sciences of the United States of America*, 101(21), 8174–9.

Goldin, P. R., McRae, K., Ramel, W. & Gross, J. J. (2008). The neural bases of emotion regulation: reappraisal and suppression of negative emotion. *Biological psychiatry*, 63(6), 577–86.

Goldin-Meadow, S. & Beilock, S. L. (2010). Action's influence on thought: The case of gesture. *Perspectives on Psychological Science*, 5(6), 664–74.

Goldin-Meadow, S., Cook, S. W. & Mitchell, Z. A. (2009). Gesturing gives children new ideas about math. *Psychological Science*, 20(3), 267–72.

Goodwin, B. (2001). *How the leopard changed its spots*. Princeton, NJ: Princeton University Press.

Goroshit, M. & Hen, M. (2016). Teachers' empathy: Can it be predicted by self-efficacy? *Teachers and Teaching*, 22(7), 805–18.

Goswami, U. (1992). *Analogical reasoning in children*. Hove: Lawrence Erlbaum.

Gotlieb, R., Hyde, E., Immordino-Yang, M. H. & Kaufman, S. B. (2016). Cultivating the social–emotional imagination in gifted education: Insights from educational neuroscience. *Annals of the New York Academy of Sciences*, 1377(1), 22–31.

Gould, S. J. (1981). *The mismeasure of man*. New York: W. W. Norton.

Gould, S. J. (1991). Exaptation: A crucial tool for an evolutionary psychology. *Journal of Social Issues*, 47(3), 43–65.

Goupil, L. & Kouider, S. (2016). Behavioral and neural indices of metacognitive sensitivity in preverbal infants. *Current Biology*, 26, 3038–45.

Goupil, L., Romand-Monnier, M. & Kouider, S. (2016). Infants ask for help when they know they don't know. *Proceedings of the National Academy of Sciences*, 113(13), 3492–6.

Graham, G. (2019). Behaviorism. In E. N. Zalta (ed.), *The Stanford Encyclopedia of Philosophy*. Retrieved from: https://plato.stanford.edu/archives/spr2019/entries/behaviorism

Grant, C. (2017). *The contribution of education to economic growth*. Brighton: Institute of Development Studies.

Greene, J. D. (2015). The cognitive neuroscience of moral judgment and decision making. In J. Decety & T. Wheatley (eds), *The moral brain: A multidisciplinary perspective*. Cambridge, MA: The MIT Press, pp. 35–48.

Greene, J. D., Sommerville, R. B., Nystrom, L. E., Darley, J. M. & Cohen, J. D. (2001). An fMRI investigation of emotional engagement in moral judgment. *Science*, 293, 2105–8.

Gruber, M., Gelman, B. D. & Ranganath, C. (2014). States of curiosity modulate hippocampus-dependent learning via the dopaminergic circuit. *Neuron*, 84(2),486–96.

Gurunandan, K., Arnaez-Telleria, J., Carreiras, M. & Paz-Alonso, P. (2020). Converging evidence for differential specialization and plasticity of language systems. *Journal of Neuroscience*, 40(50), 9715–24.

Guyer, A. E., McClure-Tone, E. B., Shiffrin, N. D., Pine, D. S. & Nelson, E. E. (2009). Probing the neural correlates of anticipated peer evaluation in adolescence. *Child Development*, 80, 1000–15.

Habermas, J. (1972). *Knowledge and human interests* (trans. J. Shapiro). Portsmouth, NH: Heinemann.

Hadamard, J. (1945). *An essay on the psychology of invention in the mathematical field*. Princeton, NJ: Princeton University Press.

Hadjikhani, N., Joseph, R. M., Snyder, J. & Tager-Flusberg, H. (2006). Anatomical differences in the mirror neuron system and social cognition network in autism. *Cerebral Cortex*, 16(9), 1276–82.

Hajcak, G., Dunning, J. P. & Foti, D. (2009). Motivated and controlled attention to emotion: Time-course of the late positive potential. *Clinical Neurophysiology*, 120(3), 505–10.

Hamlin, J. K., Wynn, K. & Bloom, P. (2010). Three-month-olds show a negativity bias in their social evaluations. *Developmental Science*, 13(6), 923–9.

Hamston, J., Weston, J., Wajsenberg, J. & Brown, D. (2010). *Giving voice to the impacts of values education: The final report of the Values in Action schools project*. Melbourne: Education Services Australia.

Hansen, D. V., Lui, J. H., Parker, P. R. & Kriegstein, A. R. (2010). Neurogenic radial glia in the outer subventricular zone of human neocortex. *Nature*, 464(7288), 554–61.

Hare, T. A., Tottenham, N., Davidson, M. C., Glover, G. H. & Casey, B. J. (2008). Biological substrates of emotional reactivity and regulation in adolescence during an emotional go–no go task. *Biological Psychiatry*, 63, 927–34.

Hattie, J. (2008). *Visible learning: A synthesis of over 800 meta-analyses relating to achievement*. London: Routledge.

Hattie, J. (2012). *Visible learning for teachers: Maximising impact on learning*. London: Routledge.

Hattie, J. & Timperley, H. (2007). The power of feedback. *Review of Educational Research*, 77(1), 81–112.

Hay, D. F., Nash, A. & Pederson, J. (1981). Responses of six-month-olds to the distress of their peers. *Child Development*, 52, 1071–5.

Heath, S. B. (1997). Culture: Contested realm in research on children and youth. *Applied Developmental Science*, 1(3), 113–23.

Heaven, D. (2020). Why faces don't always tell the truth about feelings. *Nature*, 578, 502–4.

Hebb, D. O. (1949). *The organization of behavior: A neuropsychological theory*. New York: John Wiley & Sons.

Hodges, N. J. & Franks, I. M. (2002). Modelling coaching practice: The role of instruction and demonstration. *Journal of Sports Sciences*, 20(10), 793–811.

Hoffman, M. L. (1975). Developmental synthesis of affect and cognition and its interplay for altruistic motivation. *Developmental Psychology*, 11, 607–22.

Hofstadter, D. R. (1995). *Fluid concepts and creative analogies*. New York: Basic Books.

Hofstadter, D. (2001). Analogy as the core of cognition. In D. Gentner, K. J. Holyoak & B. N. Kokinov (eds), *The analogical mind: Perspectives from cognitive science*. Cambridge, MA: MIT Press, pp. 499–538.

Holroyd, C. B., Nieuwenhuis, S., Yeung, N., Nystrom, L., Mars, R. B., Coles, M. G. & Cohen, J. D. (2004). Dorsal anterior cingulate cortex shows fMRI response to internal and external error signals. *Nature Neuroscience*, 7(5), 497–8.

Horky, P. (2013). *Plato and Pythagoreanism*. New York: Oxford University Press.

Howard-Jones, P. A. (2014). Neuroscience and education: Myths and messages. *Nature Reviews Neuroscience*, 15, 817–24.

Howe, M. L. & Lewis, M. D. (2005). The importance of dynamic systems approaches for understanding development. *Developmental Review*, 25(1), 247–51.

Huang, W., Bhaduri, A., Velmeshev, D., Wang, S., Wang, L., Rottkamp, C. A., Alvarez-Buylla, A., Rowitch, D. H. & Kriegstein, A. R. (2020). Origins and proliferative states of human oligodendrocyte precursor cells. *Cell*, 182(3), 594–608.e11.

Huang, X., Li, W., Sun, B., Chen, H. & Davis, M. H. (2012). The validation of the interpersonal reactivity index for Chinese teachers from primary and middle schools. *Journal of Psychoeducational Assessment*, 30(2), 194–204.

Iacoboni, M. (2005). Understanding others: Imitation, language, and empathy. In S. Hurley & N. Chater (eds), *Perspectives on imitation: From neuroscience to social science: Vol. 1. Mechanisms of imitation and imitation in animals*. Cambridge, MA: MIT Press, pp. 77–99.

Immordino-Yang, M. H. (2016). *Emotions, learning, and the brain: Exploring the educational implications of affective neuroscience*. New York: W.W. Norton.

Immordino-Yang, M. H. & Damasio, A. (2016). We feel, therefore we learn: The relevance of affective and social neuroscience to education. In M. H. Immordino-Yang (ed.), *Emotions, learning, and the brain: Exploring the educational implications of affective neuroscience*. New York: W.W. Norton, pp. 27–42.

Imms, W., Mahat, M., Byers, T. & Murphy, D. (2017). *Type and use of innovative learning environments in Australasian schools*. ILETC Survey No. 1. Melbourne: University of Melbourne. Retrieved from www.iietc.com.au/publications/reports

Ingold, T. (2008). The social child. In A. Fogel, B. J. King & S. G. Shanker (eds), *Human Development in the twenty-first century: Visionary ideas from systems scientists*. Cambridge: Cambridge University Press, pp. 112–18.

Inhelder, B. & Piaget, J. (1958). *An essay on the construction of formal operational structures. The growth of logical thinking: From childhood to adolescence* (A. Parsons & S. Milgram, trans.). New York: Basic Books.

International Mind, Brain and Education Society (IMBES) (2018). Homepage. Retrieved from: https://imbes.org

Izard, V., Sann, C., Spelke, E. S. & Streri, A. (2009). Newborn infants perceive abstract numbers. *Proceedings of the National Academy of Sciences of the United States of America*, 106, 10382–5.

James, W. (1890). *The principles of psychology*. New York: Henry Holt.

Jamieson, J. P., Mendes, W. B., Blackstock, E. & Schmader, T. (2010). Turning the knots in your stomach into bows: Reappraising arousal improves performance in the GRE. *Journal of Experimental Social Psychology*, 46, 208–12.

Jelbert, S. A., Hosking, R. J., Taylor, A. H. & Gray, R. D. (2018). Mental template matching is a potential cultural transmission mechanism for New Caledonian crow tool manufacturing tradition. *Nature Scientific Reports*, 8, 8956. https://doi.org/10.1038/s41598-018-27405-1

Johnson, C. E. (2000). Children's phoneme identification in reverberation and noise. *Journal of Speech, Language & Hearing Research*, 43(1), 144–57.

Jones, P. (2016). Language and social determinism in the Vygotskian tradition: A response to Ratner (2015). *Language and Sociocultural Theory*, 3(1), 3–10.

Kaas, J. H. (2012). Somatosensory system. In J. K. Mai & G. Paxinos (eds), *The human nervous system* (3rd ed.). Amsterdam: Academic Press, pp. 1074–109.

Kandel, E. R. (2006). *In search of memory: The emergence of a new science of mind*. New York: W.W. Norton.

Karim, M. (2003). *Imam Ghazali's book of religious learnings*. New Delhi: Islamic Book Service.

Karpicke, J. & Blunt, J. R. (2011). Retrieval practice produces more learning than elaborative studying with concept mapping. *Science*, 331(6018), 772–5.

Kauffman, S. (1995). *At home in the universe: The search for laws of self-organization and complexity*. Oxford: Oxford University Press.

Kauffman, S. (2008). *Reinventing the sacred*. New York: Basic Books.

Kearney, H. (1971). *Science and change, 1500–1700*. London: Weidenfeld & Nicolson.

Kennerley, S. W., Walton, M. E., Behrens, T. E., Buckley, M. J. & Rushworth, M. F. (2006). Optimal decision making and the anterior cingulate cortex. *Nature Neuroscience*, 9(7), 940–7.

Kettle, K. L. & Häubl, G. (2010). Motivation by anticipation: Expecting rapid feedback enhances performance. *Psychological Science*, 21(4), 545–7.

Khalsa, S. S., Adolphs, R., Cameron, O. G., Critchley, H. D., Davenport, P. W., … Interoception Summit Participants (2016). Interoception and mental health: A roadmap. *Cognitive Neuroscience and Neuroimaging*, 3(6), 501–13.

Kiefer, M. & Pulvermüller, F. (2012). Conceptual representations in mind and brain: Theoretical development, current evidence and future directions. *Cortex*, 48, 805–25.

Kiefer, M. & Trumpp, N. M. (2012). Embodiment theory and education: The foundations of cognition in perception and action. *Trends in Neuroscience and Education*, 1(1), 15–20.

Kim, M., Decety, J., Wu, L. Baek, S. & Sankey, D. (2021). Neural computations in children's third-party interventions are modulated by their parents' moral values. *Science of Learning*, 6(38). https://doi.org/10.1038/s41539-021-00116-5

Kim, M. & Sankey, D. (2009). Towards a dynamic systems approach to moral development and moral education: A response to the JME special issue. *Journal of Moral Education*, 38(3), 283–98.

Kim, M. & Sankey, D. (2010). The dynamics of emergent self-organisation: Reconceptualising child development in teacher education. *Australian Journal of Teacher Education*, 35(4). http://dx.doi.org/10.14221/ajte.2010v35n4.6

Kim, M. & Sankey, D. (2018). Philosophy, neuroscience and pre-service teachers' beliefs in neuromyths: A call for remedial action. *Educational Philosophy and Theory*, 50(13), 1214–27.

Kimel, S. Y., Mischkowski, D., Kitayama, S. & Uchida, Y. (2017). Culture, emotions, and the cold shoulder: Cultural differences in the anger and sadness response to ostracism. *Journal of Cross-Cultural Psychology*, 48(9), 1307–19.

King, R. B. & Bernardo, A. B. I. (2016). *The psychology of Asian learners: A Festschrift in honour of David Watkins*. Singapore: Springer.

Kiriazis, J. & Slobodchikoff, C. N. (2006). Perceptual specificity in the alarm calls of Gunnison's prairie dogs. *Behavioural Process*, 73(1), 29–35.

Kitayama, S., Duffy, S. & Uchida, Y. (2007). Self as cultural mode of being. In S. Kitayama & D. Cohen (eds), *Handbook of cultural psychology*. New York: Guilford Press, pp. 136–74.

Kitayama, S. & Park, J. (2010). Cultural neuroscience of the self: Understanding the social grounding of the brain. *Social Cognitive and Affective Neuroscience*, 5(2–3), 111–29.

Kitayama, S. & Salvador, C. E. (2017). Culture embrained: Going beyond the nature–nurture dichotomy. *Perspectives on Psychological Science*, 12(5), 841–54.

Kitayama, S. & Uskul, A. (2011). Culture, mind, and the brain: Current evidence and future directions. *Annual Review of Psychology*, 62, 419–49.

Klatte, M., Hellbrück, J., Seidel, J. & Leistner, P. (2010). Effects of classroom acoustics on performance and well-being in elementary school children: A field study. *Environment and Behavior*, 24(5), 659.

Klis, M. & Kossewska, J. (1996). *Empathy in the structure of personality of special educators*. Krakow: Pedagogical University.

Koestler, A. (1964). *The act of creation*. New York: Macmillan.

Kohlberg, L. (1981). *Essays on moral development. Vol I: The philosophy of moral development*. San Francisco: Harper & Row.

Kolb, D. A. (1984). *Experiential learning: Experience as the source of learning and development*. Englewood Cliffs, NJ: Prentice-Hall.

Kontra, C., Goldin-Meadow, S. & Beilock, S. L. (2012). Embodied learning across the life span. *Topics in Cognitive Sciences*, 4, 731–9.

Kontra, C., Lyons, D. J., Fischer, S. M. & Beilock, S. L. (2015). Physical experience enhances science learning. *Psychological Science*, 26(6), 737–49.

Kuhn, T. S. (1962). *The structure of scientific revolutions*. Chicago: University of Chicago Press.

Kuhn, T. S. (1970). *The structure of scientific revolutions* (2nd ed.). Chicago: University of Chicago Press.

Lagercrantz, H., Hanson, M. A., Ment, L. R. & Peebles, D. M. (2011). *The newborn brain: Neuroscience and clinical applications* (2nd ed.). Cambridge: Cambridge University Press.

Lakatos, I. (1970). Falsification and the methodology of scientific research programmes. In I. Lakatos & A. Musgrave (eds), *Criticism and the growth of knowledge*. Cambridge: Cambridge University Press.

Lakoff, G. & Johnson, M. (1999). *Philosophy in the flesh: The embodied mind and its challenge to Western thought*. New York: Basic Books.

Lambert, D. & Sankey, D. (1994). Classrooms as ecosystems. *Journal of Teacher Development*, 3(3), 175–82.

Lapsley, D. K. (1996). *Moral psychology*. Boulder, CO: Westview Press.

Lapsley, D. K. & Hill, P. L. (2008). On dual processing and heuristic approaches to moral cognition. *Journal of Moral Education*, 37(3), 313–32.

Le Van Quyen, M. (2011). The brainweb of cross-scale interactions. *New Ideas in Psychology*, 29, 57–63.

Leone, D. P., Srinivasan, K., Chen, B., Alcamo, E. & McConnell, S. K. (2008). The determination of projection neuron identity in the developing cerebral cortex. *Current Opinion in Neurobiology*, 18(1), 28–35.

Lepora, N. F. & Pezzulo, G. (2015). Embodied choice: How action influences perceptual decision making. *PLoS Computational Biology*, 11, e1004110.

Leydesdorff, L. (2000). Luhmann, Habermas, and the theory of communication. *Systems Research and Behavioural Science*, 17(3), 273–88.

Lieberman, M. D., Jarcho, J. M., Berman, S., Naliboff, B. D., Suyenobu, B. Y., Mandelkern, M. & Mayer, E. A. (2004). The neural correlates of placebo effects: A disruption account. *NeuroImage*, 22(1), 447–55.

Lindsey, R. V., Shroyer, J. D., Pashler, H. & Mozer, M. (2014). Improving students' long-term knowledge retention through personalizer review. *Psychological Science*, 25, 639–47.

Liu, D., Sabbagh, M. A., Gehring, W. J. & Wellman, H. M. (2009). Neural correlates of children's theory of mind development. *Child Development*, 80(2), 318–26.

Lovat, T. & Dally, K. (2018). Testing and measuring the impact of character education on the learning environment and its outcomes. *Journal of Character Education*, 14(2), 1–22.

Lovat, T. & Toomey, R. (2010). The evolution of student wellbeing pedagogy and its implications for teacher education. In R. Toomey, T. Lovat, N. Clement & K. Dally (eds), *Teacher education and values pedagogy: A student wellbeing approach*. Terrigal: David Barlow Publishing, pp. 1–14.

Luck, S. J. (2005). *An introduction to the event-related potential technique*. Cambridge, MA: MIT Press.

Luria, A. R. (1973). *The man with a shattered world: the history of a brain wound*. London: Jonathan Cape.

Lynn, R. (2006). *Race differences in intelligence: An evolutionary analysis*. Augusta, GA: Washington Summit Publishers.

MacLean, P. (1990). *The triune brain in evolution: Role in paleocerebral functions*. Washington, DC: Plenum Press.

Mahat, M., Bradbeer, C., Byers, T. & Imms, W. (2018). *Innovative learning environments and teacher change: Defining key concepts*. Melbourne: University of Melbourne.

Mainhard, T., Pennings, H. J. M., Wubbels, T. & Brekelmans, M. (2012). Mapping control and affiliation in teacher–student interaction with state space grids. *Teaching and Teacher Education*, 28(7), 1027–37.

Maloney, E. A. & Beilock, S. L. (2012). Math anxiety: Who has it, why it develops, and how to guard against it. *Trends in Cognitive Sciences*, 16(8), 404–6.

Malti, T., Ongley, S. F., Peplak, J., Chaparro, M. P., Buchmann, M., Zuffianò, A. & Cui, L. (2016). Children's sympathy, guilt, and moral reasoning in helping, cooperation, and sharing: A 6-year longitudinal study. *Child Development*, 87(6), 1783–95.

Mangels, J. A., Butterfield, B., Lamb, J., Good, C. & Dweck, C. S. (2006). Why do beliefs about intelligence influence learning success? A social cognitive neuroscience model. *Social Cognitive and Affective Neuroscience*, 1(2), 75–86.

Marshall, L. E. & Marshall, W. L. (2011). Empathy and antisocial behaviour. *The Journal of Forensic Psychiatry and Psychology*, 22(5), 742–59.

Martin, G. B. & Clark, R. D. (1987). Distress crying in neonates: Species and peer specificity. *Developmental Psychology*, 18, 3–9.

Maslin, K. T. (2001). *An introduction to the philosophy of mind*. Cambridge: Polity Press.

Masten, C. L., Gillen-O'Neel, C. & Spears Brown, C. (2010). Children's intergroup empathic processing: The roles of novel ingroup identification, situational distress, and social anxiety. *Journal of Experimental Child Psychology*, 106, 115–28.

Matsumoto, D., Yoo, S. H. & Nakagawa, S. (2008). Culture, emotion regulation, and adjustment. *Journal of Personality and Social Psychology*, 94(6), 925–37.

McCloskey, M. (1992). Cognitive mechanisms in numerical processing: Evidence from acquired dyscalculia. *Cognition*, 44, 107–57.

McCrink, K. & Wynn, K. (2004). Large-number addition and subtraction by 9-month-old infants. *Psychological Science*, 15, 776–81.

Mealings, K. T., Buchholz, J. M., Demuth, K. & Dillon, H. (2015). Investigating the acoustics of a sample of open plan and enclosed kindergarten classrooms in Australia. *Applied Acoustics*, 100, 95–105.

Mehler, J. & Bever, T. G. (1967). Cognitive capacity of very young children. *Science*, 158, 141–2.

Michaelson, J., Abdallah, S., Steuer, N., Thompson, S. & Marks, N. (2009). *National accounts of well-being: Bringing real wealth onto the balance sheet*. London: New Economics Foundation.

Michel, C. (2021). Overcoming the modal/amodal dichotomy of concepts. *Phenomenology and the Cognitive Sciences*, 20, 655–77.

Midgley, M. (1987). *Evolution as a religion: Strange hopes and stranger fears*. London: Methuen.

Miller, A. K. (2015). *Freedom to fail: How do I foster risk-taking and innovation in my classroom?* New York: ASCD Arias.

Milner, B. (1962). Les troubles de la memoire accompagnant des lesions hippocampiques bilaterales. In P. Passouant (ed.), *Physiologie de l'hippocampe*. Paris: Centre National de la Recherche Scientifique, pp. 257–72.

Milner, B., Corkin, S. & Teuber, H. L. (1968). Further analysis of the hippocampal amnesic syndrome: 14-year follow-up study of H.M. *Neuropsychologia*, 6(3), 215–34.

Ministerial Council on Education, Employment, Training and Youth Affairs (MCEETYA) (2001). *New framework for vocational education in schools: A comprehensive guide about pathways for young Australians in transition*. Melbourne: MCEETYA.

Mitchell, M. & Hofstadter, D. R. (1995). Perspectives on copycat: Comparisons with recent work. In D. R. Hofstadter (ed.), *Fluid concepts and creative analogies*. New York: Basic Books, pp. 275–300.

Mithen, S. (1996). *The prehistory of the mind: A search for the origins of art, religion and science*. London: Thames and Hudson.

Morais, J., Cary, L., Alegria, J. & Bertelson, P. (1979). Does awareness of speech as a sequence of phones arise spontaneously? *Cognition*, 7, 323–31.

Moriguchi, Y., Ohnishi, T., Mori, T., Matsuda, H. & Komaki, G. (2007). Changes of brain activity in the neural substrates for theory of mind in childhood and adolescence. *Psychiatry and Clinical Neurosciences*, 61, 355–63.

Moshman, D. (2011). *Adolescent rationality and development: Cognition, morality and identity* (3rd ed.). New York: Psychology Press.

Mulkay, M. (1979). *Science and the sociology of knowledge*. London: Allen & Unwin.

Murata, A., Moser, J. S. & Kitayama, S. (2013). Culture shapes electrocortical responses during emotion suppression. *Social Cognitive and Affective Neuroscience*, 8(5), 595–601.

Myoga, M. H. (2020). The superior colliculus: Auditory inputs integrate with other senses to coordinate orientation movements. In B. Fritzsch (ed.), *The senses: A comprehensive reference* (2nd ed.). New York: Elsevier, pp. 601–22.

Narvaez, D. (2016). *Embodied morality: Protectionism, engagement, and imagination*. New York: Palgrave Macmillan.

National Academies of Sciences, Engineering, and Medicine (NASEM) (2018). *How people learn II: Learners, contexts, and cultures*. Washington, DC: National Academies Press.

National Aphasia Association (2021). Aphasia definitions. Retrieved from: www.aphasia.org/aphasia-definitions

National Research Council (2001). *Knowing what students know: The science and design of educational assessment*. Washington, DC: National Academies Press.

New Zealand Ministry of Education (2018). Innovative learning environments. Retrieved from: http://elearning.tki.org.nz/Teaching/Innovative-learning-environments

Newell, K. M. (1986). Constraints of the development of coordination. In M. G. Wade & H. Whiting (eds), *Motor development in children: Aspects of coordination and control*. Dordrecht: Martinus Nijhoff, pp. 341–60.

Newell, K. M. & Jordan, K. (2007). Constraints and movement organization: A common language. In W. E. Davis & G. D. Broadhead (eds), *Ecological task analysis and movement*. Champaign, IL: Human Kinetics, pp. 5–13.

Niyozov, S. & Pluim, G. (2009). Teachers' perspectives on the education of Muslim students: A missing voice in Muslim education research. *Curriculum Inquiry*, 39, 637–77.

Noë, A. (2004). *Action in perception*. Cambridge, MA: MIT Press.

NSW Department of Education (2020). Culture and diversity. Retrieved from: https://education.nsw.gov.au/teaching-and-learning/curriculum/multicultural-education/culture-and-diversity

Office of the eSafety Commissioner (2018). *State of play-youth, kids and digital dangers*. Canberra: Australian Government.

Ohira, H., Nomura, M., Ichikawa, N., Isowa, T., Iidaka, T., Sato, A., Fukuyama, S., Nakajima, T. & Yamada, J. (2006). Association of neural and physiological responses during voluntary emotion suppression. *NeuroImage*, 29(3), 721–33.

Ojose, B. (2008). Applying Piaget's theory of cognitive development to mathematics instruction. *The Mathematics Educator*, 18(1), 26–30.

Organisation for Economic Co-operation and Development (OECD) (2002). *Understanding the brain: Towards a new learning science*. Paris: OECD.

Organisation for Economic Co-operation and Development (OECD) (2007). *Understanding the brain: The birth of a learning science*. Paris: OECD.

Organisation for Economic Co-operation and Development (OECD) (2013). *PISA 2012 results. Ready to learn: Students' engagement, drive and self-beliefs (Volume III)*. Paris: OECD.

Organisation for Economic Co-operation and Development (OECD) (2015). *PISA in focus: Does math make you anxious?* Paris: OECD.

Organisation for Economic Co-operation and Development (OECD) (2016). *PISA 2015 results (Volume I): Excellence and equity in education*. Paris: OECD.

Organisation for Economic Co-operation and Development (OECD) (2019a). *PISA 2018 results (Volume I): What students know and can do*. Paris: OECD.

Organisation for Economic Co-operation and Development (OECD) (2019b). *PISA 2018 results (Volume III): What School Life Means for Students' Lives*. Paris: PISA, OECD Publishing. https://doi.org/10.1787/acd78851-en.

Organisation for Economic Co-operation and Development (OECD) (2020). Students' self-efficacy and fear of failure. InPISA, *Results (Volume III): What school life means for students' lives*. Paris: OECD.

Ouyang, M., Dubois, J., Yu, Q., Mukherjee, P. & Huang, H. (2019). Delineation of early brain development from foetuses to infants with diffusion MRI and beyond. *NeuroImage*, 185, 836–50.

Overton, W. F. (2007). A coherent metatheory for dynamic systems: Relational organicism-contextualism. *Human Development*, 50, 154–9.

Park, D., Ramirez, G. & Beilock, S. L. (2014). The role of expressive writing in math anxiety. *Journal of Experimental Psychology: Applied*, 20, 103–11.

Partanen, E., Kujala, T., Näätänen, R., Liitola, A., Sambeth, A. & Huotilainen, M. (2013). Learning-induced neural plasticity of speech processing before birth. *Proceedings of the National Academy of Sciences of the United States of America*, 110(37), 15145–50.

Pashler, H., McDaniel, M., Rohrer, D. & Bjork, R. (2008). Learning styles: Concepts and evidence. *Psychological Science in the Public Interest*, 9, 105–19.

Pembrey, M. E., Bygren, L. O., Kaati, G., Edvinsson, S., Northstone, K., Sjöström, M., Golding, J. & ALSPAC Study Team (2006). Sex-specific, male-line transgenerational responses in humans. *European Journal of Human Genetics*, 14(2), 159–66.

Penfield, W. (1958). *The excitable cortex in conscious man*. Liverpool: Liverpool University Press.

Peters, S., der Meulen, V., Zanolie, K. & Crone, E. A. (2017). Predicting reading and mathematics from neural activity for feedback learning. *Developmental Psychology*, 53(1), 149–59.

Pew Research Center (2020). *The global God divide*. Retrieved from: www.pewresearch.org/global/wp-content/uploads/sites/2/2020/07/PG_2020.07.20_Global-Religion_FINAL.pdf

Pezzulo, G. & Cisek, P. (2016). Navigating the affordance landscape: Feedback control as a process model of behavior and cognition. *Trends in Cognitive Sciences*, 20(6), 414–24.

Piaget, J. (1949). La pédagogie moderne. *Gazette de Lausanne et Journal suisse (Lausanne)*, 152(63), 10.

Piaget, J. (1965). *The child's conception of number*. New York: W.W. Norton.

Ping, R., Church, R. B, Decatur, M., Larson, S. W., Zinchenko, E. & Goldin-Meadow, S. (2021). Unpacking the gestures of chemistry learners: What the hands tell us about correct and incorrect conceptions of stereochemistry. *Discourse Processes*, 58(3), 213–32.

Plowden Report (1967). *Children and their primary schools – a report of the Central Advisory Council for Education (England)*. London: HMSO.

Polanyi, M. (1958). *Personal knowledge: Towards a post-critical philosophy*. Chicago: University of Chicago Press.

Polanyi, M. (1983). *The tacit dimension*. Gloucester, MA: Peter Smith.

Popper, K. (1978 [1963]). *Conjectures and refutations: The growth of scientific knowledge*. New York: Routledge & Kegan Paul.

Popper, K. (1994). *All life is problem solving*. London: Routledge.

Popper, K. R. & Eccles, J. C. (1977). *The self and its brain: An argument for interactionism*. Cham: Springer-Verlag.

Preston, S. D. & de Waal, F. B. M. (2002). Empathy: Its ultimate and proximate bases. *The Behavioral and Brain Sciences*, 25(1), 1–71.

Price, C. & Devlin, J. (2003). The myth of the visual word form area. *Neuroimage*, 19, 473–81.

Prigogine, I. (1997). *The end of certainty: Time, chaos, and the new laws of nature*. New York: The Free Press.

Pring, R. (2001). Education as a moral practice. *Journal of Moral Education*, 30(2), 101–12.

Ramirez, G., Gunderson, E. A., Levine, S. C. & Beilock, S. L. (2013). Math anxiety, working memory and math achievement in early elementary school. *Journal of Cognitive Development*, 14(2), 187–202.

Rees, G., Bradshaw, J., Goswami, H. & Keung, A. (2010). *Understanding children's well-being: A national survey of young people's well-being*. York: The Children's Society. Retrieved from: www.york.ac.uk/inst/spru/research/pdf/Understanding.pdf

Reeve, R. A., Reynolds, F., Paul, J. & Butterworth, B. L. (2018). Culture-independent prerequisites for early arithmetic. *Psychological Science*, 29(9), 1383–92.

Reid, V. M., Dunn, K., Young, R. J., Amu, J., Donovan, T. & Reissland, N. (2017). The human fetus preferentially engages with face-like visual stimuli. *Current Biology*, 27(12), 1825–8.

Repacholi, B. M. & Gopnik, A. (1997). Early reasoning about desires: Evidence from 14- and 18-month-olds. *Developmental Psychology*, 33, 12–21.

Rest, J. (1974). *Development in judging moral issues*. Minneapolis, MN: University of Minnesota Press.

Rest, J., Narvaez, D., Bebeau, M. J. & Thoma, S. J. (1999). *Postconventional moral thinking: A neo-Kohlbergian approach*. Mahwah, NJ: Lawrence Erlbaum.

Rhodes, M. & Chalik, L. (2013). Social categories as markers of intrinsic interpersonal obligations. *Psychological Science*, 24, 999–1006.

Richardson, K. (2000). *Developmental psychology: How nature and nurture interact*. Mahwah, NJ: Lawrence Erlbaum.

Rogowsky, B., Calhoun, B. M. & Tallal, P. (2015). Matching learning styles to instructional method: Effects on comprehension. *Journal of Educational Psychology*, 107(1), 64–78.

Rosenzweig, E. Q., Wigfield, A. & Eccles, J. S. (2019). Expectancy-value theory and its relevance for student motivation and learning. In K. A. Renninger & S. E. Hidi (eds), *The Cambridge handbook of motivation and learning*. Cambridge: Cambridge University Press, pp. 617–44.

Rothbart, M. K., Ahadi, S. A. & Hershey, K. L. (1994). Temperament and social behavior in childhood. *Merrill-Palmer Quarterly*, 40, 21–39.

Roth-Hanania, R., Davidov, M. & Zahn-Waxler, C. (2011). Empathy development from 8 to 16 months: Early signs of concern for others. *Infant Behavior and Development*, 34, 447–58.

Rouault, M., McWilliams, A., Allen, M. & Fleming, S. (2018). Human metacognition across domains: Insights from individual differences and neuroimaging. *Personality Neuroscience*, 1, E17. https://doi.org/10.1017/pen.2018.16

Rowan, A. J. & Tolunsky, E. (2003). *Primer of EEG with a mini-atlas*. Philadelphia, PA: Elsevier.

Sabbagh, M. A., Bowman, L. C., Evraire, L. & Ito, J. M. B. (2009). Neurodevelopmental correlates of theory of mind in preschool children. *Child Development*, 80, 1147–62.

Saltmarsh, S., Chapman, A., Campbell, M. & Drew, C. (2015). Putting 'structure within the space': Spatially un/responsive pedagogic practices in open-plan learning environments. *Educational Review*, 67(3), 315–27.

Sandseter, E. B. H. (2009). Affordances for risky play in preschool: The importance of features in the play environment. *Early Childhood Education Journal*, 36, 439–46.

Sanger, M. N. (2008). What we need to prepare teachers for the moral nature of their work. *Journal of Curriculum Studies*, 40(2), 169–85.

Sankey, D. (1996). Beyond the ideology of school-based teacher training. *Journal of Primary Education*, 5(1–2), 67–77.

Sankey, D. (1999). *Classrooms as safe places to be wrong*. Retrieved from www.gov.uk/government/uploads/system/uploads/attachment_data/file/18171 8/DFE-RR049.pdf

Sankey, D. (2006). The neuronal, synaptic self: Having values and making choices. *Journal of Moral Education*, 35(2), 27–42.

Sankey, D. (2011). Future horizons: Moral learning and the socially embedded synaptic self. *Journal of Moral Education*, 40(3), 417–25.

Sankey, D. (2016). The neurobiology of trust and schooling. *Educational Philosophy and Theory*, 50(2), 183–92.

Sankey, D. & Kim, M. (2013). A dynamic systems approach to moral and spiritual development. In J. Arthur & T. Lovat (eds), *The Routledge international handbook of education, religion and values*. London: Routledge, pp. 182–93.

Sankey, D. & Kim, M. (2016). Cultivating moral values in an age of neuroscience. In C. W. Joldersma (eds), *Neuroscience and education: A philosophical appraisal*. New York: Routledge, pp. 111–27.

Sawyer, R. K. (2005). *Social emergence: Societies as complex systems*. Cambridge: Cambridge University Press.

Schouten, A., Boiger, M., Kirchner-Häusler, A., Uchida, Y. & Mesquita, B. (2020). Cultural differences in emotion suppression in Belgian and Japanese couples: A social functional model. *Frontiers in Psychology*, 11, 1048. https://doi.org/10.3389/fpsyg.2020.01048

Schwartz, O. (2019). Don't look now: why you should be worried about machines reading your emotions. *The Guardian*, 6 March. Retrieved from: www.theguardian.com/technology/2019/mar/06/facial-recognition-software-emotional-science

Scott, D. & Bhaskar, R. (2015). *Roy Bhaskar: A theory of education*. London: Springer.

Scriven, M. (1967). The methodology of evaluation. In R. E. Stake (ed.), *Curriculum evaluation*. Chicago: Rand McNally, pp. 39–83.

Segool, N., Carlson, J., Goforth, A., von der Embse, N. & Barterian, J. (2013). Heightened test anxiety among young children: Elementary school students' anxious responses to high-stakes testing. *Psychology in the School*, 50(5), 489–99.

Shaffer, D. & Kipp, K. (2007). *Developmental psychology* (8th ed.). Belmont, CA: Cengage Learning.

Shenk, D. (2010). *The genius in all of us: Why everything you have been told about genetics, talent and IQ is wrong*. New York: Doubleday.

Sherrington, C. (1906). *The integrative action of the nervous system*. New Haven, CT: Yale University Press.

Sheya, A. & Smith, L. (2019). Development weaves brains, bodies and environments into cognition. *Language, Cognition and Neuroscience*, 34(10), 1266–73.

Smith, J. D., Couchman, J. J. & Beran, M. J. (2012). The highs and lows of theoretical interpretation in animal-metacognition research. *Philosophical Transactions of The Royal Society B*, 367, 1297–309.

Smith, L. B. (2009). Dynamic systems, sensorimotor processes, and the origins of stability and flexibility. In J. Spencer, M. S. C. Thomas & J. L. McClelland (eds), *Toward a unified theory of development: Connectionism and dynamic systems theory re-considered*. Oxford: Oxford University Press, pp. 67–85.

Smith, L. (2013). It's all connected: Pathways in visual object recognition and early noun learning. *The American Psychologist*, 68(8), 618–29.

Solé, R. & Goodwin, B. (2000). *Signs of life: How complexity pervades biology*. New York: Basic Books.

Soon, C. S., Brass, M., Heinze, H. J. & Haynes, J. D. (2008). Unconscious determinants of free decisions in the human brain. *Nature Neuroscience*, 11(5), 543–5.

South Australian Department of Education (2021). Applying interoception skills. Retrieved from: www.education.sa.gov.au/schools-and-educators/curriculum-and-teaching/curriculum-programs/applying-interoception-skills-classroom

Soylu, F., Lester, F. K. & Newman, S. D. (2018). You can count on your fingers: The role of fingers in early mathematical development. *Journal of Numerical Cognition*, 4(1), 107–35.

Spencer, J., Perone, S. & Buss, A. (2011). Twenty years and going strong: A dynamic systems revolution in motor and cognitive development. *Child Development Perspectives*, 5(4), 260–6.

Squire, L. R. (2009). The legacy of patient H. M. for neuroscience. *Neuron*, 61(1), 6–9.

Squire, L. R. & Kandel, E. (2009). *Memory: From mind to molecules* (2nd ed.). Greenwood Village, CO: Roberts & Company.

Stanley, T. (2010). *Pythagoras: His life and teachings*. Lake Worth, FL: Ibis Press.

Stojiljković, S., Djigić, G. & Zlatković, B. (2012). Empathy and teachers' roles. *Procedia – Social and Behavioral Sciences*, 69(24), 960–6.

Stojiljković, S., Todorović, J., Digić, G. & Dosković, Z. (2014). Teachers' self-concept and empathy. *Procedia – Social and Behavioral Sciences*, 116, 875–9.

Tamir, C., Connaughton, A. & Salazar, A. M. (2020). *The global God divide: People's thoughts on whether belief in God is necessary to be moral vary by economic development, education and age*. Washington, DC: Pew Research Center. Retrieved from: www.pewresearch.org/global/2020/07/20/the-global-god-divide

Tavor, I., Botvinik-Nezer, R., Bernstein-Eliav, M., Tsarfaty, G. & Assaf, Y. (2019). Short-term plasticity following motor sequence learning revealed by diffusion magnetic resonance imaging. *Human Brain Mapping*, 41, 442–52.

Taylor, C. & Dewsbury, B. M. (2018). On the problem and promise of metaphor use in science and science communication. *Journal of Microbiology & Biology Education*, 19(1). https://doi.org/10.1128/jmbe.v19i1.1538

Thagard, P. (2014). Cognitive science. In E. N. Zalta (ed.), *The Stanford Encyclopedia of Philosophy*. Stanford, CA: Stanford University Press.

Thelen, E. & Smith, L. B. (1994). *A dynamic systems approach to the development of cognition and action*. Cambridge, MA: MIT Press.

Thelen, E. & Smith, L. B. (2006). Dynamic systems theories. In W. Damon (editor-in-chief) & R.M. Lerner (ed.), *Handbook of child psychology. Volume 1: Theoretical models of human development* (6th ed.). New York: Wiley, pp. 258–312.

Toomey, R., Lovat, T., Clement, N. & Dally, K. (2010). *Teacher education and values pedagogy: A student wellbeing approach*. Terrigal: David Barlow Publishing.

Trumpp, N. M., Kliese, D., Hoenig, K., Haarmaier, T. & Kiefer, M. (2013). Losing the sound of concepts: Damage to auditory association cortex impairs the processing of sound-related concepts. *Cortex*, 49, 474–86.

Tsakiris, M. & Critchley, H. (2016). Interoception beyond homeostasis: Affect, cognition and mental health. *Philosophical Transactions of Royal Society B*, 371. http://doi.org/10.1098/rstb.2016.0002

Turing, A. (1952). The chemical basis of morphogenesis. *Philosophical Transactions of the Royal Society B*, 237(641), 37–72.

Tversky, A. & Kahneman, D. (1992). Advances in prospect theory: Cumulative representation of uncertainty. *Journal of Risk and Uncertainty*, 5, 297–323.

Ulrich, B. D., Ulrich, D. A. & Angulo-Kinzler, R. M. (1998). The impact of context manipulations on movement patterns during a transition period. *Human Movement Science*, 46, 327–46.

UNESCO (2020). *Learning for empathy: A teacher exchange and support program.* Retrieved from: https://en.unesco.org/fieldoffice/bangkok/project/learning-for-empathy

UNESCO Institute for Statistics (2017). *More than one-half of children and adolescents are not learning worldwide.* UIS Fact Sheet No. 46. Montreal: UIS. Retrieved from: http://uis.unesco.org/sites/default/files/documents/fs46-more-than-half-children-not-learning-en-2017.pdf

University of Southampton (n.d.). *Emergence and self-organization.* Retrieved from: www.complexity.soton.ac.uk/theory/_Emergence_and_Self-Organization.php

US Department of Education Office of Vocational and Adult Education (2020). *The teaching excellence in adult literacy (TEAL).* Retrieved from: https://teal.ed.gov/tealguide/metacognitive

Van Geert, P. (1994). *Dynamic systems of development: Change between complexity and chaos.* New York: Harvester Wheatsheaf.

Van Geert, P. (2008). Nonlinear complex dynamical systems in developmental psychology. In S. Guastello et al. (eds), *Chaos and complexity in psychology: The theory of non-linear dynamical systems.* Cambridge: Cambridge University Press, pp. 242–81.

Van Langen, M. A. M., Wissink, I. B., Van Vugt, E. S., Van der Stouwe, T. & Stams, G. J. J. M. (2014). The relation between empathy and offending: A meta-analysis. *Aggression and Violent Behaviour*, 19(2), 179–89.

Van Rinsveld, A., Dricot, L., Guillaume, M., Rossion, B. & Schiltz, C. (2017). Mental arithmetic in the bilingual brain: Language matters. *Neuropsychologia*, 101, 17–29.

Victoria State Government Education and Training (2020). Promote mental health: social and emotional learning. Retrieved from: www.education.vic.gov.au/school/teachers/health/mentalhealth/Pages/socialemotion.aspx

Vygotsky, L. (1978a). Interaction between learning and development. In M. Gauvain & M. Cole (eds), *Readings on the development of children*. New York: Scientific American Books, pp. 34–40.

Vygotsky, L. S. (1978b). *Mind in society: The development of higher psychological processes.* Cambridge, MA: Harvard University Press.

Vygotsky, L. S. (1986). *Thought and language.* Cambridge, MA: MIT Press.

Wang, W., Xu, K., Niu, H. & Miao, X. (2020). Emotion recognition of students based on facial expressions in online education based on the perspective of computer simulation. *Complexity.* https://doi.org/10.1155/2020/4065207

Weisberg, S. M. & Newcombe, N. S. (2017). Embodied cognition and STEM learning: overview of a topical collection in CR:PI. *Cognitive Research: Principles and Implications*, 2(38). https://doi.org/10.1186/s41235-017-0071-6

Wellcome Trust (2014). Education grants awarded. Retrieved from: https://wellcome.org/what-we-do/our-work/education-grants-awarded#education-and-neuroscience-initiative-(2014)-b21e

Western Australian Department of Education (2018). *Domain guide: Emotional maturity.* Retrieved from: www.education.wa.edu.au/dl/0de8xq

Wicha, N. Y., Dickson, D. S. & Martinez-Lincoln, A. (2018). Arithmetic in the bilingual brain. In D. B. Berch, D. C. Geary & K. M. Koepke (eds), *Language and culture in mathematical cognition.* New York: Academic Press, pp. 145–72.

Williams, K. D., Cheung, C. K. T. & Choi, W. (2000). Cyberostracism: Effects of being ignored over the internet. *Journal of Personality & Social Psychology*, 79, 748–62.

Wilson, M. A. & McNaughton, B. L. (1994). Reactivation of hippocampal ensemble memories during sleep. *Science*, 265(5172), 676–9.

Wilson, R. A. & Lucia, F. (2017). Embodied cognition. In E. N. Zalta (ed.), *The Stanford encyclopedia of philosophy*. Stanford, CA: Stanford University.

Woltering, S. & Lewis, M. D. (2009). Developmental pathways of emotion regulation in childhood: A neuropsychological perspective. *Mind, Brain and Education*, 3(3), 160–9.

Wynn, K. (1992). Addition and subtraction by human infants. *Nature*, 358, 749–50.

Yasnitsky, A. (2019). *Questioning Vygotsky's legacy: Scientific psychology or heroic cult*. London: Routledge.

Yates, L. & Hobson, H. (2020). Continuing to look in the mirror: A review of neuroscientific evidence for the broken mirror hypothesis, EP-M model and STORM model of autism spectrum conditions. *Autism*, 24(8), 1945–59.

Zahavi, D. (2009). Is the self a social construct? *An Interdisciplinary Journal of Philosophy*, 52(6), 551–73.

Zambo, D. (2013). The practical and ethical concerns of using neuroscience to teach young children and help them self-regulate. In L. H. Wasserman & D. Zambo (eds), *Early childhood and neuroscience: Links to development and learning*. Dordrecht: Springer, pp. 7–21.

Zambo, D. & Zambo, R. (2011). Teachers' beliefs about neuroscience and education. *Teaching Educational Psychology*, 7(2), 25–41.